Elements of Pennsylvanian Stratigraphy, Central Appalachian Basin

Edited by

Charles L. Rice
U.S. Geological Survey
National Center, MS 926
Reston, Virginia 22092

SPECIAL PAPER

294

1994

NORTHWEST MISSOURI STATE
UNIVERSITY LIBRARY
MARYVILLE, MO 64468

Copyright © 1994, The Geological Society of America, Inc. (GSA). All rights reserved. GSA grants permission to individual scientists to make unlimited photocopies of one or more items from this volume for noncommercial purposes advancing science or education, including classroom use. Permission is granted to individuals to make photocopies of any item in this volume for other noncommercial, nonprofit purposes provided that the appropriate fee ($0.25 per page) is paid directly to the Copyright Clearance Center, 27 Congress Street, Salem, Massachusetts 01970, phone (508) 744-3350 (include title and ISBN when paying). Written permission is required from GSA for all other forms of capture or reproduction of any item in the volume including, but not limited to, all types of electronic or digital scanning or other digital or manual transformation of articles or any portion thereof, such as abstracts, into computer-readable and/or transmittable form for personal or corporate use, either noncommercial or commercial, for-profit or otherwise. Send permission requests to GSA Copyrights.

Copyright is not claimed on any material prepared wholly by government employees within the scope of their employment.

Published by The Geological Society of America, Inc.
3300 Penrose Place, P.O. Box 9140, Boulder, Colorado 80301

Printed in U.S.A.

GSA Books Science Editor Richard A. Hoppin

Library of Congress Cataloging-in-Publication Data

Elements of Pennsylvanian stratigraphy, central Appalachian Basin /
   edited by Charles L. Rice.
      p.   cm. — (Special paper / Geological Society of America ;
   294)
   Includes bibliographical references and index.
   ISBN 0-8137-2294-2
   1. Geology, Stratigraphic—Pennsylvanian. 2. Geology—Appalachian
Basin. 3. Paleontology—Pennsylvanian. 4. Paleontology-
-Appalachian Basin. I. Rice, Charles L. II. Series: Special
papers (Geological Society of America) ; 294.
QE673.E47   1994
551.7'52'0974—dc20   94-32979
                                                              CIP

Front cover: Limestone concretions (about 3 ft thick) of the Cannelton limestone (White, 1885; see Blake et al., this volume) at its type locality on U.S. Highway 60 just east of Cannelton, West Virginia. Photograph taken in March of 1986 by C. L. Rice.
Back cover: Beta-form quartz from the tonstein of the Fire Clay coal bed. Photograph by H. E. Belkin.

10 9 8 7 6 5 4 3 2 1

Contents

1. *Introduction* 1
   Charles L. Rice

2. *Revision of Nomenclature and Correlations of Some Middle Pennsylvanian Units in the Northwestern Part of the Appalachian Basin, Kentucky, Ohio, and West Virginia* 7
   Charles L. Rice, Robert M. Kosanke, and Thomas W. Henry

3. *Key Rock Units and Distribution of Marine and Brackish Water Strata in the Pottsville Group, Northeastern Ohio* 27
   Ernie R. Slucher and Charles L. Rice

4. *Revised Stratigraphy and Nomenclature for the Middle Pennsylvanian Kanawha Formation in Southwestern West Virginia* 41
   Bascombe M. Blake, Jr., Alan F. Keiser, and Charles L. Rice

5. *Palynostratigraphy of Selected Middle Pennsylvanian Coal Beds in the Appalachian Basin* 55
   Cortland F. Eble

6. *Facies Analysis of Middle Pennsylvanian Marine Units, Southern West Virginia* 69
   Ronald L. Martino

7. *The Pennsylvanian Fire Clay Tonstein of the Appalachian Basin— Its Distribution, Biostratigraphy, and Mineralogy* 87
   Charles L. Rice, Harvey E. Belkin, Thomas W. Henry, Robert E. Zartman, and Michael J. Kunk

8. *High-Precision $^{40}Ar/^{39}Ar$ Age Spectrum Dating of Sanidine from the Middle Pennsylvanian Fire Clay Tonstein of the Appalachian Basin* 105
   Michael J. Kunk and Charles L. Rice

9. *Glossary of Pennsylvanian Stratigraphic Names, Central Appalachian Basin* 115
   Charles L. Rice, John K. Hiett, and Elizabeth D. Koozmin

Introduction

Charles L. Rice
U.S. Geological Survey, National Center, MS 926, Reston, Virginia 22092

The chapters in this volume concern the Pennsylvanian stratigraphy of the central Appalachian basin. The volume is meant to serve as an addendum to the works of Harold R. Wanless (1939, 1946, 1975), which give (in combination) the only comprehensive descriptions of the stratigraphy for that area. Syntheses of Pennsylvanian stratigraphy have been made for individual states (see for example, U.S. Geological Survey, 1979), but, in many cases, they are provincial in character and tend to either gloss over regional problems or treat these problems from the narrow perspective of the individual state. Wanless's regional stratigraphic framework generally ignored state boundaries and, therefore, tended to expose stratigraphic problems that were obscure in the context of local stratigraphy. The following chapters examine some of the problems raised by Wanless's syntheses and attempt to update the nomenclature, the physical stratigraphy, and the biostratigraphy of the region. Much of material introduced here is based on detailed geologic mapping completed in the last 30 yr by federal and state agencies and on associated stratigraphic and biostratigraphic studies.

The central Appalachian basin, which extends from New York to Alabama (Fig. 1), contains a variety of terrigenous sedimentary rocks that were deposited in a foreland basin northwest of the ancestral Appalachian mountains along the southeastern margin (present direction) of the North American craton. The deepest part of the eroded basin is preserved in southern West Virginia and southwestern Virginia. From that area, the Pennsylvanian strata thin toward the north, northwest, and west onto the craton. The thinnest parts of the Pennsylvanian deposits are in Pennsylvania, Ohio, and northwestern West Virginia, areas that also contain the youngest Pennsylvanian rocks. The asymmetrical nature of the basin, its geographic extent, and the concomitant differences of facies have made analyses of the Pennsylvanian stratigraphy singularly difficult.

The three studies by Wanless (1939, 1946, 1975) are the only major attempts to unify the Pennsylvanian stratigraphy of the Appalachian basin on a bed-by-bed basis. When Wanless began his studies, the key beds of the Pottsville and Allegheny Formations (Lower and Middle Pennsylvanian) had been traced across Pennsylvania into Ohio. Many of the county and other geologic reports concerning the Pennsylvanian deposits had been completed in Ohio. Additionally, all of the county geologic reports for the coal-bearing areas of West Virginia had been completed. Government geologists had reported on all parts of the coal basin, and much of the discovery work in the main parts of the Appalachian coal field had been undertaken by coal companies. There was a wealth of stratigraphic detail in coal company and published reports that were couched in a colorful and varied nomenclature. That terminology reflected the ruggedness of the well-dissected Cumberland plateau, which contains the coal field, as well as the essentially isolated nature of many coal-mining districts within the plateau.

As a result of their county geologic reports, both Ohio (Bownocker and Dean, 1929) and West Virginia (White, 1914) had established standard Pennsylvanian sections into which all stratigraphic units within their political boundaries were expected to fit. However, it was a matter of conjecture for Wanless and other workers as to how the Lower and Middle Pennsylvanian sections of those two states fit together because of differences of facies and the fact that the stratigraphic intervals exposed in West Virginia are much thicker than the comparable intervals in Ohio. There was also a wide separation of outcrop areas of correlative units between Ohio and West Virginia (see Fig. 1).

In the stratigraphically thicker and southern part of the basin, the topography is particularly rugged, and the coal districts of eastern Kentucky, southwestern Virginia, and eastern Tennessee are partitioned by major mountain or ridge divides, some of which are associated with faults that caused significant displacement of strata. Considering the repetitive nature of the Pennsylvanian section and its general lack of recognizable and widespread marker beds, it is remarkable how well Wanless was able to synthesize those varied and isolated sections into a workable stratigraphic framework that is both detailed and comprehensive.

In the early part of the 20th century, many stratigraphers had recognized the value of the extensive Pennsylvanian marine limestones and shales as regional stratigraphic marker beds. The marine units in the southern part of the central Appalachian basin—unlike the thin but relatively easily traced

Rice, C. L., 1994, Introduction, *in* Rice, C. L., ed., Elements of Pennsylvanian Stratigraphy, Central Appalachian Basin: Boulder, Colorado, Geological Society of America Special Paper 294.

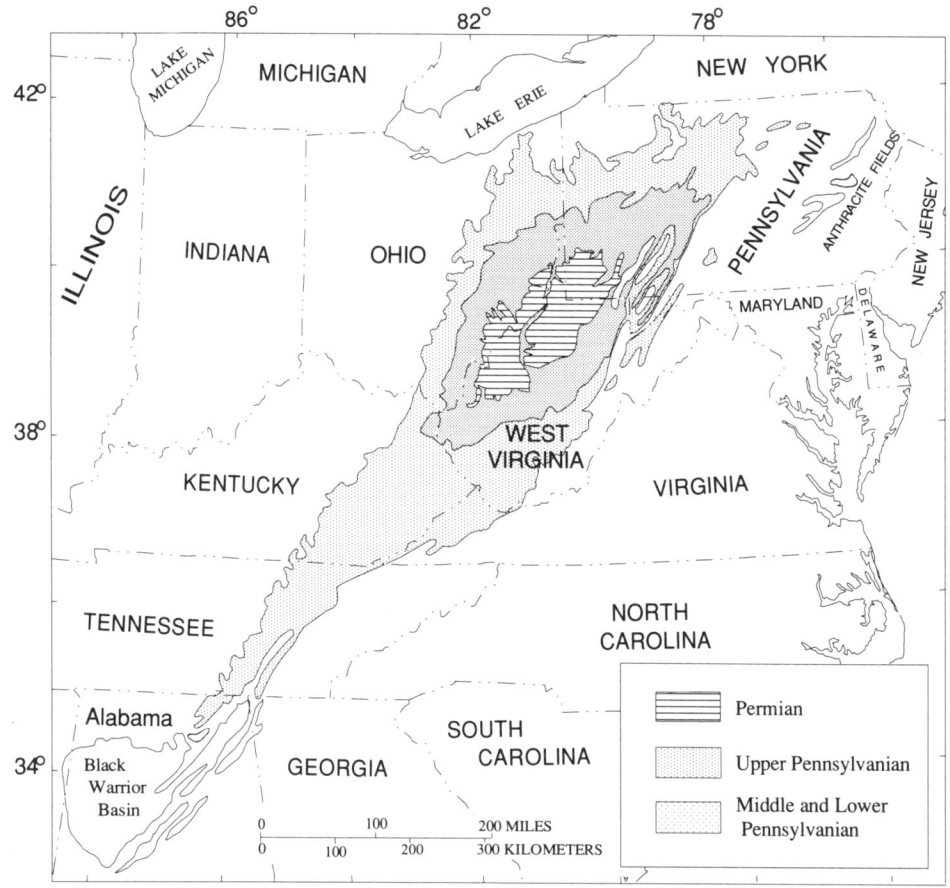

Figure 1. Generalized geology of the central Appalachian basin (patterned) showing the distribution of Pennsylvanian and Permian strata.

marine limestones of the Pottsville and Allegheny Formations of Pennsylvania and Ohio—are mostly shales that seldom crop out. In the late 1950s, when detailed (7.5-min quadrangle) geologic mapping began in the middle and southern parts of the basin in Kentucky, there was considerable doubt whether the marine shales could be traced widely enough to be useful as marker beds. Until that time, the marine units had been simply described as fossiliferous beds or shales that, in effect, had no defined tops or bottoms. Furthermore, some geologists argued that most of the Pennsylvanian section was marine and that marine and brackish water fossils could be found somewhere over every coal bed. The uncertainty of the regional correlation of the marine units is apparent in many of Wanless's stratigraphic descriptions.

Subsequent geologic mapping has demonstrated the continuity of the major marine units and their utility as major elements of the stratigraphic framework in the Appalachian basin as a whole. Many of the more important marine units have been described and given formal names. More importantly, the marine units have begun to be viewed as packages of structured deposits whose parts (or facies) can be attributed to specific depositional environments within transgressive and regressive regimes. Studies of sedimentation models, trace fossils, and fossil assemblages (such as the papers by Martino and by Slucher and Rice) have produced a clearer understanding of the different terrestrial and marine facies and their extent. These investigations doubtlessly will lead to future revisions of the descriptions of many formal Pennsylvanian stratigraphic units.

The correlation charts in Figure 2 show how important key marine units are for analysis of regional Pennsylvanian stratigraphy. The upper part of Figure 2 is generalized from Wanless's (Tables 5–6, 8, 9, 1946) southern Appalachian study and shows his reliance on the Kendrick and Magoffin marine units for the development of a stratigraphy framework for the middle and upper parts of the Pennsylvanian section. In that chart, Wanless indicated his uncertainty in some usages of the names Magoffin and Kendrick by placing them in quotes. His caution was warranted because the Magoffin marine zone of northeastern Tennessee, which he correctly identified in some Tennessee sections, is miscorrelated in the upper part of Figure 2; that marine unit is probably equivalent to the much younger Stoney Fork Member (formerly Lost Creek limestone) of the Breathitt Formation in Kentucky.

Wanless (1946) recognized the discrepancies in many of

Introduction

the contacts of the Hance, Mingo, Catron, Hignite, and Bryson Formations between the Black Mountains and Log Mountains areas on the Cumberland overthrust sheet (upper part, Fig. 2), probably because of their stratigraphic positions with respect to the Kendrick and Magoffin marine units. The contacts between those formations were based primarily on tops and bottoms of coal beds, some of which he recognized as being miscorrelated. His inability to identify the Kendrick shale in the Log Mountains area, however, led him to accept the miscorrelation of the Wallins Creek coal with the Poplar Lick coal, which underlies the Kendrick (lower chart, Fig. 2).

The lower part of Figure 2 incorporates data generated mainly from the detailed geologic quadrangle mapping in Kentucky, Tennessee, and Virginia in the last 30 yr. It shows that the main elements of the stratigraphic framework are the Fire Clay coal bed and equivalents and three marine shales, the Betsie Shale Member, the Kendrick Shale Member, and the Magoffin Member, all of the Breathitt Formation and equivalent strata. The Kendrick Shale Member does extend into Tennessee on the Cumberland overthrust sheet (Rice and Newell, 1990), but its precise position in the Graves Gap Formation (Fig. 2) on the autochthonous plate in Tennessee is not clear at this time. The projected position of the Stoney Fork Member (Fig. 2) with respect to strata on the Cumberland overthrust sheet in Kentucky and Virginia also is unclear because the stratigraphic section on the overthrust sheet is exceedingly thick and the Stoney Fork does not extend into that area.

Numerous formational subdivisions have been made of the Middle Pennsylvanian rocks, as is apparent in Figure 2, although there is little or no lithologic basis for such divisions. Often, boundaries were placed at the tops or bottoms of locally thick sandstones, which were not persistent and could not be mapped regionally. In many cases, the contacts were placed at the tops or bottoms of single coal beds or other key beds, as in the example of the formations on the Cumberland overthrust sheet (upper chart, Fig. 2); these beds are commonly discontinuous or misidentified. Curiously, most Pennsylvanian stratigraphers have not used the extensive marine units in the Middle Pennsylvanian for subdivision. The marine units are generally the most reliable stratigraphic markers or key beds, but they cannot be traced across the entire basin. Furthermore, they do not, in themselves, have the lithic characteristics that meet the criteria of a formation or a formation boundary. In spite of the above facts, the lower part of Figure 2 shows that present-day subdivisions of the Pennsylvanian continue to utilize the tops and bottoms of locally extensive coal and sandstone beds as formation contacts.

Wanless's (1975) major synthesis, which was edited and published after his death on June 3, 1970, made only minor changes in the regional stratigraphy that he had developed (actually in greater detail) in his earlier works. Geological Society of America Memoir 13 (Wanless, 1946), for instance, contains extensive and detailed measured sections for large areas in Kentucky and Tennessee. Some stratigraphic units were miscorrelated, but the lithologic descriptions of long trail sections are unmatched (other than by some modern drillhole descriptions) for the information that they contain. Unfortunately, Wanless never had occasion to question the consistency of the stratigraphy of the extensive mapping done by the state geological surveys in Ohio and West Virginia in the early part of the century; that stratigraphy has only recently been subject to reexamination. Several chapters in this volume deal with a revision of the Pennsylvanian stratigraphic framework of West Virginia (Blake et al.) and Ohio (Slucher and Rice) and the resolution of Pennsylvanian stratigraphic correlations in and between those states and Kentucky (Rice, Kosanke, and Henry).

Only a meager amount of biostratigraphic data was available to Wanless, and his regional correlations apparently were based on personal familiarity with Pennsylvanian fossils and on such general paleontologic studies as those of Read and Mamay (1964). Plant spores of Middle Pennsylvanian coal beds were just beginning to be recognized in the second half of this century as useful for correlation purposes and are only now beginning to be used with confidence for regional and interregional stratigraphic analyses. Several chapters in this volume deal both directly (Eble) and indirectly (Rice, Kosanke, and Henry) with that aspect of the Pennsylvanian stratigraphy.

Although fusulinids were recognized to be a very powerful biostratigraphic tool, they were not studied in enough detail in the Appalachian basin to aid in the correlation of marine units until the work of Douglass (1987). Douglass suggested a correlation of the "southern Vanport" limestone of southern Ohio and northeastern Kentucky with the Columbiana Member of the Allegheny Formation of northern Ohio; the Columbiana overlies the Vanport of Pennsylvania and northern Ohio. This correlation and Douglass's suggested correlation of the Stoney Fork Member of the Breathitt Formation in southeastern Kentucky with the Boggs (Blunt Run) limestone of central Ohio resulted in a profound change of the regional stratigraphic framework. Correlation of the marine units of northern Ohio with those of southeastern Kentucky had always been a problem for physical stratigraphers, primarily because of the great differences of thickness between comparable sections. Douglass's fusulinid studies now are supported by recent investigations of conodont biostratigraphy in the Pennsylvanian of the Appalachian basin (Wardlaw et al., 1993). The preliminary result of those investigations is a zonation of the Middle and Upper Pennsylvanian conodonts by which many of the marine limestones of that age can be characterized and correlated in the Appalachian basin. The consequences of those studies are discussed in the revision of nomenclature and Middle Pennsylvanian stratigraphy for adjoining areas of Ohio, Kentucky, and West Virginia (Rice, Kosanke, and Henry).

The occurrence of widespread thin partings in coal beds has fascinated many Pennsylvanian stratigraphers, who have commonly used them for coal-bed correlation. This usage is particularly true of the flint clay parting (tonstein) of the Fire Clay coal bed and equivalent beds in the central Appalachian

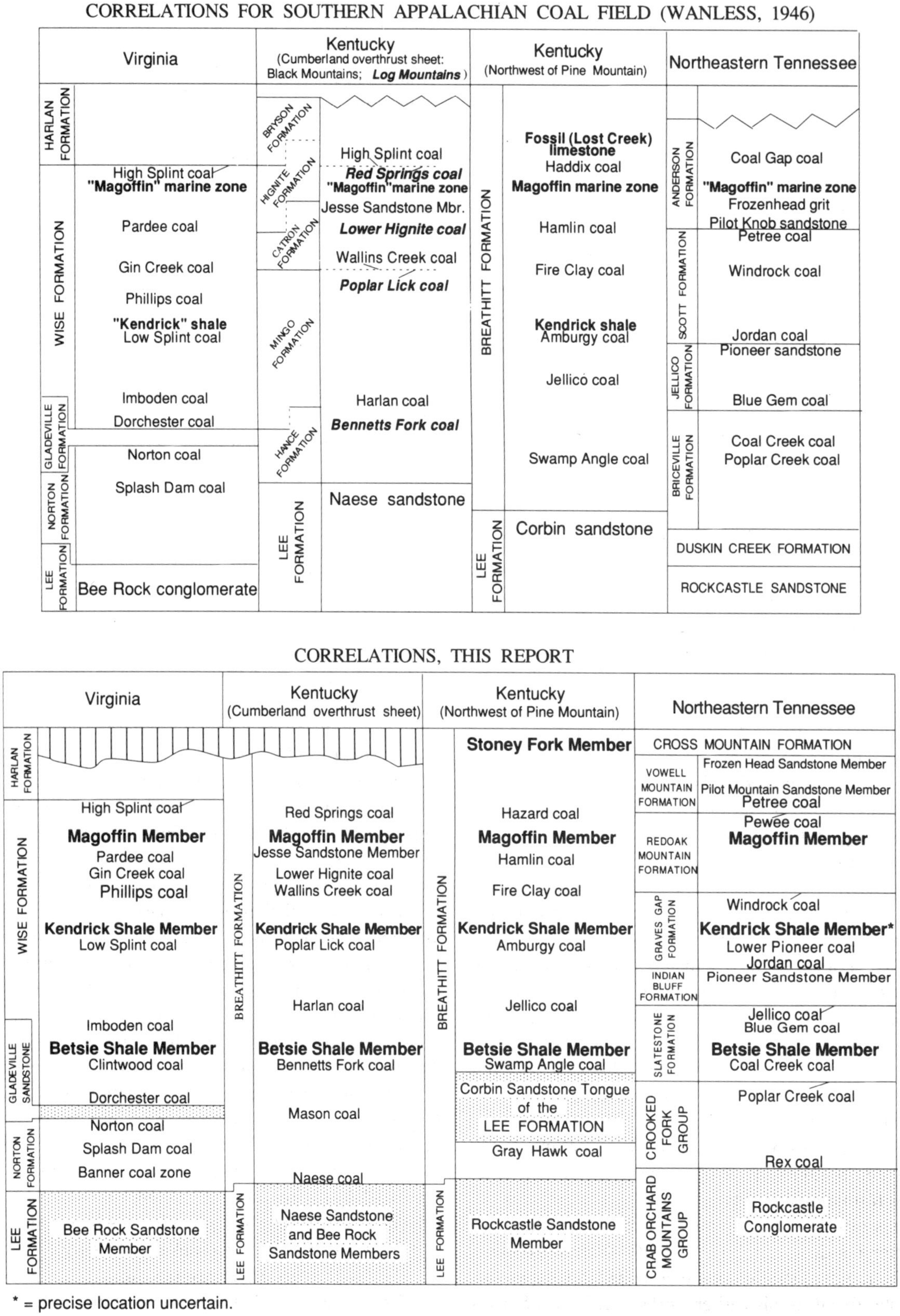

CORRELATIONS FOR SOUTHERN APPALACHIAN COAL FIELD (WANLESS, 1946)

Virginia	Kentucky (Cumberland overthrust sheet: Black Mountains; *Log Mountains*)	Kentucky (Northwest of Pine Mountain)	Northeastern Tennessee
HARLAN FORMATION	BRYSON FORMATION	Fossil (Lost Creek) limestone / Haddix coal	ANDERSON FORMATION — Coal Gap coal
WISE FORMATION — High Splint coal / "Magoffin" marine zone / Pardee coal / Gin Creek coal / Phillips coal / "Kendrick" shale / Low Splint coal	HIGNITE FORMATION — High Splint coal / *Red Springs coal* / "Magoffin" marine zone / Jesse Sandstone Mbr. / *Lower Hignite coal* / Wallins Creek coal / *Poplar Lick coal* / CATRON FORMATION / MINGO FORMATION	BREATHITT FORMATION — Magoffin marine zone / Hamlin coal / Fire Clay coal / Kendrick shale / Amburgy coal / Jellico coal	"Magoffin" marine zone / Frozenhead grit / Pilot Knob sandstone / SCOTT FORMATION — Petree coal / Windrock coal / JELLICO FORMATION — Jordan coal / Pioneer sandstone / Blue Gem coal
GLADEVILLE FM — Imboden coal / Dorchester coal	HANCE FORMATION — Harlan coal / *Bennetts Fork coal*	Swamp Angle coal	BRICEVILLE FORMATION — Coal Creek coal / Poplar Creek coal
NORTON FORMATION — Norton coal / Splash Dam coal	LEE FORMATION — Naese sandstone	LEE FORMATION — Corbin sandstone	DUSKIN CREEK FORMATION
LEE FORMATION — Bee Rock conglomerate			ROCKCASTLE SANDSTONE

CORRELATIONS, THIS REPORT

Virginia	Kentucky (Cumberland overthrust sheet)	Kentucky (Northwest of Pine Mountain)	Northeastern Tennessee
HARLAN FORMATION		**Stoney Fork Member**	CROSS MOUNTAIN FORMATION / VOWELL MOUNTAIN FORMATION — Frozen Head Sandstone Member / Pilot Mountain Sandstone Member / Petree coal
WISE FORMATION — High Splint coal / **Magoffin Member** / Pardee coal / Gin Creek coal / Phillips coal / **Kendrick Shale Member** / Low Splint coal	BREATHITT FORMATION — Red Springs coal / **Magoffin Member** / Jesse Sandstone Member / Lower Hignite coal / Wallins Creek coal / **Kendrick Shale Member** / Poplar Lick coal / Harlan coal	BREATHITT FORMATION — Hazard coal / **Magoffin Member** / Hamlin coal / Fire Clay coal / **Kendrick Shale Member** / Amburgy coal / Jellico coal	REDOAK MOUNTAIN FORMATION — Pewee coal / **Magoffin Member** / GRAVES GAP FORMATION — Windrock coal / **Kendrick Shale Member*** / Lower Pioneer coal / Jordan coal / INDIAN BLUFF FORMATION — Pioneer Sandstone Member / SLATESTONE FORMATION — Jellico coal / Blue Gem coal
Imboden coal / **Betsie Shale Member** / Clintwood coal / GLADEVILLE SANDSTONE / Dorchester coal	**Betsie Shale Member** / Bennetts Fork coal / Mason coal	**Betsie Shale Member** / Swamp Angle coal / Corbin Sandstone Tongue of the LEE FORMATION / Gray Hawk coal	**Betsie Shale Member** / Coal Creek coal / CROOKED FORK GROUP — Poplar Creek coal / Rex coal
NORTON FORMATION — Norton coal / Splash Dam coal / Banner coal zone	LEE FORMATION — Naese coal / Naese Sandstone and Bee Rock Sandstone Members	LEE FORMATION — Rockcastle Sandstone Member	CRAB ORCHARD MOUNTAINS GROUP — Rockcastle Conglomerate
LEE FORMATION — Bee Rock Sandstone Member			

* = precise location uncertain.

Figure 2. Evolution of the stratigraphic framework for the Pennsylvanian of the southern Appalachian coal field. Italicized names in upper chart are coals at formational boundaries in the Log Mountains area of the Cumberland overthrust sheet.

basin. Wanless (1939) considered this parting of the Fire Clay coal as the most useful marker bed in the Middle Pennsylvanian of the Appalachian region. He even correlated the Fire Clay coal with the Upper DeLong coal of the Eastern Interior (Illinois) basin, which has a similarly persistent clay parting (Wanless, 1952). The extent of these thin partings and their persistence across large areas suggested to many workers that they are airborne volcanic ash deposits.

Only in the last decade, however, have we been able to disaggregate the flint clays and analyze the accessory minerals to show that they consist mostly of primary igneous crystals. It now appears that the tonsteins of the acid peat mires are the equivalents of volcanic bentonites that were deposited in marine or freshwater lake environments. The details of the Fire Clay tonstein—its biostratigraphic position, mineralogy, distribution, and possible source—are discussed by Rice, Belkin, et al. A second work (Kunk and Rice) concerns the basinwide $^{40}Ar/^{39}Ar$ isotopic dating of sanidine from the Fire Clay tonstein. This stratigraphic unit is the first in the Carboniferous of North America to be dated by isotopic means. The precision of its date makes it suitable for numerical refinement of the geologic time scale for the continent.

Important components of Wanless's 1939 and 1946 papers were the glossaries of Pennsylvanian stratigraphic names in which he identified for each unit the type areas, local stratigraphic position, and probable regional correlatives—information essential to coherent nomenclatural usage. Those glossaries have been combined and updated in the final chapter of this volume (Rice, Hiett, and Koozmin) in order to provide a new and more comprehensive listing of formal and informal Pennsylvanian stratigraphic units and their correlatives for the central Appalachian basin.

Most of the chapters in this volume deal with the contentious subject of correlation of Pennsylvanian units. The difficulty of tracing coal beds and sandstone and marine units from one area to another has suggested to some workers that the established time-stratigraphic framework and its "layer-cake" model have lost their usefulness for Pennsylvanian stratigraphic analyses. New depositional models that emphasized the time-transgressive nature of sedimentary units proposed a much more complex stratigraphic framework than what had been envisioned previously. Although much has been learned from recent studies of depositional models, not the least is that we need a better knowledge of physical stratigraphy. We have improved the Pennsylvanian stratigraphic framework by detailed geologic mapping supported by focused biostratigraphic investigations and by analyses of much drillhole data. Wanless (1939, 1946) recognized the need to study many of the subjects treated in the chapters of this volume and listed them in sections entitled Problems Deserving Further Study. He specifically mentioned the need for biostratigraphic studies of invertebrate fossils and plant fossils and particularly the study of plant spores from coal maceration. He also identified geographic areas requiring further detailed study such as the tristate area of Ohio, Kentucky, and West Virginia. Some of our more important stratigraphic problems have been in areas that we have too long avoided or taken for granted. As is apparent from Wanless's suggestions of further studies and the conclusions reached in the following papers, we have just begun to acquire the stratigraphic data needed to effectively analyze the regional depositional and tectonic patterns of the Pennsylvanian in the central Appalachian basin.

The following is a list of people who assisted in the review of the chapters comprising this volume: Bascombe M. Blake, Jr., Donald R. Chesnut, Jr., Aureal T. Cross, Raymond C. Douglass, Avery A. Drake, Jr., J. Thomas Dutro, Jr., Jack B. Epstein, John C. Ferm, Romeo M. Flores, Robert A. Gastaldo, Gregory S. Gohn, Norman C. Hester, Alan F. Keiser, Robert P. Koeppen, Robert M. Kosanke, Marvin A. Lanphere, Edward T. Luther, Walter L. Manger, Lucy McCartan, Robert C. McDowell, Glen K. Merrill, Molly F. Miller, Robert B. Mixon, Jack E. Nolde, Russel A. Peppers, David S. Powars, Harold B. Rollins, Arthur P. Schultz, Victor M. Seiders, Ernie R. Slucher, Lawrence W. Snee, Bryon D. Stone, John F. Sutter, Carl C. Swisher, Robert E. Weems, Eugene G. Williams, and Robert E. Zartman.

REFERENCES CITED

Bownocker, J. A., and Dean, E. S., 1929, Analysis of the coals of Ohio: Ohio Division of Geological Survey Bulletin 34, 360 p.

Douglass, R. C., 1987, Fusulinid biostratigraphy and correlations between the Appalachian and Eastern Interior Basins: U.S. Geological Survey Professional Paper 1451, 95 p., 20 plates.

Read, C. B., and Mamay, S. H., 1964, Upper Paleozoic floral zones and floral provinces of the United States: U.S. Geological Survey Professional Paper 454-K, p. K1–K35.

Rice, C. L., and Newell, W. L., 1990, Geologic map of part of the Jellico East quadrangle, Campbell and Claiborne Counties, Tennessee: U.S. Geological Survey Geologic Quadrangle Map GQ-1674, scale 1:24,000.

U.S. Geological Survey, 1979, The Mississippian and Pennsylvanian (Carboniferous) Systems in the United States: U.S. Geological Survey Professional Paper 1110-A-L, 376 p.

Wanless, H. R., 1939, Pennsylvanian correlations in the Eastern Interior and Appalachian coal fields: Geological Society of America Special Paper 17, 130 p.

Wanless, H. R., 1946, Pennsylvanian geology of a part of the southern Appalachian coal field: Geological Society of America Memoir 13, 162 p.

Wanless, H. R., 1952, Studies of field relations of coal beds, in Proceedings, 2nd Conference on the origin and constitution of coal, 1952: Halifax, Nova Scotia Department of Mines and Nova Scotia Research Foundation, p. 148–180 (with discussion).

Wanless, H. R., 1975, Appalachian region, in McKee, E. D., and Crosby, E. J., coords., Paleotectonic investigations of the Pennsylvanian system in the United States, pt. 1, Introduction and regional analyses of the Pennsylvanian System: U.S. Geological Survey Professional Paper 853, p. 69–96.

Wardlaw, B. R., Rice, C. L., and Stamm, R. G., 1993, Preliminary analysis of conodont occurrences in Pennsylvanian strata of Ohio and Kentucky: U.S. Geological Survey Open-File Report 93-312, 9 p.

White, I. C., 1914, Introduction, in Krebs, C. E., and Teets, D. D., Jr., Kanawha County: West Virginia Geological and Economic Survey [County Report], p. xvii–xxviii.

MANUSCRIPT ACCEPTED BY THE SOCIETY FEBRUARY 1, 1994

Printed in U.S.A.

Geological Society of America
Special Paper 294
1994

Revision of nomenclature and correlations of some Middle Pennsylvanian units in the northwestern part of the Appalachian basin, Kentucky, Ohio, and West Virginia

Charles L. Rice
U.S. Geological Survey, National Center, MS 926, Reston, Virginia 22092
Robert M. Kosanke and Thomas W. Henry
U.S. Geological Survey, DFC, Box 25046, MS 919, Denver, Colorado 80225

ABSTRACT

Studies of the physical stratigraphy and analyses of the Middle Pennsylvanian flora and fauna of some coal beds and marine units of the Breathitt Formation in Kentucky, the Pottsville and Allegheny Formations in Ohio, and the Kanawha Formation and Charleston Sandstone in West Virginia show a need for the revision of stratigraphic nomenclature of the Pottsville Formation and the lower part of the Allegheny Formation and equivalent strata. Attempts to project single stratigraphic elements from one region to another have resulted historically in multiple miscorrelations. A major marine unit (previously misidentified as the Vanport limestone of the Breathitt and Allegheny Formations) is here named the Obryan Member of the Breathitt Formation in northeastern Kentucky and of the Allegheny Formation in southern Ohio. The Obryan is characterized by the fusulinid *Beedeina ashlandensis* Douglass and is correlated with the Columbiana Member of the Allegheny Formation in central Ohio, which also contains that fusulinid. This correlation and the correlation of the Boggs Limestone Member of the Pottsville Formation in Ohio with the Stoney Fork Member of the Breathitt Formation in Kentucky are supported by analyses of Middle Pennsylvanian conodonts. A preliminary zonation of conodonts for strata of the Pottsville and Allegheny Formations shows that the major marine units of these formations in Ohio are biostratigraphically distinct. The Obryan Member is locally absent, but its position is marked by overlying clay beds in many parts of Kentucky, Ohio, and West Virginia. The Vanport Limestone Member of the Allegheny Formation (as identified in Pennsylvania and central Ohio) is here correlated with the Zaleski Flint Member (Allegheny Formation) in southern Ohio. The Kilgore Flint Member (new name, Breathitt Formation), the informal Limekiln limestone and informal Flint Ridge flint (Breathitt Formation) in Kentucky, and the informal Kanawha black flint (Kanawha Formation) in West Virginia are correlated with the Putnam Hill Limestone Member (Allegheny Formation) in Ohio. These latter chert deposits are shoreward (southward and southeastward) facies of a marine unit deposited mostly in restricted estuaries and bays. The chert deposits appear to result from a widespread episode of silicification of fossiliferous marine siltstones and limestones that locally affected underlying peats and silts. The Kilgore

Rice, C. L., Kosanke, R. M., and Henry, T. W., 1994, Revision of nomenclature and correlations of some Middle Pennsylvanian units in the northwestern part of the Appalachian basin, Kentucky, Ohio, and West Virginia, *in* Rice, C. L., ed., Elements of Pennsylvanian Stratigraphy, Central Appalachian Basin: Boulder, Colorado, Geological Society of America Special Paper 294.

and Obryan Members and their equivalents are used as the two principal stratigraphic marker beds for analyses of Middle Pennsylvanian sections extending across northeastern Kentucky from central Ohio to central West Virginia. Clay units and coal beds that overlie the Obryan Member contain flint clay beds (tonsteins) that are, in part, the product of volcanic ash falls. The range zones of selected palynomorphs from northeastern Kentucky and southeastern Ohio corroborate some of the correlations proposed herein.

INTRODUCTION

The Kilgore Flint Member and the Obryan Member (new names) are the two youngest named rock units containing marine fossils in the upper part of the Breathitt Formation (Lower and Middle Pennsylvanian) in Kentucky. They and their equivalents crop out in an area mainly in the northern and western flanks of a broad east-northeastward–plunging syncline south and east of Ashland, Kentucky (Fig. 1). Because of lateral discontinuities and structure, these outcrops are isolated from other areas that contain correlative strata farther south and southeast in Kentucky and in West Virginia. The key marine beds of the middle and lower parts of the Breathitt Formation—such as the Kendrick Shale Member, the Magoffin Member, and the Stoney Fork Member—are discontinuous or pinch out northeastward in Kentucky and do not provide much-needed stratigraphic control for that area. The lack of persistent marine marker beds and coal beds in northeastern Kentucky and western West Virginia on the southern side of the syncline makes correlation of the younger Middle Pennsylvanian units difficult in a southward-thickening section, particularly between isolated hilltops where the Kilgore and Obryan (or equivalent strata) might occur. In addition, the tri-state area of Ohio, Kentucky, and West Virginia is one of regional facies change in the Pennsylvanian between the more marine strata of the Pottsville and Allegheny Formations of central Ohio and the less marine equivalent strata of northeastern Kentucky and western West Virginia. Consequently, there has been a lack of attention to some long-standing regional correlation problems that have broad implications with respect to the distribution of facies and the geometry of the basin.

Miscorrelations of principal units, particularly the continued use of the name "Vanport limestone member" (introduced by Phalen, 1908; see Rice, 1977) for a marine unit probably younger than the Vanport Limestone Member of the Allegheny Formation of central and northern Ohio, have added to the stratigraphic confusion. Reinvestigation of the physical stratigraphy, as well as of the macrofaunas and microfaunas of marine units of this part of the Pennsylvanian section, and an examination of the palynology of associated coal beds suggest a need for formal, well-defined names for the upper Middle Pennsylvanian marine units in eastern Kentucky. These studies also show a need for a reevaluation of regional correlations of Middle Pennsylvanian marine units and related strata in Ohio, Kentucky, and West Virginia. Furthermore, many of the commonly used names for stratigraphic units in the tristate area were first defined in Pennsylvania; these usages also require examination for their appropriateness.

The Kendrick Shale Member and the Magoffin Member of the Breathitt Formation in Kentucky and their equivalents—the Dingess and Winifrede Shale Members of the Kanawha Formation in West Virginia—are widely distributed marine units that are key elements of the Middle Pennsylvanian stratigraphic framework in the central Appalachian basin. Because these units have been miscorrelated with named marine limestones of the Pottsville Formation in Ohio (see, for example, Wanless, 1939, 1975), their relative stratigraphic position is also discussed in light of the biostratigraphic data presented in this chapter.

The stratigraphic nomenclature for the Pennsylvanian of Ohio was established long before the publication of the North American Stratigraphic Code of 1983 (North American Commission on Stratigraphic Nomenclature, 1983) or any of its predecessors. Consequently, the Ohio nomenclature does not follow that code and contains some widely recognized stratigraphic marker beds such as the Upper and Lower Mercer limestone units, which, according to the code, are improperly named for formal members. Because of these and other nomenclatural problems in Ohio that are beyond the scope of our considerations (and for the sake of consistency and simplicity), this report in general follows the usage of Wanless (1975) where those important marine marker beds are listed as formal members of the Pottsville and Allegheny Formations. The accompanying glossary (Rice, Hiett, and Koozmin, this volume) designates these units as informal, unranked units, following the usage of the State of Ohio.

PREVIOUS WORK

The correlation and distribution of marine units in Middle Pennsylvanian strata in Kentucky and adjacent areas of Ohio and West Virginia have not been studied comprehensively since the investigations of Phalen (1908, 1912) in the Kenova 30-min quadrangle (Kentucky–West Virginia–Ohio) and those of Stout (1916) in southern Ohio (Fig. 2). White (1878) had given the name "Vanport limestone" to a limestone previously called the "Ferriferous limestone" from outcrops near Vanport, Pennsylvania (Fig. 1). Although many of the limestones in this part of the Pennsylvanian section in Ohio and Pennsylvania are directly overlain by thin beds of iron ore, the name "Ferriferous" was applied commonly to a single easily identified unit in western Pennsylvania and in central and northern Ohio. Following White's usage, Phalen (1908, 1912)

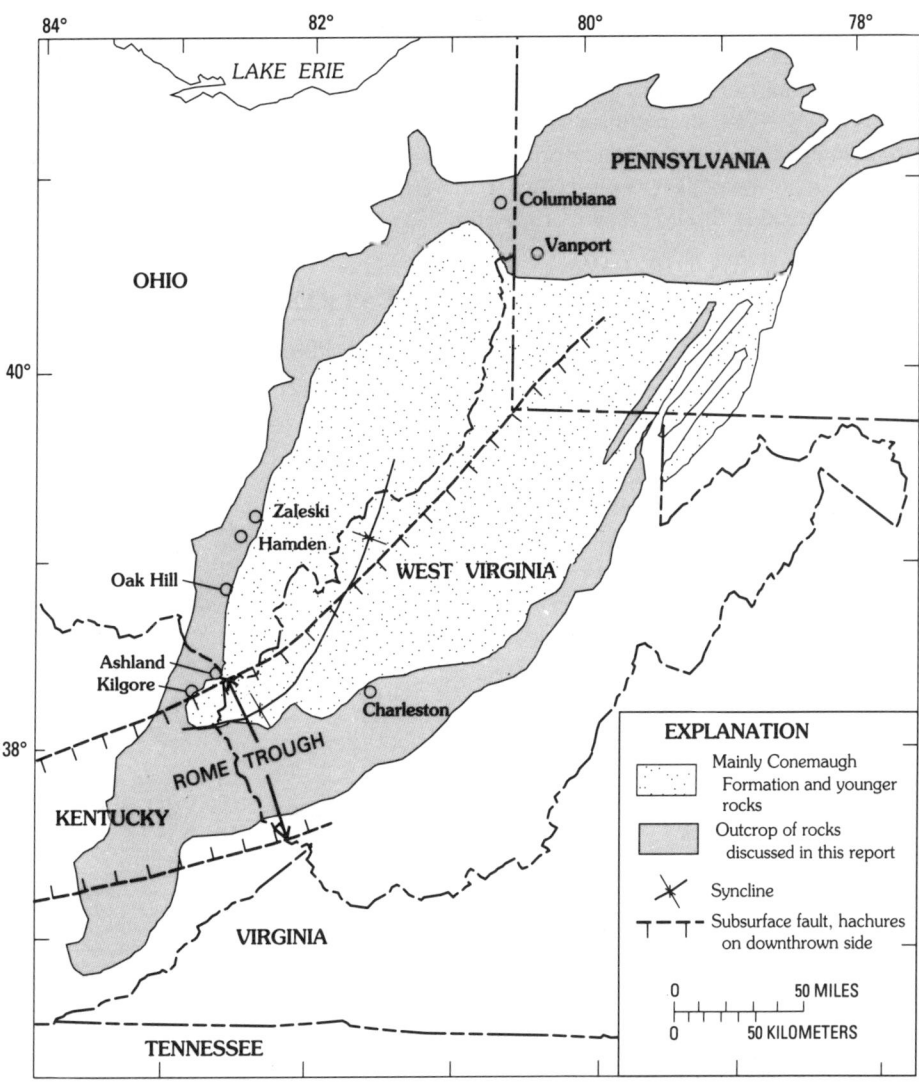

Figure 1. Distribution of strata studied in this report and major structural features. Location of Rome trough from Harris (1975).

applied the name Vanport limestone to a limestone called the "Ferriferous" or "Hanging Rock" limestone in southern Ohio and Kentucky by earlier workers (Orton, 1878). In southern Ohio and northeastern Kentucky, that limestone is fairly continuous and varies in thickness from 1 to 3 m. Because of its continuity and thickness, the limestone was an important source of iron ore and flux for the iron industry, a source of lime for the production of mortar, and a source of rock for road metal in southern Ohio in the last century.

The Vanport limestone was the only marine marker bed identified in the Allegheny and Pottsville Formations in the Kenova Quadrangle by Phalen (1908, 1912). Stout (1916) mapped the limestone bed and followed Phalen's usage of the name in southern Ohio. But Stout's section, shown in part in the upper half of Figure 2, included many marine units that were recognized in central Ohio, where they are extensive and well defined; he projected those southward into the Kenova area, where many are typically represented by either a brackish water or freshwater facies. Figure 2 shows a stratigraphic synthesis by Wanless (1939, 1975) for a part of the Middle Pennsylvanian for Ohio (probably based on the more complete sections of central Ohio) and for Kentucky and West Virginia. Many of the names of coal and sandstone beds that were important elements of the stratigraphic sections of earlier workers in the tristate area are not regionally continuous and are not shown in Figure 2 for the sake of simplicity; some of those names are listed by Webb (1963).

Because both the Hamden and the Boggs iron ores of southern Ohio are brackish water or nonmarine deposits in their type sections (Morningstar, 1922), the use of those names for marine equivalents in the northern part of Ohio probably is not appropriate because these units have not been, and probably cannot be, traced from one to the other. Thus, in central and northern Ohio, Sturgeon and DeLong (1964) gave the name "Columbiana Member" to a marine limestone and shale unit, which was generally (but, as we show, mistakenly) believed to be the marine equivalent of the Hamden ore. The "Boggs limestone member," however, is a well-recognized and documented marine limestone in the Pottsville Formation in central Ohio (Stout, 1918), in which, in many places, significant replacement of marine fossils by iron carbonate has occurred (R. C. Douglass, oral communication, 1978).

Other historical problems illustrated by Figure 2 include Stout's (1916) first placement of a black flint marine unit at the top of the Pottsville Formation, just below the Brookville coal bed (see also Morningstar, 1922). He later named that unit the

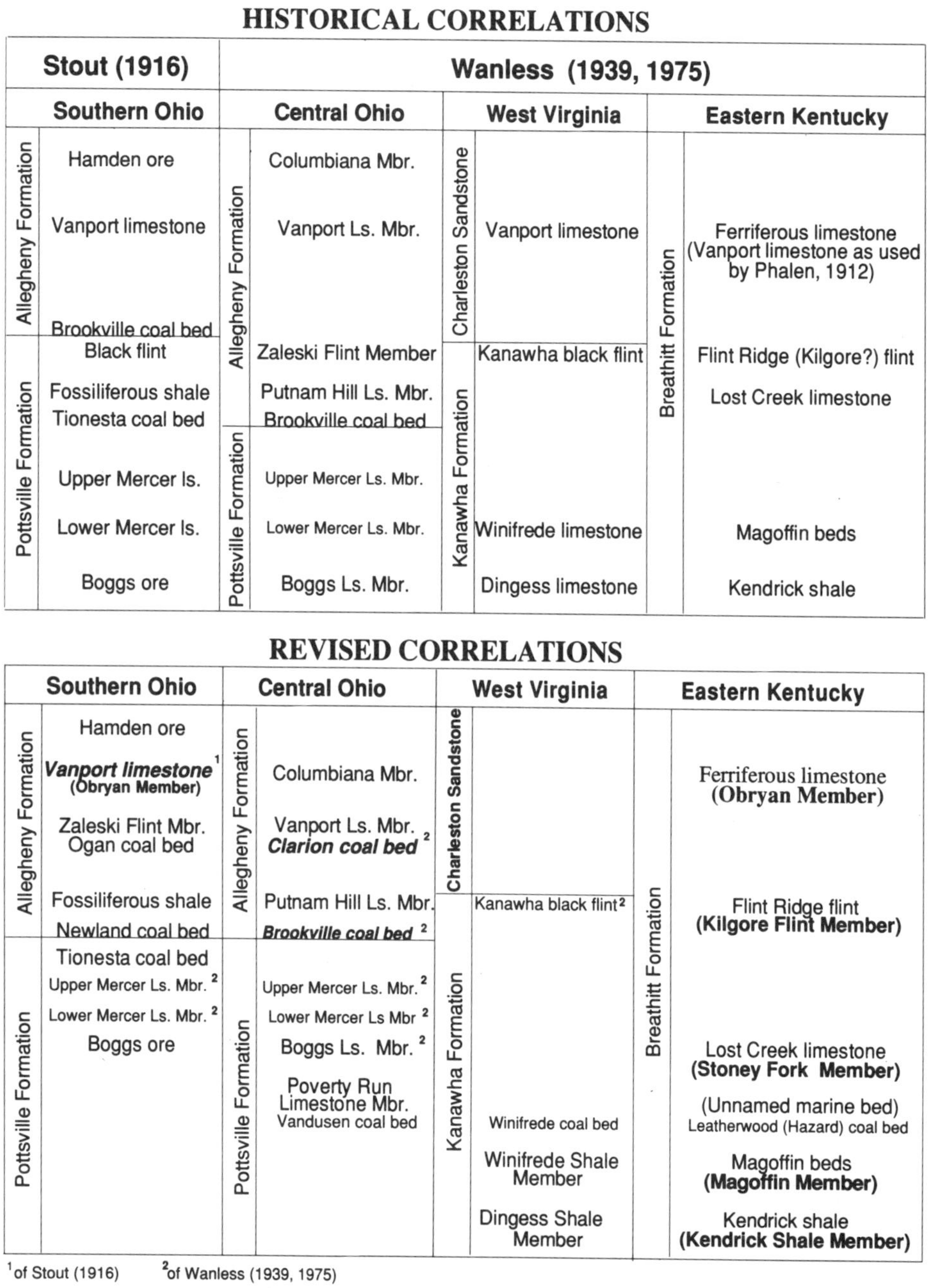

HISTORICAL CORRELATIONS

Stout (1916)	Wanless (1939, 1975)		
Southern Ohio	Central Ohio	West Virginia	Eastern Kentucky
Allegheny Formation: Hamden ore; Vanport limestone; Brookville coal bed	**Allegheny Formation**: Columbiana Mbr.; Vanport Ls. Mbr.; Zaleski Flint Member; Putnam Hill Ls. Mbr.; Brookville coal bed	**Charleston Sandstone**: Vanport limestone; **Kanawha Formation**: Kanawha black flint	**Breathitt Formation**: Ferriferous limestone (Vanport limestone as used by Phalen, 1912); Flint Ridge (Kilgore?) flint; Lost Creek limestone
Pottsville Formation: Black flint; Fossiliferous shale; Tionesta coal bed; Upper Mercer ls.; Lower Mercer ls.; Boggs ore	**Pottsville Formation**: Upper Mercer Ls. Mbr.; Lower Mercer Ls. Mbr.; Boggs Ls. Mbr.	Winifrede limestone; Dingess limestone	Magoffin beds; Kendrick shale

REVISED CORRELATIONS

Southern Ohio	Central Ohio	West Virginia	Eastern Kentucky
Allegheny Formation: Hamden ore; *Vanport limestone*[1] (Obryan Member); Zaleski Flint Mbr.; Ogan coal bed; Fossiliferous shale; Newland coal bed	**Allegheny Formation**: Columbiana Mbr.; Vanport Ls. Mbr.; *Clarion coal bed*[2]; Putnam Hill Ls. Mbr.; *Brookville coal bed*[2]	**Charleston Sandstone**; **Kanawha Formation**: Kanawha black flint[2]	**Breathitt Formation**: Ferriferous limestone (**Obryan Member**); Flint Ridge flint (**Kilgore Flint Member**)
Pottsville Formation: Tionesta coal bed; Upper Mercer Ls. Mbr.[2]; Lower Mercer Ls. Mbr.[2]; Boggs ore	**Pottsville Formation**: Upper Mercer Ls. Mbr.[2]; Lower Mercer Ls Mbr[2]; Boggs Ls. Mbr.[2]; Poverty Run Limestone Mbr.; Vandusen coal bed	Winifrede coal bed; Winifrede Shale Member; Dingess Shale Member	Lost Creek limestone (**Stoney Fork Member**); (Unnamed marine bed); Leatherwood (Hazard) coal bed; Magoffin beds (**Magoffin Member**); Kendrick shale (**Kendrick Shale Member**)

[1] of Stout (1916)   [2] of Wanless (1939, 1975)

Figure 2. Evolution of stratigraphic names for a part of the Middle Pennsylvanian in adjacent areas of Ohio, West Virginia, and Kentucky. Names in bold italics are incorrectly but commonly used in areas shown. Stratigraphic names in Ohio attributed to Wanless (1939, 1975) are mostly derived from county reports of Stout (1916, 1918, 1927). Formal names of marine units used in this report for eastern Kentucky are shown in bold in parentheses in lower chart. Only selected coal-bed names are shown for simplicity. Ls. = Limestone; Mbr. = Member.

"Zaleski flint member" and placed it in the overlying Allegheny Formation (Stout, 1927, p. 170, footnote). That change requires an extensive reevaluation of Stout's measured sections in southern Ohio (Stout, 1916), which provide the only published record of the entire Pennsylvanian section for that area. The confusion concerning the position of the Zaleski Flint Member is further complicated by the miscorrelation of the "Ferriferous" limestone by Phalen (1908) and Stout (1916) and by the suggestions of Cavaroc and Ferm (1968, Fig. 4) that the Zaleski Flint Member and Putnam Hill Limestone Member may be equivalent units, and that there may be two or even three (Ferm, 1970) different "Vanport" limestones.

Wanless (1975) correlated the informal Kanawha black flint of central West Virginia with the informal Flint Ridge flint and informal Kilgore flint of eastern Kentucky (Fig. 2) and the Zaleski Flint Member of Ohio. The Kilgore flint had earlier been referred to as the "black flint" of the Princess reserve district by Ferm (1963). Tentative correlations made by Huddle and Englund (1966) and repeated by Rice et al. (1979) and Outerbridge (1989) equated the Kanawha black flint of West Virginia with the informal marine Lost Creek limestone (now called the Stoney Fork Member) in southeastern Kentucky. The later correlation of Huddle and Englund (1966) was not noted by Wanless (1975) in his synthesis of Appalachian Pennsylvanian stratigraphy.

Merrill (1968) studied the conodonts of the Allegheny Formation of Ohio and Pennsylvania and also concluded that the "southern Vanport" was different from, but stratigraphically closely related to, the "northern Vanport." Wardlaw et al. (1993) showed that conodont faunas can be used for the correlation of many Pennsylvanian limestones in the Middle and Upper Pennsylvanian of the central Appalachian basin. Kosanke (1973), Helfrich (1981), and Phillips et al. (1985) reported on the palynomorphs of the coals of the upper part of the Breathitt Formation in the Princess reserve district of northeastern Kentucky, and Gray (1967) studied similar coal beds in the Allegheny Formation in Ohio and Pennsylvania. Both Kosanke and Phillips et al. discussed their findings in terms of regional and interregional correlations of key coal beds.

STRATIGRAPHY OF THE MARINE UNITS

The lower half of Figure 2 shows our resolution of some of the Pennsylvanian stratigraphic problems for the central Appalachian basin, based in large part on the fusulinid biostratigraphy developed by Douglass (1987), who correlated the informal Lost Creek and Ferriferous limestones of eastern Kentucky with the Boggs Limestone Member and the Columbiana Member of central Ohio, respectively. These correlations and others indicated in Figure 2 are shown in other parts of this report to be supported by detailed analyses of miospores from coal beds and conodonts from marine limestones.

Our discussion also includes the Kendrick Shale Member and the Magoffin Member of the Breathitt Formation (and equivalents in West Virginia) for two reasons: because of the correlation of these units with the Mercer limestones and Boggs limestone in Ohio by Wanless (1939, 1975), and because of the importance of these units in the stratigraphic framework of the central Appalachian basin. Many of the names of key marine units of the Breathitt Formation in Kentucky used in this report have been formalized since the reports of Wanless (1975). Those indicated in parentheses in the last column of the lower half of Figure 2 are used herein in place of the informal names.

Kilgore Flint Member and related strata

In northeastern Kentucky, Webb (1963) named the black flint bed of Ferm (1963) the "Kilgore flint." The unit contains marine fossils and locally overlies the Princess No. 5 coal bed. The name was used informally by Ferm et al. (1971) and herein is formalized as the Kilgore Flint Member of the Breathitt Formation. Type section for the Kilgore (Fig. 3) is here designated in the roadcut on the northwestern exit ramp at the intersection of U.S. Highway 60 and U.S. Interstate Highway I-64 near Coalton in the Rush 7.5-min quadrangle, about 1.6 km north of Kilgore, Kentucky, for which the member is named.

The member crops out in only a small area of about 41 km^2 in northeastern Kentucky (Fig. 4); it has a maximum thickness of about 3.7 m and consists of chert and silicified siltstone, both of which are fossiliferous (Ferm, 1963; Whaley, 1969). However, the member is associated locally in Kentucky and Ohio with some silicification of the underlying coal bed (Ferm, 1963; Webb, 1963) and, in Kentucky, with a broad,

Figure 3. Type section of the Kilgore Flint Member of the Breathitt Formation on the northwest exit ramp at the intersection of U.S. Highway 60 and U.S. Interstate I-64 near Coalton, Boyd County, Kentucky. The Kilgore Flint Member is the light colored unit just above top of car. The equivalent of the Obryan Member of the Breathitt Formation is reported by Ferm et al. (1971) to be a thin fossiliferous shale just above coal bed in middle part of outcrop and below the thin, ledgy siltstone beds. Princess No. 6 coal bed is at top of Hitchins clay unit at top of outcrop.

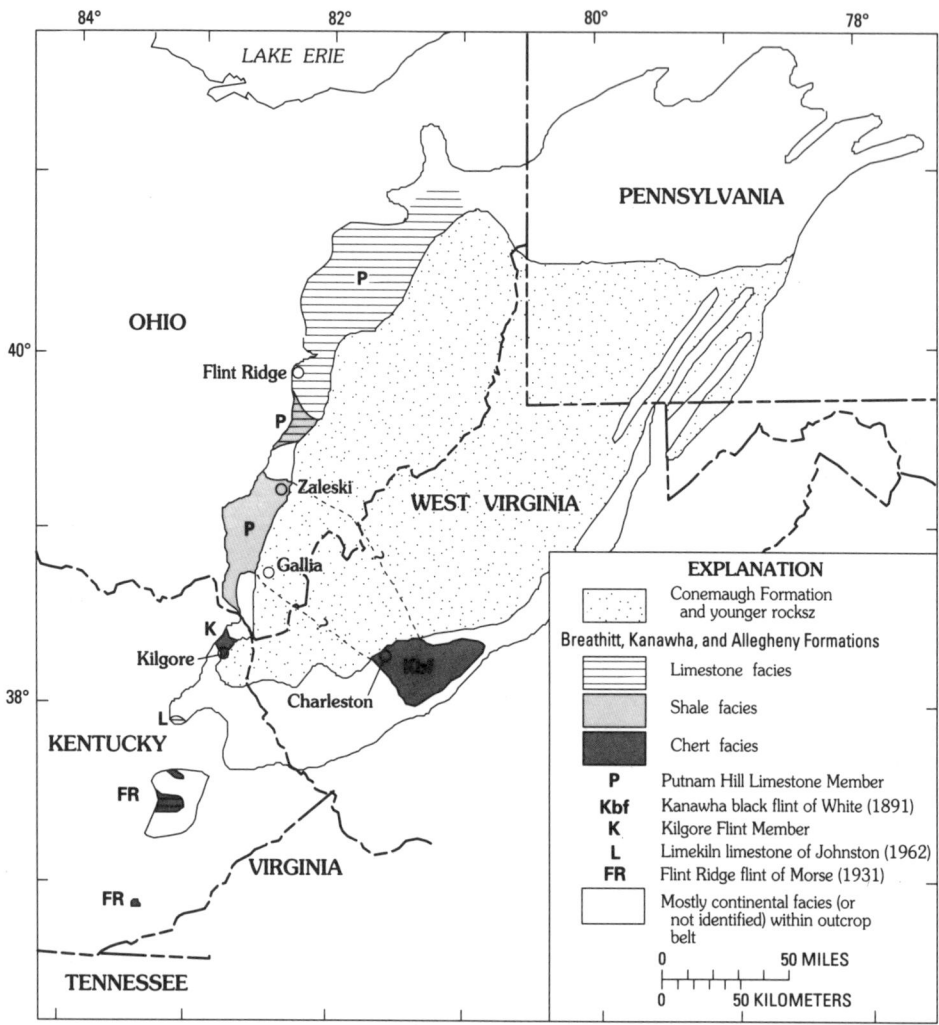

Figure 4. Generalized distribution of facies in outcrops of the Kilgore Flint Member of the Breathitt Formation and equivalent marine units in Kentucky, Ohio, and West Virginia. Putnam Hill Limestone Member of the Allegheny Formation has not been identified separately in northernmost Ohio or in Pennsylvania and may be included with strata of the Vanport Limestone Member of the Allegheny Formation. Data sources concerning distribution: for the Kilgore Flint Member and equivalent strata in Kentucky, U.S. Geological Survey geologic quadrangle maps; for the Putnam Hill Limestone Member of the Allegheny Formation in Ohio, county and regional reports of the Ohio Geological Survey; for the Kanawha black flint of White (1891) in West Virginia, Reppert (1979).

thin zone of silicified siltstone containing plant fossils that occurs just below or at about the same horizon as the Kilgore (see Carlson, 1971). These silicic rocks extend discontinuously as much as 40 km south and southeast of the type section of the marine facies of the Kilgore in Kentucky (see, for example, Ward, 1978). Additionally, petrified wood and silicified charcoal have been reported by Outerbridge (1964) from the equivalent horizon in areas as much as 70 km south of Kilgore, Ky.

Correlation of the Kilgore Flint Member with the informal Flint Ridge flint of Morse (1931) in Kentucky, first proposed by Wanless (see Fig. 2), here follows the correlation of Rice and Smith (1980), which was based on analyses of the palynology of coal beds from above and below the two units (R. M. Kosanke, written communication, 1976) as well as a review of the physical stratigraphy of all pertinent 7.5-min geologic quadrangle (GQ) maps in the intervening area. Kosanke indicated that a 15-cm-thick core sample of the first coal about 6 m above the main body of the Flint Ridge flint (see Fig. 4 for location) contained both *Schopfites* and *Thymospora pseudothiessenii*; these palynomorphs first occur stratigraphically in the coal beds between the Kilgore Flint Member and the younger Obryan Member in northeastern Kentucky. On the basis of these palynomorphs and the occurrence of *Zoster-*

osporites triangularis in the underclay of the coal bed (first described in the seatrock of the Princess No.5B coal bed), Kosanke suggested that the coal bed was equivalent to the Princess No. 5B (see Fig. 13). Additionally, palynomorphs from a 20-cm-thick rider coal bed from the highwall of a strip bench of the uppermost Skyline coal bed (a correlative of the Princess No. 5), about 9 m below the Flint Ridge flint included none of the above palynomorphs but others more representative of the Princess No. 5 coal bed. Thus, the palynomorphs of these two coal beds appear to bracket the Flint Ridge flint in the same way as those of the Princess No. 5 and 5B coal beds bracket the Kilgore Flint Member in northeastern Kentucky.

Figure 4 shows the discontinuous nature of the outcrop areas of the Kilgore Flint Member and related marine units in the central Appalachian basin. In many of the areas in the southern part of the basin, the units were either not deposited (most probable) or subsequently were eroded. Marine faunas of the equivalent units in Kentucky, as well as the Kanawha black flint of White (1891) in West Virginia, are restricted ecologically and apparently do not contain fusulinids or biostratigraphically diagnostic macrofauna. The main body of the Flint Ridge flint in Kentucky is in a narrow, west-trending basin about 8 km wide that, like the basin of the Kanawha black flint, may be controlled in part by movement of fault blocks within the east-northeast–trending Rome trough (McGuire and Howell, 1963), shown in Figure 1. The maximum thickness of the Flint Ridge flint is about 9 m, and it consists of partly silicified fossiliferous limestone, which thins laterally to generally less than 0.3 m of sparsely fossiliferous chert (Hinricks, 1978; Rice, 1975). The Kanawha black flint is probably an estuarine or shallow bay deposit, which is as much as 3 m thick and grades laterally to a thin orbiculoid shale facies that also is silicified locally (Cavaroc and Ferm, 1968; Reppert, 1979).

Webb (1963) mapped the Kilgore Flint Member in Kentucky and correlated it with a fossiliferous shale and siltstone in Ohio that overlies the Newland coal bed (see lower part of Fig. 2), which he mistakenly called Brookville or Tionesta after Stout (1916, 1927). Webb correlated the Newland (Brookville) with the Princess No. 5 coal bed. The marine unit in Ohio contains an abundant fauna (see Morningstar, 1922) in the area of and south of Zaleski, Ohio; its composition is 30 to 50% $CaCO_3$, mostly from fossil shells (Stout, 1927). Although Stout (see Morningstar, 1922) initially called this unit the McArthur Member of the Pottsville Formation, he later correctly indicated that it was the Putnam Hill Limestone Member of the Allegheny Formation (Stout, 1927, p. 170). The unit is also highly fossiliferous where it goes below drainage about 5 km west-northwest of Gallia (Stout, 1916) and between that area and Zaleski, Ohio (see Fig. 4). Although Webb (1963) indicated that the shale facies of the Putnam Hill in southern Ohio was divided into a western facies containing calcareous brachiopods and an eastern facies containing phosphatic brachiopods, our reconnaissance indicates that the calcareous brachiopod facies dominates in the region south and southeast of Zaleski. Given our correlation of the Kanawha black flint of West Virginia with the Putnam Hill Limestone Member, we project this region to be the mouth of the bay in which the Kanawha black flint was deposited (see Fig. 4).

The Putnam Hill Limestone Member is best developed and and contains fusulinids in north-central Ohio, where it is a limestone bed that ranges from less than 1 m to as much as 4 m thick. In northeastern Ohio, it apparently merges with the overlying Vanport Limestone Member about 50 km west of the Ohio-Pennsylvania state line (see Fig. 4) (DeLong and White, 1963; see also Zimmerman, 1966; Ferm, 1970) and is not identified separately to the east in Ohio or in Pennsylvania (Lamborn, 1954). The "Clarion" coal bed of Stout (1918), which overlies the Putnam Hill and underlies the Vanport in central Ohio, pinches out eastward and may have no coal bed correlative in Pennsylvania. Similarly, the Ogan coal bed, which overlies the Putnam Hill fossiliferous shale in southern Ohio, pinches out south of Zaleski (Stout, p. 175, 1927; Webb, 1963) near the pinchout of the overlying Zaleski Flint Member (see Figs. 5, 6, 13). The Ogan coal bed probably has no correlative in Kentucky or West Virginia.

The informal Limekiln limestone of Johnston (1962) in northeastern Kentucky (Fig. 4) contains the same species of conodonts found in the Putnam Hill and Vanport Limestone Members in Ohio (B. R. Wardlaw and R. G. Stamm, written communication, 1993). Because the Putnam Hill and Vanport cannot be differentiated on the basis of conodonts at the present time (see Fig. 7), the Limekiln is here correlated with the Kilgore Flint Member and the Flint Ridge flint, which are correlatives of the Putnam Hill. Unfortunately, we have not been able to make conodont separations from cherty strata to make a more direct comparison between the Limekiln and the Kilgore and Flint Ridge marine units.

Vanport Limestone Member and related strata

Stratigraphic data for the lower part of the Allegheny Formation are sparse and inconclusive in the area between Zaleski and Flint Ridge, Ohio. However, on the basis of stratigraphic position, we correlate the Zaleski Flint Member with the fusulinid-bearing cherts of Shawnee and Flint Ridge, Ohio (Fig. 5). These cherts are assigned to the Vanport Limestone Member of the Allegheny Formation, as it has been mapped in central and northeastern Ohio (Stout, 1918; Flint, 1951). As indicated above, this unit appears to merge with the underlying Putnam Hill Limestone Member in northeastern Ohio and western Pennsylvania. In southern Ohio, a similar merger of the Putnam Hill with the overlying Zaleski Flint Member occurs locally where the Ogan coal pinches out southwest of Hamden (Fig. 1) and where Stout (1916, p. 180) stated the flint "appears to represent the final deposit" of the fossiliferous shales (the Putnam Hill) overlying the "Tionesta" (Newland) coal bed. Thus, the Vanport Limestone Member in most of Pennsylvania is probably a single, uninterrupted marine-bay

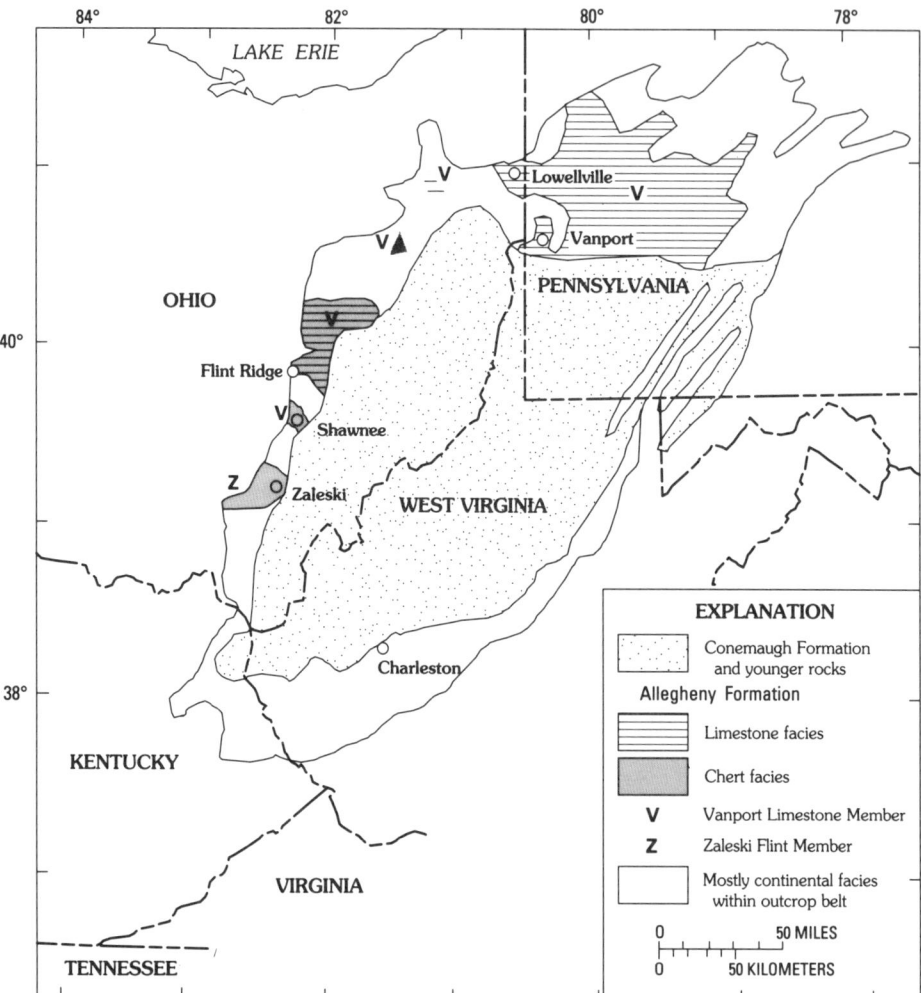

Figure 5. Generalized distribution of facies in outcrops of the Vanport Limestone Member and Zaleski Flint Member of the Allegheny Formation in Ohio and Pennsylvania. Ohio data from county and regional reports of the Ohio Geological Survey. Details of distribution for the Vanport in western Pennsylvania from Williams and Ferm (1964) and Williams and Bragonier (1974).

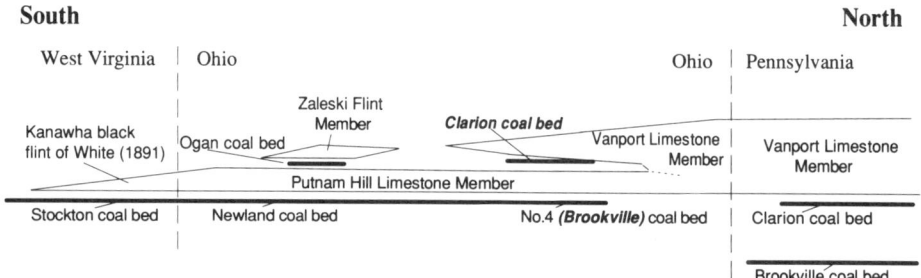

Figure 6. Stratigraphic relations of the Putnam Hill Limestone Member, Vanport Limestone Member, and the Zaleski Flint Member of the Allegheny Formation and related coal beds of Pennsylvania and Ohio with the Kanawha black flint of White (1891) and Stockton coal bed of West Virginia showing the misuse of the Clarion and Brookville (in bold italics) coal-bed names in Ohio.

deposit, whereas an earlier transgression (represented by the Putnam Hill Limestone Member, Kilgore Flint Member, Limekiln limestone, Flint Ridge flint, and Kanawha black flint) resulted in deposits that extended far south into eastern Kentucky and West Virginia (see Figs. 4, 6). That transgression was terminated by a major regressive event, which, in Ohio, included deposition of the Clarion and Ogan coal beds and associated clastic sediments (this regressive wedge corresponds in part to the "Clarion α" clastic wedge of Ferm, 1970). A later, more limited Vanport transgression in Ohio is represented by the Vanport Limestone Member of which the Zaleski Flint Member is the southernmost deposit; correlatives of this unit probably do not occur in eastern Kentucky or West Virginia because of nondeposition.

Preliminary analyses of conodonts in the Putnam Hill and Vanport Limestone Members in Ohio indicate that they have a

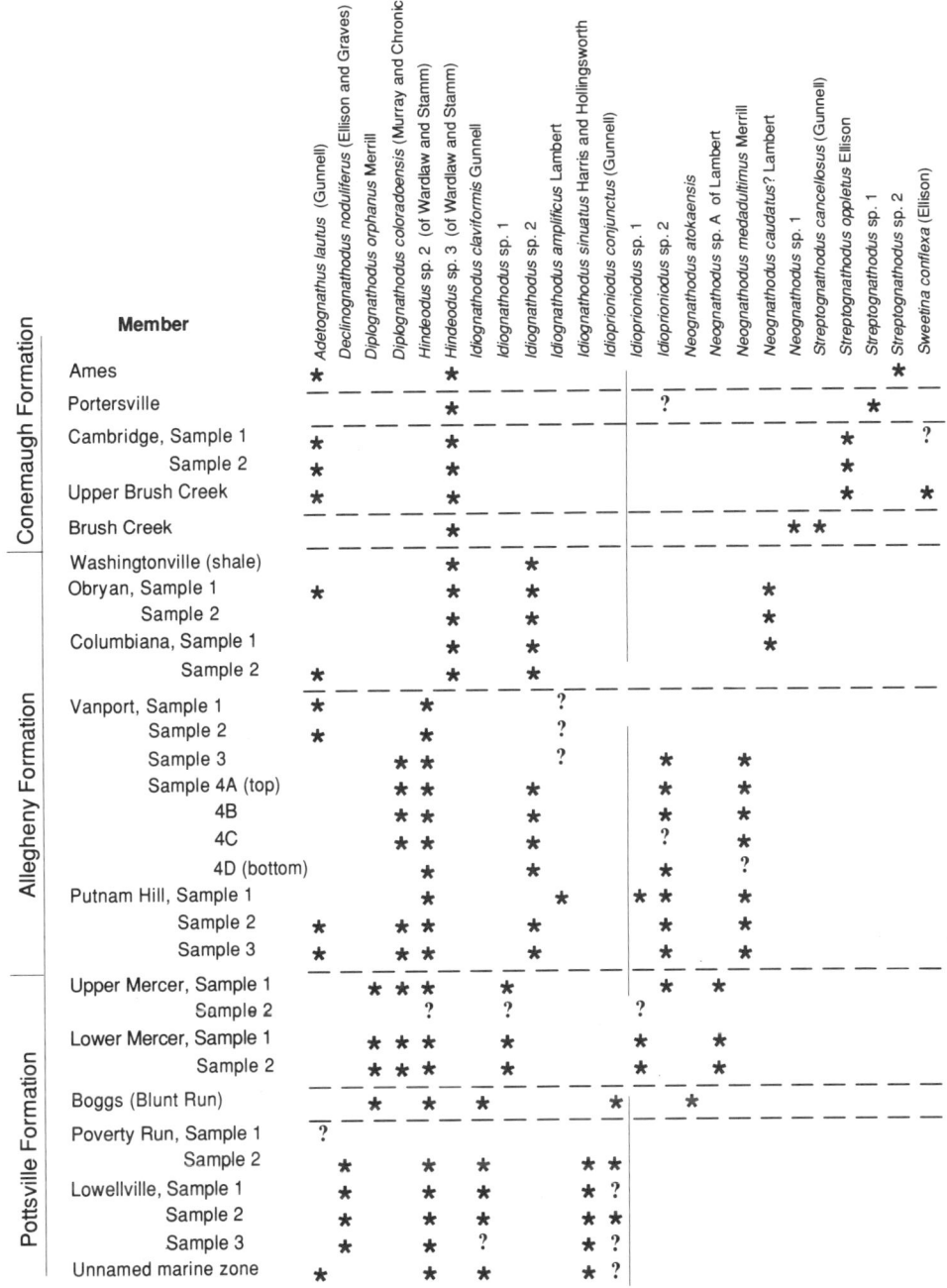

Figure 7. Preliminary analyses of conodont occurrences in 16 formal and informal marine units collected from Pennsylvanian strata in Ohio and northeastern Kentucky. All lithologies are limestone unless otherwise indicated. Dashed lines denote major changes in fauna. Question mark symbol indicates identification uncertain. Data from Wardlaw et al. (1993).

similar faunal variability (see Fig. 7) and may represent parts of the same marine unit (Wardlaw et al., 1993). The close temporal association of the Putnam Hill and Vanport in Ohio and the Vanport in Pennsylvania is indicated by the occurrence of *Beedeina leei* (Skinner) (see localities 79 and 87 of Douglass, 1987) in the Putnam Hill of Ohio and the Vanport (Mahoning limestone of Rogers, 1858) of Pennsylvania. Additionally, *B. carmani* (Thompson) was originally collected from the Vanport in Ohio (see locality 75 of Douglass, 1987) and was identified also in the Putnam Hill (see locality 71 of Douglass, 1987) (this locality 71 was inadvertently identified earlier as Vanport but now is recorded as the Putnam Hill in the records of the Ohio State Division of Geological Survey). In addition to the two fusulinids listed above, the Vanport of Ohio is also characterized by the fusulinid *Wedekindellina euthysepta* (Henbest). Douglass (1987) placed the boundary between the Atokan and the Desmoinesian Provincial Series at the base of the Putnam Hill Limestone Member on the basis of the occurrence of *B. leei* in that unit. For stratigraphic ranges of these fusulinids, see Douglass (1987, Fig. 5).

Figure 5 shows that the western facies of the Vanport Limestone Member as described in Ohio is cherty. In northern Ohio, the member is discontinuous and covered by glacial deposits in some of the area between Flint Ridge and Lowellville; in that area, it is reported to be locally cherty and as much as 3 m thick in isolated hilltop outcrops (DeLong and White, 1963). In the vicinity of Lowellville, the Vanport is as much as 6 m thick (Lamborn, 1951), and, eastward into Pennsylvania, the limestone thickens locally to more than 6 m (Williams and Ferm, 1964).

Obryan Member and related strata

The name Obryan Member is here given to a marine limestone unit of the Breathitt Formation variously identified in northeastern Kentucky as the Vanport Limestone of White (1878), the Vanport Limestone Member of the Breathitt Formation, or the Vanport Limestone Member as used by Phalen (1912). It is identical to the unit referred to in southeastern Ohio as the Vanport or "Ferriferous" limestone member of the Allegheny Formation (Stout, 1916), and, in this report, the name Obryan Member is extended into Ohio as a member of the Allegheny Formation to the vicinity of Zaleski (Fig. 8). The type locality for the Obryan Member is along a power line that parallels U.S. Highway 23 about 1.6 km east of the Obryan Cemetery, for which the member is named, and just north of Bench Mark 544 in the Ironton 7.5-min quadrangle, Greenup County, Kentucky, about 4.3 km northwest of Ashland, Kentucky (locality 61 of Douglass, 1987). The member lies about 7.6 m below the Princess No. 6 coal bed (see Dobrovolny et al., 1966) and is exposed for about 300 m at the top of the east-facing roadcut of U.S. Highway 23. A readily accessible reference section is identified by Phalen (1912) as the Vanport limestone, just above and about 6 m northeast of U.S. Highway 52 in Ohio opposite Ashland, Kentucky (locality 62 of Douglass, 1987).

At its type locality, the Obryan Member is as much as 1.8 m thick but pinches out northward along the highway roadcut. The member consists of an irregularly but massively bedded, medium gray to brown bioclastic limestone that has a micritic matrix; the upper 15 cm of the member is ferruginous and weathers moderate bright red. The Obryan is commonly very fossiliferous and contains whole and broken brachiopods (productids, spiriferids, and chonetids), rostrospiraceans, and pectens, gastropods (Webb, 1963), conodonts, and the fusulinid *Beedeina ashlandensis* Douglass. In many places along the outcrop belt, the Obryan has been recrystallized and locally silicified; where fusulinids are found, they are abundant in the lower part of the member. At the type locality, the Obryan overlies a rooted, medium gray, carbonaceous, silty to sandy claystones and is overlain by light gray rooted claystone. In the Ashland area and southward, the Obryan occurs in discontinuous lenses, generally less than 1 m thick (Fig. 9). In some areas to the west and south, the member occurs as weathered, discontinuous, fossiliferous nodules within the Hitchins clay bed (Sheppard and Ferm, 1962). Hilltop exposures of clay beds (including flint clay) associated with deposition of the Obryan Member extend at least 14 km southeast of the most southerly of the limestone deposits in northeastern Kentucky (Jenkins, 1966) (Fig. 8). The clay unit has also been traced into West Virginia below the Princess No. 6 coal bed (Connor and Flores, 1978) and, where it can be identified, provides a useful marker for regional stratigraphic analyses in place of the Obryan.

The Obryan Member is relatively continuous from Ashland, Kentucky, to the area of Zaleski, Ohio (Fig. 8). The member has been extensively mined throughout that area and is being mined presently with overlying and underlying coal and clay beds. Samples of limestone from the base of a quarry of the Obryan near its northern limit of outcrop (NC sec. 36, R.17W., T.11N.) about 5.6 km east-southeast of McArthur, Ohio (Fig.8), also contain the diagnostic fusulinid *Beedeina ashlandensis* Douglass (R. C. Douglass, oral communication, 1989). In southern Ohio and northeastern Kentucky, this horizon was used as a datum for analyses of stratigraphic intervals that included the Kilgore and Zaleski Flint Members by both Webb (1963) and Cavaroc and Ferm (1968).

North of Zaleski to the vicinity of Flint Ridge, Ohio, the Obryan Member is absent or very thin and has not been identified, as shown in Figure 8; it may be eroded locally and replaced by sandstone (see, for example, Stout, 1927). On the basis of the fusulinid *Beedeina ashlandensis,* Douglass (1987, locality 78; see also Smyth, 1957, Fig. 11, locality 75) correlated the Obryan Member with the Columbiana Member of the Allegheny Formation in the Flint Ridge area of central Ohio. In the area of Flint Ridge, the Obryan equivalent (the Columbiana Member) mistakenly was referred to as the "Hamden" limestone by Stout (1918, 1927) and Flint (1951). The Co-

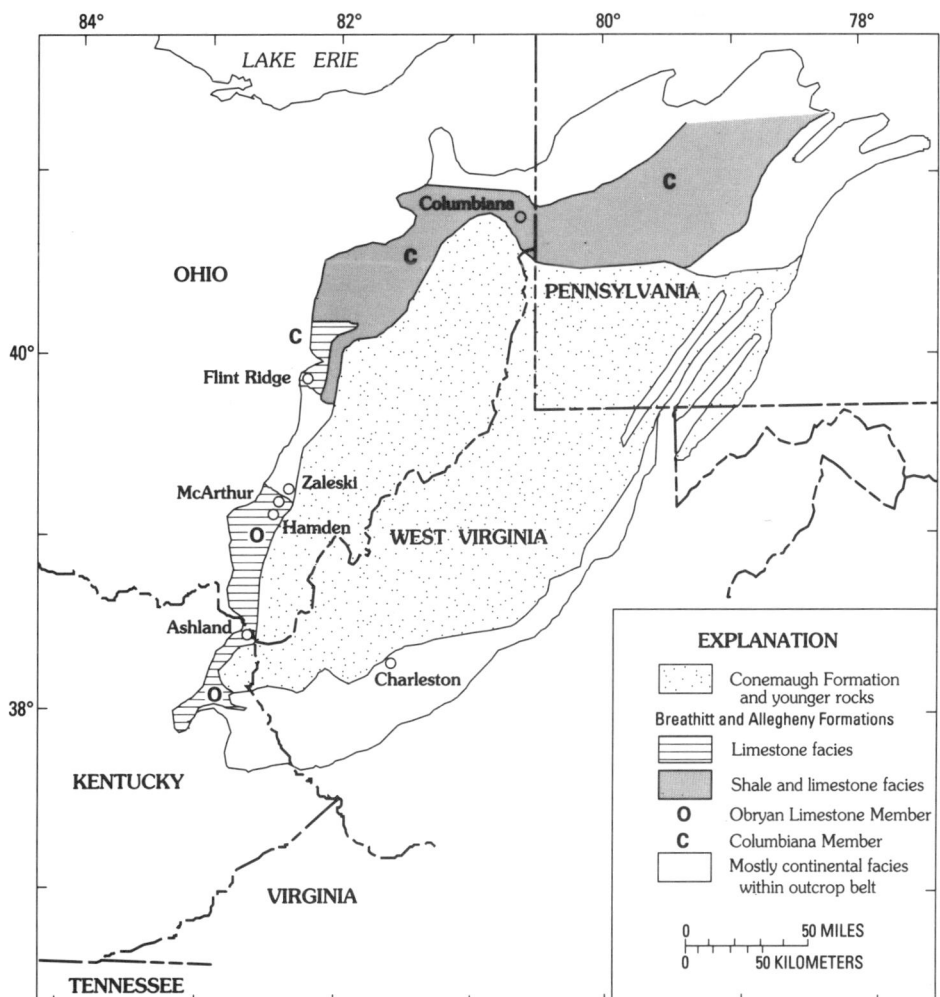

Figure 8. Generalized distribution of the Obryan Member of the Breathitt and Allegheny Formations and facies of the equivalent Columbiana Member of the Allegheny Formation in the central Appalachian basin. Data for distribution of facies in Pennsylvania and Ohio from county and regional reports of the Ohio Geological Survey and from Williams and Ferm (1964), modified by suggestions of J.C. Ferm (written communication, 1991).

lumbiana is widespread and well defined in the area east and northeast of Flint Ridge, Ohio, where it is commonly ferruginous and ranges from less than 0.3 m to as much as 1.2 m in thickness (Stout, 1918).

The Columbiana ("Hamden") Member is described by Stout in the Flint Ridge area to be nodular limestone, very irregular in shape and size, and to contain fusulinids and brachiopods. Nodular masses of the limestone are locally reported by Stout to be bedded in the body of the "Oak Hill" clay, similar to the description of the Obryan Member in northeastern Kentucky, where the Obryan is found locally as limestone and ironstone nodules within the Hitchins or Lawrence clay unit (Fig. 9). Northeast of Flint Ridge, the limestone facies of the member becomes less continuous, and the member is represented by an upward-coarsening marine sequence of shale and siltstone that commonly contains small nodules and thin lenses of limestone associated with a highly fossiliferous shale unit near its base (Stout and Lamborn, 1924). In places in northern Ohio and in Pennsylvania, the basal part of the Columbiana Member contains a dark gray fossiliferous argillaceous limestone bed as much as 0.6 m thick. The occurrence of the shale facies in Pennsylvania is described in detail by Ferm and Williams (1965).

In a study of conodonts of the marine units of the Allegheny Formation in Ohio and Pennsylvania, Merrill (1968) identified the Vanport as two distinct but closely related limestones or stratigraphic units, a younger "northern" Vanport and an older "southern" Vanport (the Obryan Member). As Figures 5 and 8 suggest, these two units are separated by an area in south-central Ohio where marine rocks and faunas do

Figure 9. Obryan Member of the Allegheny Formation in middle of roadcut in center of photograph pinches out into rooted clay shale toward right. Newland (No. 4) coal bed (at base of highway sign at left) is overlain by shale containing brackish water fossils that is equivalent to the Putnam Hill Limestone Member. Roadcut just southeast of Ironton, Ohio, on U.S. Highway 52.

not appear to be present at these stratigraphic horizons. A more recent analysis of the conodonts of the limestones of the Allegheny Formation in Ohio by Wardlaw et al. (1993) indicates that the Columbiana and Obryan Members appear to be correlatives. Four samples from the Columbiana and Obryan Members (two widely spaced samples from each) yielded new species of *Hindeodus* as well as a species of *Neognathodus* different from that of the underlying Vanport Limestone Member (Fig. 7). A single sample of the overlying Washingtonville Member of the Allegheny Formation, a calcareous shale, yielded a poor fauna similar to that of the Columbiana and Obryan Members (Wardlaw et al., 1993). Sample 2 of the Columbiana Member (Fig. 7) was collected from a large, irregular (1.2 to 1.5 m in length) limestone nodule from the Flint Ridge area in central Ohio near Smyth's (1957, locality 75) sample, which contained *Beedeina ashlandensis* Douglass.

Type areas of both the Hamden ore and the Oak Hill clay are in southeastern Ohio, where the ore deposit appears to have had a limited local distribution in the vicinity of Hamden (Fig. 1). Stout (1927, p. 304) described it as being generally poorly represented by scattered nodules of impure ore at the base of the Oak Hill clay bed; both units generally directly overlie the No. 5 coal bed of the Oak Hill area (Fig. 1). Because the Obryan Member is correlated herein with the Columbiana Member, the underlying coal in southeastern Ohio, the No. 4A coal bed, is correlated with the Lower Kittanning coal bed, which underlies the Columbiana Member of central Ohio (see Fig. 13), as well as northeastern Ohio and western Pennsylvania. Thus, the No. 5 coal bed, which is above the Obryan Member in the Oak Hill area, may either be the Middle Kittanning or one bed of a zone of coal beds associated with the Middle Kittanning. Thus, the stratigraphic position of the Hamden ore is somewhat higher than the Columbiana Member, its presumed correlative.

Regional correlation of other marine units

As Figure 2 suggests, many of the early correlations of marine units from Ohio to northeastern Kentucky and West Virginia seem to have been made by matching named limestones from one region to the other. Most macrofossils are not well enough known to be stratigraphically diagnostic, and many of the regional correlations in the Appalachian basin were "best" guesses. The identification of fusulinids in the Stoney Fork Member of the Breathitt Formation in southeastern Kentucky (Douglass, 1987) provided, for the first time, a definitive paleontologic basis for the correlation of this part of the Breathitt with other Pennsylvanian sections in eastern North America. Where these strata are present in the northern part of the Appalachian basin in Pennsylvania and Ohio or in the Eastern Interior basin, equivalent sections are very thin and contain only a few thin coal beds. *Fusulinella pria* Douglass, which occurs in the Stoney Fork, belongs to a group of rapidly evolving fusulinids that are intermediate between *Profusulinella* and *Fusulinella*. *Fusulinella pria* is similar to two other intermediate fusulinid forms that occur in the Boggs Limestone Member in Ohio (Douglass, 1987).

Additionally, two samples of the Stoney Fork Member from eastern Kentucky contain the same conodont fauna, species for species, as two samples of the Boggs Limestone Member from Ohio (including the one sample identified in Fig. 7) (B. R. Wardlaw and R. G. Stamm, written communication, 1993). Unfortunately, outcrops of the Stoney Fork and the Boggs could not be identified any closer than about 160 km (see Fig. 10), but, on the basis of their fossil content, the members are regarded here as correlative and as possible deposits in two widely separated embayments of the same broad epicontinental sea.

The correlation of the Stoney Fork Member with the Kanawha black flint as suggested by Huddle and Englund (1966) seems unlikely, given the distribution and the nature of the lithology of the Stoney Fork Member and its correlative, the Boggs Limestone Member. As Figure 10 shows, the Stoney Fork Member does not appear to extend northeastward toward West Virginia. The related Boggs is a freshwater or brackish water deposit marked only by an iron ore irregularly bedded in shale in southern Ohio; in central Ohio, the limestone is generally less than 0.6 m thick and pinches out southeastward (Morningstar, 1922). In the westernmost hilltop outcrops in areas west of Ashland, Kentucky (Fig. 10), the "Main Block Ore," which occurs about at the position of the Stoney Fork Member (see Rice and Smith, 1980), is reported to be calcareous and to contain marine fossils (Whittington and Ferm, 1965). The discontinuous "Boggs" ore in southern Ohio is at about the same stratigraphic horizon as the Main Block Ore, and the two may be correlatives.

Magoffin, only the later of which contains the primitive non-fusiform fusulinid *Millerella* (Douglass, 1987).

Other suggested correlatives for the Mercer limestones have been the Kanawha black flint with the Lower Mercer limestone (Outerbridge, 1989) and the Upper Mercer limestone (White, 1908). But the flint facies of the Mercer limestones is generally limited to central and northern Ohio, whereas the proximal-continental siderite facies extends to the south and southeast, in the direction of the Kanawha black flint. Given the facies distribution of these three units, the correlation of either of the Mercer limestones with the Kanawha black flint seems unlikely. It seems doubtful that any correlatives of the Mercer limestones will be found in eastern Kentucky or West Virginia, except, possibly, as discontinuous and largely untraceable nodular iron ore beds.

Analyses of miospores from coal beds in the basal part of the Pottsville Formation near Jackson, Ohio, indicate that the earliest occurrence of the palynomorph *Radiizonates* sp. is in the Vandusen coal bed (Fig. 2), which underlies the Poverty Run Limestone Member of the Pottsville Formation (Rice et al., 1992, chart 1). Because the first occurrence of this palynomorph is in the Winifrede coal bed in West Virginia (Kosanke, 1988) and equivalents such as the Leatherwood coal bed in eastern Kentucky (R. M. Kosanke, written communication, 1968), we correlate the Poverty Run with an unnamed marine unit that locally overlies these coal beds and their equivalents in West Virginia and Kentucky. Several thin marine and brackish water shale beds occur below the Poverty Run Limestone Member in Ohio (Rice et al., 1992), but none are sufficiently well developed to have been named. The relation of these units to the Kendrick Shale Member and the Magoffin Member in eastern Kentucky (and equivalents in West Virginia) is unknown.

REGIONAL COAL-BED CORRELATIONS

In order to supplement the biostratigraphic control for this study, the palynomorph content of selected coal beds from northeastern Kentucky and southern Ohio was studied in relation to previous studies of coal beds in Ohio and Pennsylvania (Gray, 1967), West Virginia (Kosanke, 1984, 1988), and northeastern Kentucky (Kosanke, 1973). The Princess Nos. 5, 5A, and 5B coal beds were sampled in Kentucky at the type section of the Kilgore Flint Member (Fig. 3). Samples were also collected from the Newland, Winters, No. 4A, and No. 5 coal beds in the area generally between Zaleski and McArthur, Ohio. An additional sample was collected from the No. 4 coal bed (or "Brookville") in central Ohio, east of Flint Ridge. Most of the samples were collected from active strip pits (see Fig. 11); the Princess No. 5A and No. 5B coal beds and the No. 4 coal bed (Ohio) were generally thin (18, 38, and 18 cm, respectively) and from weathered outcrops. (See Appendix 1 for descriptions and sample locations.)

All coals sampled yielded abundant palynomorphs al-

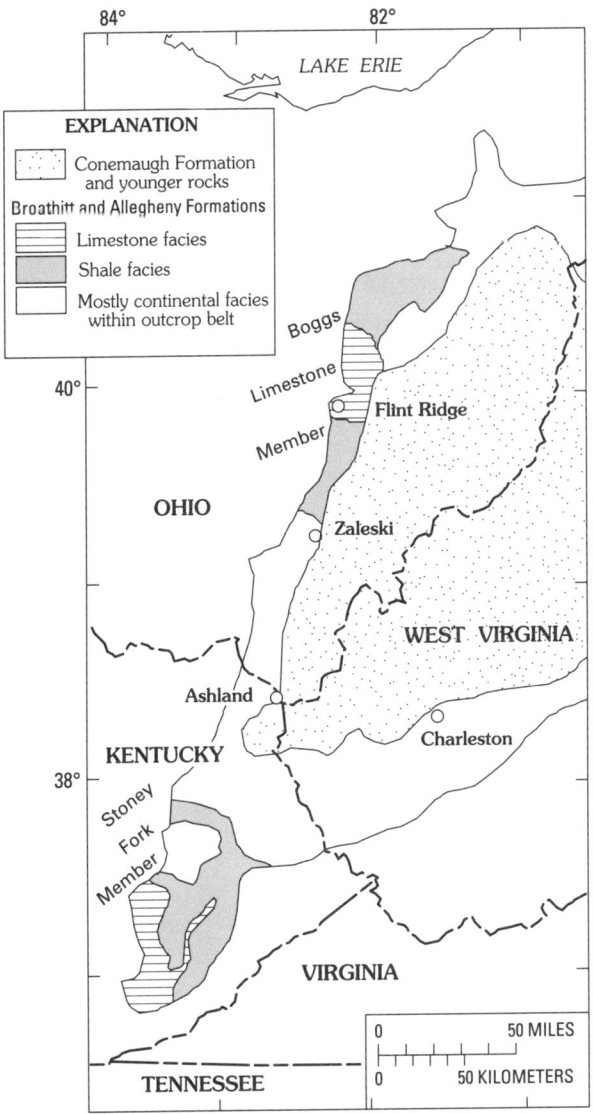

Figure 10. Generalized map of the distribution of the Stoney Fork Member of the Breathitt Formation in eastern Kentucky (data mainly from Ping and Rice, 1979) and its correlative, the Boggs Limestone Member of the Allegheny Formation in central Ohio (data mainly from Morningstar, 1922).

An early, commonly accepted set of correlations was that of the Kendrick Shale Member and Magoffin Member of eastern Kentucky and their equivalents, the Dingess and Winifrede Shale Members in West Virginia, with the Lower and Upper Mercer limestones of the Pottsville Formation in central Ohio, respectively (Wanless, 1939). Because the Mercer limestones are shown (lower half, Fig. 2) to overlie the Boggs and/or Stoney Fork stratigraphic horizon whereas the Kendrick and Magoffin underlie it, these units cannot be correlative. In addition, the middle and upper Atokan faunal assemblages of the Mercer limestones, as characterized by *Fusulinella iowensis*, clearly make these members younger than the Kendrick or

Figure 11. Strip pit of the Winters and No. 4A coal beds (not shown at base) about 5.5 km west-southwest of Zaleski, Ohio, showing where a sandstone-filled channel has replaced the Obryan Member of the Allegheny Formation at the northern limit of its distribution (see Fig. 8). Limestone nodules are found at the base of the massive sandstone at this locality. The bulldozer shown in the photo is discarding oxidized coal of the No. 5 coal bed that overlies the sandstone. The Winters and No. 4A coal beds are separated by 2 to 6 m of carbonaceous shale.

though some palynomorphs in the No. 4 and No. 5 coal beds of Ohio could have been better preserved. The abundances of selected palynomorphs of the coals studied in Ohio and northeastern Kentucky are shown in Table 1. Overall, the abundance data of the Ohio samples are similar to those of the Princess reserve district in that *Laevigatosporites* and *Lycospora* are numerically the most abundant genera followed by *Densosporites* spp. and *Torispora securis*. Figure 12 shows also a good correspondence between selected palynomorphs in correlative coal beds studied in Ohio and Kentucky. *Torispora securis* is perhaps the most significant spore for our study because the base of its range zone is in the No. 4, Newland, Princess No. 5, and Stockton coal beds, thereby supporting the correlation of the Putnam Hill Limestone Member, the Kilgore Flint Member, and the Kanawha black flint of White (1891).

Other range zones of significance in this part of the Allegheny Formation are the bases of the range zones *Thymospora pseudothiessenii* and *Schopfites* and the top of the *Radiizonates* range zone, shown in Figure 13. The start of the range zone of *T. pseudothiessenii* is in the top sample of the Winters coal bed and earlier had been identified in the underclay of the Princess No. 5B coal bed (Kosanke, 1973) (Fig. 13). As Figure 13 indicates, *T. pseudothiessenii* also is found locally in the Upper No. 5 Block coal bed in West Virginia. Preliminary analysis of the "Clarion" coal bed of central Ohio, showing that it contains both *T. pseudothiessenii* and *Schopfites* (C. Eble, oral communication, 1991) suggests that it is closely related to the overlying Kittanning coals. The position of the base of the *Schopfites* range zone shown in Figure 13 is determined by its first occurrence in the West Virginia section, the No. 6 Block coal bed. In addition to its occurrence in the Clarion coal bed of Ohio, *Schopfites* is also found in the Laurel coal bed in northeastern Kentucky (Fig. 13), and, as mentioned earlier, in the first coal above the Flint Ridge flint of Morse (1931).

The correlation of units in Figure 13 is in general supported by the analyses of palynomorphs from coal beds of Ohio, Kentucky, and West Virginia. There is some question concerning the nomenclature of coal beds of the Charleston Sandstone of West Virginia because of the complex splitting of the No. 5 Block coal bed into four or more distinct coal beds and the difficulty of identifying these in widely separated and poorly exposed sections. Therefore, it is not clear that the adjective "Lower" or "Upper" or even the number "5" or "6" always identifies the same coal bed (see Englund et al., 1979).

Another problem in correlation of coal beds is that some palynomorphs can be abundant in one maceration of a coal bed but absent in the overlying or underlying macerations (Kosanke, 1988). For example, *Radiizonates* is thought to have been produced by a herbaceous plant limited to growth in a specific environment; the palynomorph is notably absent in some samples of coal beds known to contain the spore. *Radiizonates* is reported from the Lower No. 5 Block coal bed of West Virginia, but apparently has its youngest occurrence in the Newland and Princess No. 5 coal beds in Ohio and Kentucky. Thus, the order of first occurrences in northeastern Kentucky and Ohio may differ from that in West Virginia because of the thinness of the coal beds and stratigraphic sections in Kentucky and Ohio, and also because of the differences of biofacies and depositional environments and the small percentages of stratigraphically important spores. Nevertheless, the clustering of limits of palynomorph range zones in the interval between the No. 4 coal bed of Ohio and base of the Columbiana Member is strong corroboration of the limestone correlations shown in the lower half of Figure 2 and in Figure 13 for the Allegheny Formation.

An early study by White (1900; see also White and Ashley, 1906) directly compared fossil plants from the coal beds of central West Virginia to those of the Allegheny River Valley in west-central Pennsylvania. He concluded that the proportion and range of the identical forms of flora collected from the roof shales of the Stockton coal bed of West Virginia suggested a correlation of that coal with the Clarion coal of Pennsylvania. He also stated that equivalents of the Kittanning coal beds of Pennsylvania were the coal beds found above the Kanawha black flint in West Virginia and that the Kanawha black flint was probably the "ferriferous" (Vanport) limestone of the Allegheny River Valley area in Pennsylvania. White's analysis appears to corroborate our findings that the Kanawha black flint, the Kilgore Flint Member, and the Putnam Hill Limestone Member may represent the lower part of the Vanport Limestone Member of the Allegheny River Valley in Pennsylvania and that the Stockton coal bed may be equivalent to the Clarion

TABLE 1. PERCENTAGE OF SELECTED PALYNOMORPH TAXA IN COALS FROM OHIO AND NORTHEASTERN KENTUCKY*

Taxa	Calamospora	Densosporites	Endosporites	Florinites	Laevigatosporites	Lycospora	Schopfites	Thymospora	Torispora	Triquitrites	Wilsonites
Ohio Coal Beds											
Newland (No. 4)	+	28.8	2.5	+	35.7	12.9	0	0	15.8	+	+
No. 4 (*Brookville*)	2.0	+	0	2.0	48.7	39.5	0	0	0	0	+
Winters	+	+	2.9	+	25.5	61.2	0	+	+	+	+
No. 4A	+	2.8	+	4.3	56.5	15.8	0	+	0	2.2	+
Lawrence	+	+	8.0	+	16.2	65.0	0	0	+	+	+
No. 5	1.7	+	+	+	28.2	51.7	+	2.5	3.6	+	+
No. 6 (Middle Kittanning)	1.0	+	4.0	1.0	40.0	26.0	+	13.0	2.0	2.0	+
NE Ky Coal Beds											
Princess No. 5	+	7.4	4.4	+	36.9	21.7	0	0	14.7	+	+
Princess No. 5A	+	7.0	+	9.0	46.0	9.0	0	0	+	6.0	5.5
Princess No. 5B	6.5	+	+	+	18.0	58.2	0	+	+	+	+
Princess No. 6	2.1	4.1	2.8	1.6	43.5	29.5	+	2.5	+	3.5	2.0
Princess No. 7	2.0	+	+	0	54.7	13.6	+	25.0	2.0	1.9	+

*The coal bed underlying the Putnam Hill Limestone Member of the Allegheny Formation is the Newland coal bed in southern Ohio and the No. 4 coal bed in central Ohio, where it is commonly mistakenly called Brookville (shown in italics). Data for the Princess Nos. 6 and 7 coal beds from Kosanke (1973); data for the No. 6 (Middle Kittanning) and the Lawrence coal beds from Gray (1967).
+ Indicates minor numerical importance.

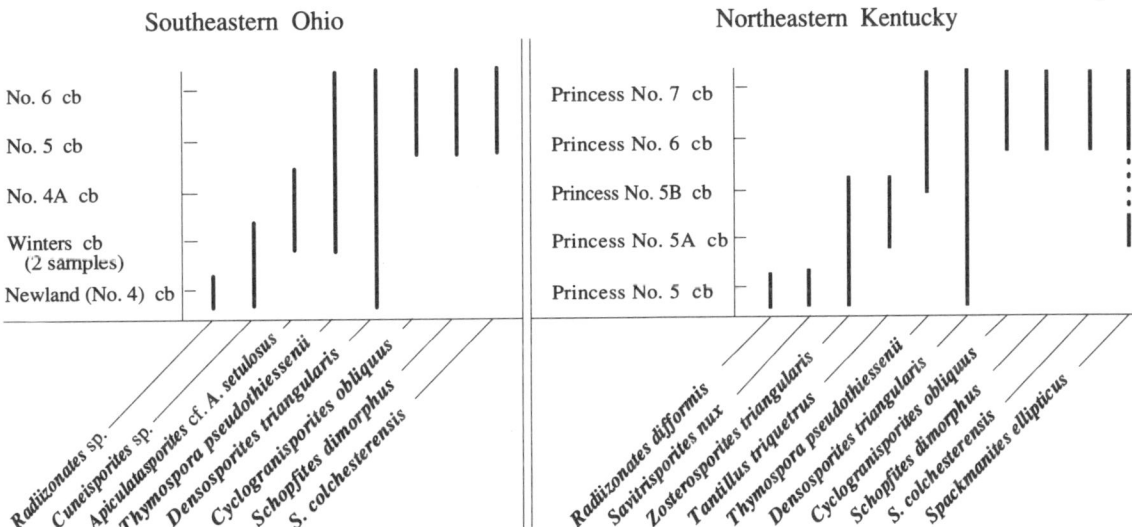

Figure 12. A comparison of selected palynomorphs in the Princess Nos. 5, 5A, 5B, 6, and 7 coal beds in northeastern Kentucky with those found in the Newland, Winters, and Nos. 4A and 5 coal beds collected from the area between McArthur and Zaleski, Ohio, and the No. 6 (Middle Kittanning) coal bed (described by Gray, 1967). Lines show the stratigraphic ranges of palynomorphs; dashed line indicates where palynomorph was not found in coal bed. cb = coal bed.

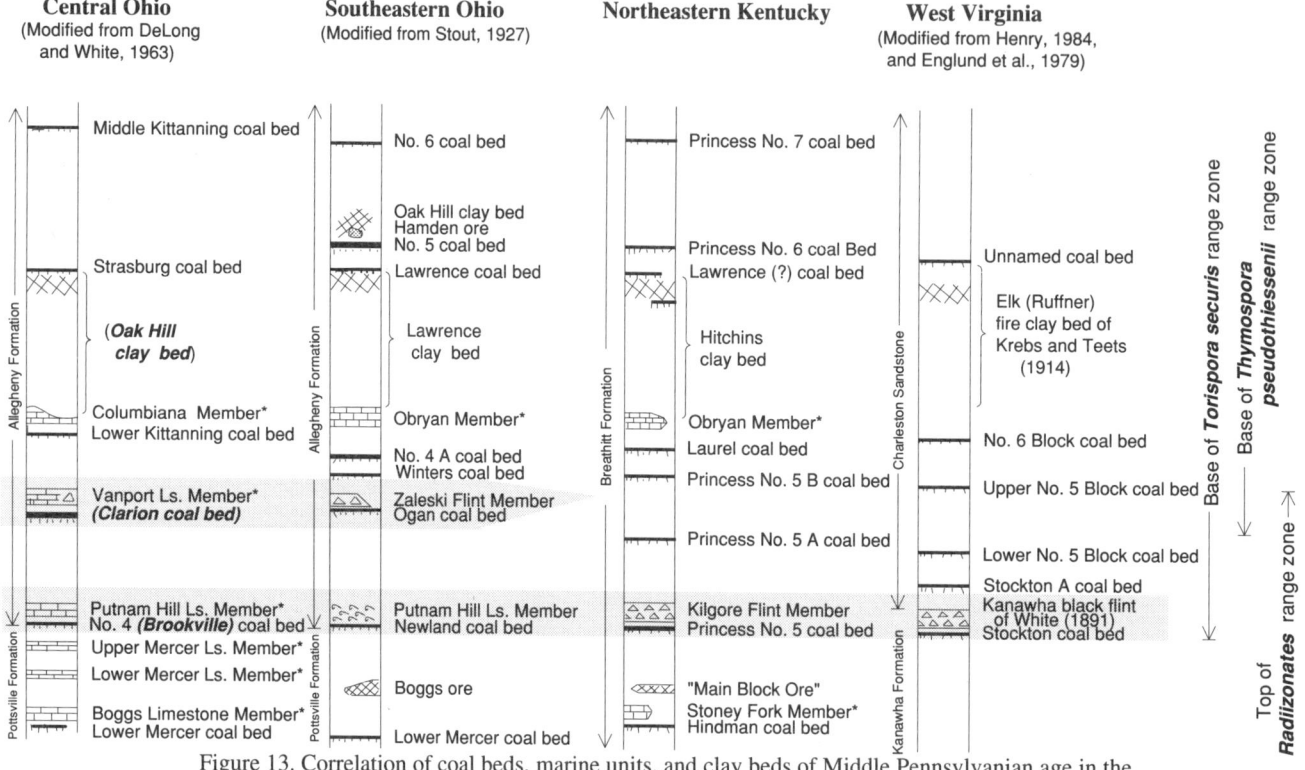

Figure 13. Correlation of coal beds, marine units, and clay beds of Middle Pennsylvanian age in the tristate area of Ohio, Kentucky, and West Virginia, based largely on physical stratigraphy and analyses of microspores, conodonts, and fusulinids. Palynomorph data in part from Kosanke (1973, 1984) and Gray (1967); fusulinid data from Douglass (1987); conodont data from Wardlaw et al. (1993) and Wardlaw and Stamm (written communication, 1993). Limestones with asterisks indicate both fusulinid and conodont data were used for identification and correlation. Palynology of the Lower Mercer and Ogan coal beds has not been studied. Names in italics are incorrectly but commonly used in areas shown. Tops and bases of range zones are shown as they occur in West Virginia, the thickest section; see text for variants in the Kentucky and Ohio sections. Crosshatching indicates flint clay; sandstone, siltstone, and shale are not shown. The tongues of the Putnam Hill Limestone and Vanport Limestone Members of the Allegheny Formation and Associated coal beds are shaded for comparison with Figure 6. Average interval thicknesses are, from left to right: central Ohio, base of Putnam Hill Limestone Member to base of Strasburg coal bed: 32 m; southeastern Ohio, base of Putnam Hill to base of No. 5 coal bed: 27 m; northeastern Kentucky, base of Kilgore Flint Member to base of Princess No. 6 coal bed: 19 m; West Virginia, base of Kanawha black flint to base of unnamed coal bed: 52 m. Ls. = limestone.

coal bed of Pennsylvania. Figure 6 illustrates these relationships as well as the probable misapplication of such coal-bed names as "Clarion," "Brookville," and even "Vanport" for that marine unit in central Ohio; these inconsistencies also have been suggested by Zimmerman (1966) and Ferm (1970).

CORRELATION OF CLAY UNITS

The commercial clay units of the upper Middle Pennsylvanian are generally not well defined as to their upper or lower contacts or even their content. They commonly contain one or more layers or discontinuous lenses of nonplastic flint clay (a conchoidally fracturing refractory clay composed mostly of well-crystallized kaolinite) associated with a thicker body of plastic clay. The Hitchins and Lawrence clay units of northeastern Kentucky and southeastern Ohio commonly have a flint clay bed that ranges from a few centimeters to less than 1 m thick at or near the top of each unit. The total thickness of the clay bodies (including plastic and flint clay) may be more than 3 m (Stout et al., 1931). Both the Hitchins and Lawrence clay "beds" are complex deposits that may be composed of at least five different stratigraphic units that are locally difficult to distinguish. They are from top to bottom: (1) the seatrock of the Princess No. 6 coal bed or equivalents, (2) the seatrock of the Lawrence coal bed, (3) a flint clay bed or zone containing several flint clay beds, (4) a seatrock of one or more unnamed coal beds below the flint clay bed, and (5) the seatrock of the coal underlying the Obryan Member.

In northeastern Kentucky at the type section of the Kilgore Flint Member, the flint clay bed of the Hitchins clay is enclosed in coal and is as much as 30 cm thick at the eastern end of the outcrop, whereas at the western end of the outcrop (about 100 m distance), it is only a few centimeters thick and is both overlain and underlain by shale. The flint clay bed represents a tonstein containing a suite of accessory minerals in which electron microprobe analyses have identified such volcanic clasts as betaform quartz, sanidine, biotite, and zircon. Also included are many nonvolcanic detrital quartz grains (20 to 25%), which are evidence that the tonstein is a secondary deposit (H. E. Belkin, oral communication, 1991). In the Ashland, Kentucky, area, the flint clay bed is thicker and more "flinty" in places where it is directly overlain and underlain by coal beds, an indication that these swampy locations were optimum (but not necessary) for deposition and preservation of the flint clay bed. The presence of other flint clay beds and partings in the coal beds of this part of the section reported by Dobrovolny et al. (1963) and Carlson (1965) suggest that volcanic ash may have been an important component of the clay deposits.

The widespread and discontinuous nature of the Lawrence and Hitchins clay units and of individual beds within these units shows that the deposits were at least locally subject to episodes of erosion and weathering; most deposits show evidence of rooting and contain thin coal beds. The close association of the clay units with the Obryan Member suggests that their deposition was in nearshore environments and that their subsequent development probably was controlled by fluctuations of a falling sea level across a broad coastal plain. Stout et al. (1931) noted that the Lawrence clay is distinct and conspicuous in a wide area in Ohio but that commercial deposits are only a few square kilometers in extent and commonly are replaced by shale or sandstone within a distance of a few hundred meters. Locally, the entire interval may be replaced by channel sandstone. Nevertheless, like other Pennsylvanian tonsteins and strata containing flint clays, these clay units belong to a deposit that probably extends across much of the northern part of the Pennsylvanian Appalachian basin. On the basis of the palynology of the coal beds of our study area and the correlation of marine units (see Fig. 13), we correlate the Hitchins clay bed of Kentucky and the Lawrence clay bed of southeastern Ohio with the Oak Hill clay bed of central Ohio and the Elk fire clay bed of West Virginia.

DISCUSSION

The Kilgore Flint Member and its equivalents in Kentucky and West Virginia seem to represent a series of parallel southeast-trending deposits that resulted from encroachment of the sea across a very broad Middle Pennsylvanian coastal plain into the interfluvial area of northwest-trending fluvial systems (Fig. 4). Perhaps some of the smaller deposits such as the Kilgore Flint Member itself or the Flint Ridge flint of Morse (1931) occupied old abandoned river channels or valleys. Silicification of many of these deposits probably occurred at a late state of deposition, or even was postdepositional, because the thicker deposits like the Flint Ridge flint in Kentucky, which is mostly limestone, show that silica replacement proceeded from the top down. The rapid increase of silicious sponges in the southern and southwestern parts of the epicontinental sea, as the abundance of sponge spicules in the cherts of these deposits indicates (Cavaroc and Ferm, 1968; Danilchik, 1977), may have been related to a local environmental condition, such as lack of turbidity in restricted slack-water areas that resulted from a late-stage contraction of the sea. The broad area of silicified shales, siltstones, coal beds, and plant materials, associated in particular with the deposits at the horizon of this marine unit in eastern Kentucky, further suggests the existence of a sea larger than what the marine deposits alone imply.

Hydrogen- and oxygen-isotope analyses of marine nodular cherts by Knauth and Epstein (1976) showed that most samples were formed in waters that had a major meteoric-water component. Knauth (1979) concluded that silicification of coastal sediments could occur as a consequence of mixing of meteoric and marine waters where shoreline sediments prograde over lagoonal or other offshore deposits, especially in high-pH evaporite conditions. Cavaroc and Ferm (1968) also suggested that siliceous spiculites were shoreline indicators in deltaic sequences. Although the distribution of the silicification may have been controlled by the position of distributaries that prograded across the coastal plain after deposition of the marine unit, the silicified marine deposits associated with the Kilgore Flint Member and the Putnam Hill Limestone Member (and their equivalent, the Kanawha black flint) appear to have occupied mainly the upper parts of ancient inlets and bays. The southwestern facies of the Vanport Limestone Member of central Ohio and the Zaleski Flint Member farther south (see Fig. 5) seem to be similarly oriented in isolated west- or northwest-trending deposits that pinch out eastward; deep drilling for coal resources east of these outcrop areas has revealed no marine chert beds in that part of the section (Couchot et al., 1980).

The stratigraphic sections in Figure 13 distinctly show the progressive change from more marine conditions and many limestone units in central Ohio to a more continental style of deposition and a single marine chert in central West Virginia. The southward thinning of comparable sections across Ohio to northeastern Kentucky indicates the presence of a basin on the stable part of the North American craton in Ohio that was separate from the Appalachian basin to the southeast at this time. Thus, the thickness of the interval between the base of the Putnam Hill and equivalents to the base of the Strasburg coal bed and equivalents in Figure 13 decreases from north to south from 32 to 27 m in Ohio, to 19 m in northeastern Kentucky. The comparable interval in the Appalachian basin in central West Virginia is 52 m thick. The practical boundary between these two basins would seem to be the northern edge of the Rome trough shown in Figure 1; however, a more general analysis of regional stratigraphy is necessary to corroborate this suggestion.

CONCLUSIONS

The greatest source of stratigraphic confusion in Pennsylvanian sections arises from attempts to construct a single nomenclature to be applied across very large areas (generally defined by state boundaries) that undoubtedly contained a variety of depositional environments at any one time. Lithic units commonly are discontinuous, and their physical appearance may depend on local depositional conditions that may be duplicated at other times in other areas. Miscorrelations of coal beds are common, given their discontinuity and tendency to split, and marine units can be so similar in physical and faunal appearance that they are readily mistaken one for another. Our study shows that it is easier and more reasonable to correlate gross lithologic units between West Virginia, Kentucky, and Ohio than it is to deal with a nomenclature not closely tied to type sections or localities. Many coal beds and marine units in southern Ohio are called by names whose type localities are in Pennsylvania, and they have been miscorrelated. It is further evident that some units have been miscorrelated between central and southern Ohio. Although we have attempted to use local names for coal beds in southeastern Ohio, the names Nos. 4, 4A, 5 and 6, which were employed by Stout (1916) in southeastern Ohio, are used in this report where a better name is not available. This usage of numbered coal-bed names is popular among the coal companies of southeastern Ohio and has been extended into other parts of Ohio. However, where the key reference unit—the Obryan Member—is not present, the coal beds may well be misidentified.

Clearly, the stratigraphy of the Pottsville and Allegheny Formations in southern Ohio needs to be reexamined by using a variety of paleontologic tools as well as physical stratigraphy. The same can be said for those strata in other parts of the outcrop belt in Ohio and Pennsylvania. Rather than isolated sampling, however, studies need to be made of sequences of strata that contain major marker beds for local and regional reference and comparison. The fusulinid, conodont, and palynologic data brought together in this report are not exhaustive; other biotic elements need to be included in the biostratigraphic framework as well. We hope that the data and correlations shown in Figures 2 and 13 will provide the stimulus for further examination of the stratigraphy of the tristate area of Ohio, Kentucky, and West Virginia, and that those studies will provide a greater insight into the tectonic and sedimentary history of the Pennsylvanian rocks of this part of the Appalachian basin.

ACKNOWLEDGMENTS

Of the many people who reviewed versions of this chapter, we give special thanks to Harold B. Rollins and John C. Ferm. In particular, the paper owes much to the penetrating critique of John Ferm and his experience and knowledge of the Pennsylvanian stratigraphy of Ohio, Pennsylvania, and Kentucky.

APPENDIX 1

Description of coal samples and sample locations

No. 4 coal bed (18 cm thick): Collected just below the Putnam Hill Limestone Member about 3 km west of the center of Zanesville, Ohio, on northern side of Interstate I-70, just east of overpass (Douglass, 1987, locality 71).

Newland coal bed (38 cm thick) and *Winters* coal bed (56 cm thick): Collected at elevations of about 223 and 230 m, respectively, from a strip pit (168 m west, 335 m north of southwestern corner of sec. 24, T.10N., R.17W.) about 6.4 km east of McArthur, Ohio. The coals were separated by very fossiliferous marine shale of the Putnam Hill Limestone Member.

Winters coal bed (1.3 m thick), *No. 4A* coal bed (1.09 m thick), and *No. 5* coal bed (84 cm thick): Collected from the strip pit shown in Figure 9 (396 m west, 457 m south of NE. corner of sec. 15, T.10N., R.17W.) about 3.2 km northwest of McArthur, Ohio.

Princess Nos. 5, 5A, and 5B coal beds (102, 18 and 38 cm thick, respectively): Collected from the roadcut that contains the type section Kilgore Flint Member (see Fig. 3), northern side of U.S. Interstate I-64 at the intersection with U.S. Highway 60 near Coalton in Boyd County, Kentucky.

REFERENCES CITED

Carlson, J. E., 1965, Geology of the Rush quadrangle, Kentucky: U.S. Geological Survey Geologic Quadrangle Map GQ-408, scale 1:24,000.

Carlson, J. E., 1971, Geologic map of the Webbville quadrangle, eastern Kentucky: U.S. Geological Survey Geologic Quadrangle Map GQ-927, scale 1:24,000.

Cavaroc, V. V., Jr., and Ferm, J. C., 1968, Siliceous spiculites as shoreline indicators in deltaic sequences: Geological Society of America Bulletin, v. 79, p. 263–272.

Connor, C. W., and Flores, R. M., 1978, Geologic map of the Louisa quadrangle, Kentucky–West Virginia: U.S. Geological Survey Geologic Quadrangle Map GQ-1462, scale 1:24,000.

Couchot, M. L., Crowell, D. L., Van Horn, R. G., and Struble, R. A., 1980, Investigations of the deep coal resources of portions of Belmont, Guernsey, Monroe, Noble, and Washington Counties, Ohio: Ohio Division of Geological Survey Report of Investigations 116, 40 p.

Danilchik, W., 1977, Geologic map of the Tiptop quadrangle, eastern Kentucky: U.S. Geological Survey Geologic Quadrangle Map GQ-1410, scale 1:24,000.

DeLong, R. M., and White, G. M., 1963, Geology of Stark County: Ohio Geological Survey Bulletin 61, 209 p.

Dobrovolny, E., Sharps, J. A., and Ferm, J. C., 1963, Geology of the Ashland quadrangle Kentucky-Ohio, and the Catlettsburg quadrangle in Kentucky: U.S. Geological Survey Geologic Quadrangle Map GQ-196, scale 1:24,000.

Dobrovolny, E., Ferm, J. C., and Eroskay, S. O., 1966, Geologic map of parts of the Greenup and Ironton quadrangles, Greenup and Boyd Counties, Kentucky: U.S. Geological Survey Geologic Quadrangle Map GQ-532, scale 1:24,000.

Douglass, R. C., 1987, Fusulinid biostratigraphy and correlations between the Appalachian and Eastern Interior basins: U.S. Geological Survey Professional Paper 1451, 95 p., 20 plates.

Englund, K. J., Arndt, H. H., and Henry, T. W., eds., 1979, Proposed Pennsylvanian System stratotype, Virginia and West Virginia, in Ninth International Congress of Carboniferous Stratigraphy and Geology, field trip no. 1: American Geological Institute Selected Guidebook 1, 138 p.

Ferm, J. C., 1963, Coal beds of the Princess Reserve District, in Huddle, J. W., Lyons, E. J., Smith, H. L., and Ferm, J. C., eds., Coal reserves of eastern Kentucky: U.S. Geological Survey Bulletin 1120, p. 32–55.

Ferm, J. C., 1970, Allegheny deltaic deposits, in Morgan, J. P., ed., Deltaic sedimentation, modern and ancient: Society of Economic Paleontologists and Mineralogists Special Publication 15, p. 246–255.

Ferm, J. C., and Williams, E. G., 1965, Characteristics of a Carboniferous marine invasion in western Pennsylvania: Journal of Sedimentary Petrology, v. 35, no. 2, p. 319–330.

Ferm, J. C., Horne, J. C., Swinchatt, J. P., and Whaley, P. W., 1971, Carboniferous depositional environments in northeastern Kentucky, in Geological Society of Kentucky, annual spring field conference guidebook, April 1971: Lexington, Kentucky Geological Survey, 30 p.

Flint, N. K., 1951, Geology of Perry County: Ohio Geological Survey Bulletin 48, 4th ser., 377 p.

Gray, L. R., 1967, Palynology of four Allegheny coals, northern Appalachian coal field: Palaeontographica, v. 121, ser. B, nos. 1-3, p. 65–86.

Harris, L. D., 1975, Oil and gas data from the Lower Ordovician and Cambrian rocks of the Appalachian basin: U.S. Geological Survey Miscellaneous Investigations Map I-917 D, scale 1:2,500,000.

Helfrich, C. T., 1981, Preliminary correlations of coals of the Princess Reserve District in eastern Kentucky, in Cobb, J. C., et al., field trip leaders, Annual Geological Society of America coal division field trip, November 5–8, 1981: Lexington, Kentucky Geological Survey, p. 106–119.

Henry, T. W., 1984, Geologic map of the Mammoth quadrangle, Kanawha and Clay Counties, West Virginia: U.S. Geological Survey Geologic Quadrangle Map GQ-1576, scale 1:24,000.

Hinricks, E. N., 1978, Geologic map of the Noble quadrangle, eastern Kentucky: U.S. Geological Survey Geologic Quadrangle Map GQ-1476, scale 1:24,000.

Huddle, J. W., and Englund, K. J., 1966, Geology and coal reserves of the Kermit and Varney area, Kentucky: U.S. Geological Survey Professional Paper 507, 83 p.

Jenkins, E. C., 1966, Geologic map of the Milo quadrangle, Kentucky, West Virginia, and part of the Webb quadrangle in Kentucky: U.S. Geological Survey Geologic Quadrangle Map GQ-543, scale 1:24,000.

Johnston, J. E., 1962, Geology of the Lenox quadrangle, Kentucky: U.S. Geological Survey Geologic Quadrangle Map GQ-181, scale 1:24,000.

Knauth, L. P., 1979, A model for the origin of chert in limestone: Geology, v. 7, p. 274–277.

Knauth, L. P., and Epstein, S., 1976, Hydrogen and oxygen isotope ratios in nodular and bedded cherts: Geochimica et Cosmochimica Acta, v. 40, p. 1095–1108.

Kosanke, R. M., 1973, Palynological studies of the coals of the Princess Reserve District in northeastern Kentucky: U.S. Geological Survey Professional Paper 839, 22 p.

Kosanke, R. M., 1984, Palynology of selected coal beds in the proposed Pennsylvanian System stratotype in West Virginia: U.S. Geological Survey Professional Paper 1318, 44 p.

Kosanke, R. M., 1988, Palynological studies of Middle Pennsylvanian coal beds of the proposed Pennsylvanian System stratotype in West Virginia: U.S. Geological Survey Professional Paper 1455, 73 p.

Krebs, C. E., and Teets, D. D., Jr., 1914, Kanawha County: West Virginia Geological Survey County Reports, 679 p.

Lamborn, R. E., 1951, Limestones of eastern Ohio: Geological Survey of Ohio Bulletin 49, 4th ser., 377 p.

Lamborn, R. E., 1954, Geology of Coshocton County: Ohio Geological Survey Bulletin 53, 245 p.

McGuire, W. H., and Howell, P., 1963, Oil and gas possibilities of the Cambrian and Lower Ordovician in Kentucky: Lexington, Kentucky, Spindletop Research Center, 216 p.

Merrill, G. K., 1968, Allegheny (Pennsylvanian) conodonts [Ph.D. thesis]: Baton Rouge, Louisiana State University Agricultural and Mechanical College, 184 p.

Morningstar, H., 1922, Pottsville fauna of Ohio: Ohio Geological Survey Bulletin 25, 4th ser., 312 p.

Morse, W. C., 1931, The Pennsylvanian invertebrate fauna of Kentucky, in Paleontology of Kentucky: Kentucky Geological Survey, ser. 6, v. 36, p. 293–348.

North American Commission on Stratigraphic Nomenclature, 1983, North American stratigraphic code: Bulletin of the American Association of Petroleum Geologists, v. 67, p. 841–875.

Orton, E., 1878, Supplemental report on the geology of the Hanging Rock District, in Geology and paleontology: Ohio Geological Survey Report, v. 3, p. 883–941.

Outerbridge, W. F., 1964, Geology of the Offutt quadrangle, Kentucky: U.S. Geological Survey Geologic Quadrangle Map GQ-348, scale 1:24,000.

Outerbridge, W. F., 1989, Correlation of the Charleston Sandstone of the proposed Pennsylvanian stratotype with strata in eastern Kentucky, western West Virginia, and southern Ohio: U.S. Geological Survey Miscellaneous Field Studies Map MF-Z110, 1 sheet.

Phalen, W. C., 1908, Economic geology of the Kenova quadrangle, Kentucky, Ohio, and West Virginia: U.S. Geological Survey Bulletin 349, 158 p.

Phalen, W. C., 1912, Description of the Kenova quadrangle [Kentucky–West Virginia–Ohio]: U.S. Geological Survey Geologic Atlas, Folio 184, scale 1:125,000.

Phillips, T. L., Peppers, R. A., and DiMichele, W. A., 1985, Stratigraphic and interregional changes in Pennsylvanian coal-swamp vegetation—Environmental inferences: International Journal of Coal Geology, v. 5, nos. 1–2, p. 43–109.

Ping, R. C., and Rice, C. L., 1979, The Stoney Fork Member of the Breathitt Formation, in Sohl, N. F., and Wright, W. B., eds., Changes in stratigraphic nomenclature by the U.S. Geological Survey, 1978: U.S. Geological Survey Bulletin 1482-A, p. A70–A76.

Reppert, R. S., 1979, Kanawha Black Flint—Its occurrence and extent in West Virginia, in Englund, K. J., Arndt, H. H., and Henry, T. W., eds., Proposed Pennsylvanian System stratotype, Virginia and West Virginia, in Ninth International Congress of Carboniferous Stratigraphy and Geology, field trip no. 1: American Geological Institute Selected Guidebook 1, p. 109–111.

Rice, C. L., 1977, The Vanport Limestone Member as used by Phalen (1912) in northeastern Kentucky, in Sohl, N. F., and Wright, W. B., eds., Changes in stratigraphic nomenclature by the U.S. Geological Survey, 1976: U.S. Geological Survey Bulletin 1435-A, p. A142–A143.

Rice, C. L., and Smith, J. H., 1980, Correlation of coal beds, coal zones, and key stratigraphic units in the Pennsylvanian rocks of eastern Kentucky: U.S. Geological Survey Miscellaneous Field Studies Map MF-1188, 1 sheet.

Rice, C. L., Kehn, T. M., and Douglass, R. C., 1979, Pennsylvanian correlations between the Eastern Interior and Appalachian basins, in Palmer, J. R., and Dutcher, R. R., eds., Depositional and structural history of the Illinois basin pt. 2, Invited papers: Champaign, Illinois State Geological Survey, p. 103–105.

Rice, C. L., Martino, R. L., and Slucher, E. R., 1992, Regional aspects of Pottsville and Allegheny stratigraphy and depositional environments,

Ohio and Kentucky: U.S. Geological Survey Open-File Report 92-558, 67 p.

Rice, D. D., 1975, Geologic map of the Helton quadrangle, southeastern Kentucky: U.S. Geological Survey Geologic Quadrangle Map GQ-175, scale 1:24,000.

Rogers, H. D., 1858, The geology of Pennsylvania; A government survey: Philadelphia, J. B. Lippincott, 2 v., 1046 p.

Sheppard, R. A., and Ferm, J. C., 1962, Geology of the Argillite quadrangle, Kentucky: U.S. Geological Survey Geologic Quadrangle Map GQ-175, scale 1:24,000.

Smyth, P., 1957, Fusulinids from the Pennsylvanian rocks of Ohio: Ohio Journal of Science, v. 57, no. 5, p. 257–283.

Stout, W., 1916, Geology of southern Ohio: Ohio Geological Survey Bulletin 20, 4th ser., 723 p.

Stout, W., 1918, Geology of Muskingum County: Ohio Geological Survey Bulletin 21, 4th ser., 402 p.

Stout, W., 1927, Geology of Vinton County: Ohio Geological Survey Bulletin 31, 4th ser., 402 p.

Stout, W., and Lamborn, R. E., 1924, Geology of Columbiana County: Ohio Geological Survey Bulletin 28, 4th ser., 408 p.

Stout, W., Shaw, M. C., Bole, G. A., and Schaaf, D., 1931, The Lawrence clay of Lawrence County: Ohio Geological Survey Bulletin 36, 4th ser., 123 p.

Sturgeon, M. T., and DeLong, R. M., 1964, Revision of some stratigraphic names between the Lower and Middle Kittanning coals in eastern Ohio: Ohio Journal of Science, v. 64, no. 1, p. 41–43.

Wanless, H. R., 1939, Pennsylvanian correlations in the Eastern Interior and Appalachian coal fields: Geological Society of America Special Paper 17, 130 p.

Wanless, H. R., 1975, Appalachian region, in McKee, E. D., et al., eds., Paleotectonic investigations of the Pennsylvanian System in the United States: U.S. Geological Survey Professional Paper 853-C, 62 p.

Ward, D. E., 1978, Geologic map of the Adams quadrangle, Lawrence County, Kentucky: U.S. Geological Survey Geologic Quadrangle Map GQ-1489, scale 1:24,000.

Wardlaw, B. R., Rice, C. L., and Stamm, R. G., 1993, Preliminary analysis of conodont occurrences in Pennsylvanian strata of Ohio and Kentucky: U.S. Geological Survey Open-File Report 93-312, 9 p.

Webb, J. E., 1963, Allegheny sedimentary geology in the vicinity of Ashland, Kentucky [Ph.D. thesis]: Baton Rouge, Louisiana State University, 168 p.

Whaley, P. W., 1969, A litho-genetic model for rocks of a lower delta plain sequence [Ph.D. thesis]: Baton Rouge, Louisiana State University, 135 p.

White, D., 1900, Relative ages of the Kanawha and Allegheny Series as indicated by the fossil plants: Geological Society of America Bulletin, v. 11, p. 145–178.

White, D., and Ashley, G. H., 1906, Correlation of coals, in Ashley, G. H., and Glenn, L. C., Geology and mineral resources of part of the Cumberland Gap coal field: U.S. Geological Survey Professional Paper 49, p. 206–212.

White, I. C., 1878, Beaver River District of the bituminous coal-fields: Pennsylvania Geological Survey Report Q, 337 p.

White, I. C., 1891, Stratigraphy of the bituminous coal field of Pennsylvania, Ohio, and West Virginia: U.S. Geological Survey Bulletin 65, 212 p.

White, I. C., 1908, Supplementary coal report: West Virginia Geological Survey, ser. 2, v. A, 720 p.

Whittington, C. L., and Ferm, J. C., 1965, Geology of the Oldtown quadrangle, Kentucky: U.S. Geological Survey Geologic Quadrangle Map GQ-353, scale 1:24,000.

Williams, E. G., and Bragonier, W. A., 1974, Controls of Early Pennsylvanian sedimentation in western Pennsylvania, in Briggs, ed., Carboniferous of the southeastern United States: Geological Society of America Special Paper 148, p. 135–152.

Williams, E. G., and Ferm, J. C., 1964, Sedimentary facies in the lower Allegheny rocks of western Pennsylvania: Journal of Sedimentary Petrology, v. 34, p. 610–614.

Zimmerman, R. K., 1966, Aspects of early Allegheny depositional environments in eastern Ohio [Ph.D. thesis]: Baton Rouge, Louisiana State University, 123 p.

MANUSCRIPT ACCEPTED BY THE SOCIETY FEBRUARY 1, 1994

Geological Society of America
Special Paper 294
1994

Key rock units and distribution of marine and brackish water strata in the Pottsville Group, northeastern Ohio

Ernie R. Slucher
Ohio Geological Survey, 4383 Fountain Square Drive, Columbus, Ohio 43224
Charles L. Rice
U.S. Geological Survey, National Center, MS 926, Reston, Virginia 22092

ABSTRACT

Core drilling in poorly exposed Lower and Middle Pennsylvanian strata of the Pottsville and Allegheny Groups in northeastern Ohio has led to a better understanding of the character and distribution of key Pennsylvanian marine units. Two unknown and three previously known marine and marine-influenced units are identified in strata of this area in the Pottsville Group. Distribution and facies maps of these marine units in northeastern Ohio suggest pre-Pennsylvanian erosional paleotopography and periodic movement on subsurface faults locally influenced depositional patterns. Additionally, the traditional placements of the Lowellville marine unit and the Quakertown coal in the stratigraphic column for Ohio appear to be in error. The new data suggest the need for an extensive revision of the geologic column for the Lower and Middle Pennsylvanian rocks in Ohio and the need to develop a more representative stratigraphic nomenclature and framework for these rocks. Core drilling referenced in this investigation was done as part of the recent statewide mapping and core-drilling program conducted by the Ohio Geological Survey.

INTRODUCTION

Rocks containing marine and brackish water invertebrate fossils of Pennsylvanian age have been recognized as key elements of the stratigraphic section in Ohio since the early 1800s. Hildreth (1828) first noted the occurrence of beds of fossiliferous limestone and flint in the vicinity of Zanesville, Ohio (Fig. 1A). He later discussed (Hildreth, 1833, 1836) their distribution, correlation, and origin and used them to determine regional dip of Pennsylvanian strata in the Muskingum River Valley. Many of the Pottsville and Allegheny marine units (such as the informal Lower Mercer, Upper Mercer, Putnam Hill, and Vanport limestones) are commonly cherty. In his stratigraphic analyses, Hildreth (1836) was able to utilize subsurface data from the reports of well diggers of the region, who recognized limestone and "flint rock" as stratigraphic marker beds useful in their exploration for brine.

Local and regional investigations in northeastern Ohio and western Pennsylvania by Newberry (1878), Chance (1879), and White (1879) demonstrated the utility of marine limestone units in developing a stratigraphic framework for the lower coal measures. Orton (1883, 1884) also recognized that limestone units were the single most important element for stratigraphic analyses, particularly when several could be identified at different stratigraphic positions in the same Pennsylvanian section.

The regional and county reports of Stout (1916, 1918, 1927) of the geology along the west-central and southern outcrop belt of the Pottsville Group (Fig. 1A) incorporated many stratigraphic conclusions of earlier workers. These culminated in the first detailed stratigraphic column of Pennsylvanian rocks in Ohio by Bownocker and Dean (1929), shown in Figure 2.

Although minor revisions have been made since to the nomenclature of the Allegheny Group and Conemaugh Group (Sturgeon and Merrill, 1949; Sturgeon and DeLong, 1964; Murphy, 1973), the Pottsville series (as defined by Bownocker and Dean, 1929) shown in Figure 2 remains the officially recognized stratigraphic section for this interval in Ohio. Intra- and interbasinal correlation studies commonly have used this

Slucher, E. R., and Rice, C. L., 1994, Key rock units and distribution of marine and brackish water strata in the Pottsville Group, northeastern Ohio, *in* Rice, C. L., ed., Elements of Pennsylvanian Stratigraphy, Central Appalachian Basin: Boulder, Colorado, Geological Society of America Special Paper 294.

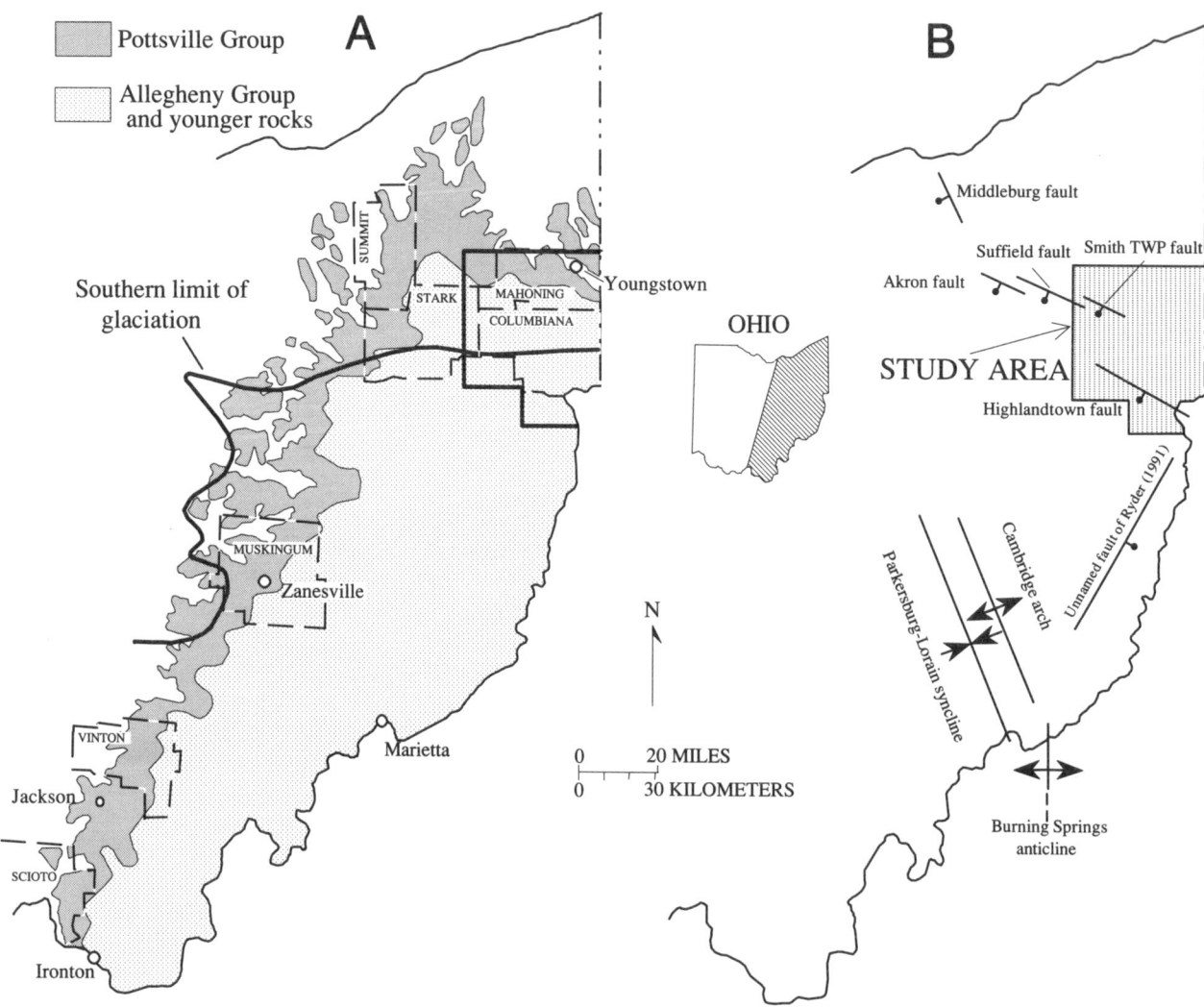

Figure 1. A, Generalized geologic map, and B, map of structural features for eastern Ohio. Bar and ball symbol indicates downthrown side of fault. Names of counties discussed in text are shown on geologic map A.

stratigraphic column to relate marine units in Ohio with strata of similar age in other parts of the Appalachian basin and in the Eastern Interior (Illinois) basin (e.g., see Wanless, 1939, 1975; Moore et al., 1944). However, more recent stratigraphic and paleontologic investigations of the Pottsville and Allegheny Groups in Ohio (Merrill, 1974, 1991; Ferm et al., 1979; Douglass, 1979, 1987; Hansen, 1986) suggest that the key beds of the Pennsylvanian stratigraphic section are not as well known as previously supposed, either in the details of their deposition and distribution or in their correlation.

Recognizing the need for better stratigraphic information and the need for modern detailed geologic maps, the Ohio Geological Survey began a new bedrock mapping initiative in 1982, to produce a new geologic map of the state. Because most of the Pennsylvanian nomenclature used in Ohio had been extended into the state from earlier geologic investigations in western Pennsylvania (Larsen, 1991), initial mapping of Pennsylvanian rocks was begun in northeastern Ohio. This mapping allowed reexamination of details of the Pottsville and Allegheny stratigraphic framework and nomenclature in the area in which they were first developed in Ohio. Although Pennsylvanian rocks underlie approximately the eastern third of Ohio, as is indicated in Figure 1A, rocks of the Pottsville Group are mostly in the subsurface and crop out in a relatively narrow belt along the western and northern edge of the Appalachian basin. Additionally, large portions of the study area of this report, including all of the northern area where Pottsville rocks crop out, were subjected to repeated glaciation during the Pleistocene. A few deeply entrenched streams that have locally eroded through the glacial sediments provide limited exposures of bedrock in areas of relatively steep topography along major drainages.

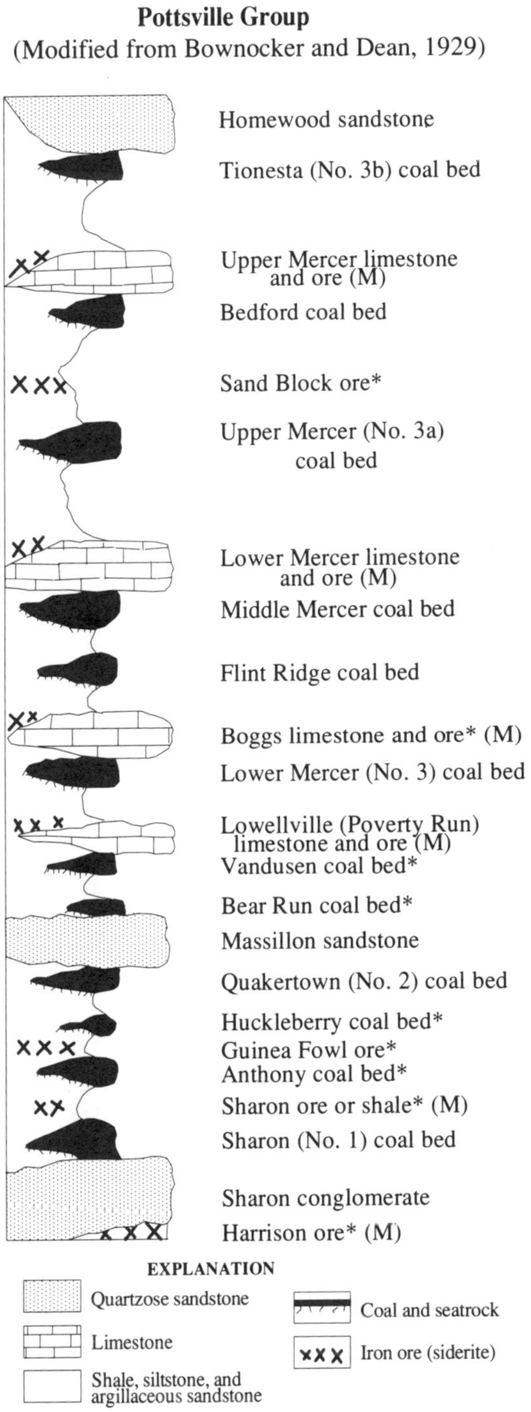

Pottsville Group
(Modified from Bownocker and Dean, 1929)

Homewood sandstone
Tionesta (No. 3b) coal bed
Upper Mercer limestone and ore (M)
Bedford coal bed
Sand Block ore*
Upper Mercer (No. 3a) coal bed
Lower Mercer limestone and ore (M)
Middle Mercer coal bed
Flint Ridge coal bed
Boggs limestone and ore* (M)
Lower Mercer (No. 3) coal bed
Lowellville (Poverty Run) limestone and ore (M)
Vandusen coal bed*
Bear Run coal bed*
Massillon sandstone
Quakertown (No. 2) coal bed
Huckleberry coal bed*
Guinea Fowl ore*
Anthony coal bed*
Sharon ore or shale* (M)
Sharon (No. 1) coal bed
Sharon conglomerate
Harrison ore* (M)

EXPLANATION
- Quartzose sandstone
- Limestone
- Shale, siltstone, and argillaceous sandstone
- Coal and seatrock
- xxx Iron ore (siderite)

Figure 2. Current standard geologic column for the Pottsville Group in Ohio. The group has an average thickness of 256 ft (78 m) in Ohio (Bownocker and Dean, 1929). Asterisk symbols denote units described and named from exposures in southern Ohio; (M) indicates units contain marine fossils.

Pottsville rocks may be obscured in unglaciated regions by outwash terraces that are remnants of former erosional levels in areas of major drainage (Stout and Lamborn, 1924; White and Totten, 1985). Thus, exposures of the strata being investigated are limited to surface mine operations, sparse roadcuts, and rare deeply incised stream valleys. To facilitate the new mapping project and to establish detailed stratigraphic sections for correlation throughout the state, the Ohio Geological Survey conducted an extensive core-drilling program in northeastern Ohio. The interpretations presented in this chapter are based in large part on data derived from the 45 core holes drilled in northeastern Ohio (see Fig. 4A).

Examination of the new core data shows that there are several marine units in the Pottsville of northeastern Ohio not previously reported. Although some Pottsville and Allegheny informal marine units such as the Lower Mercer, Upper Mercer, and Vanport limestones are persistent and are easily recognized in exposures in that area, correlation of other marine units is problematic, particularly in the case of units recognized in other areas of the state such as the Poverty Run or Boggs limestones or other named units in the Pottsville. Through examination of the character and distribution of the Pottsville marine units in northeastern Ohio, this work has led to a better understanding of their relationship to previously described Pottsville stratigraphy.

PREVIOUS WORK

White (1879) demonstrated the value of limestone horizons in stratigraphic studies of the lower part of the Pennsylvanian section in the northern part of the Appalachian basin by tracing the informal Lower and Upper Mercer limestones and the overlying Vanport limestone from western Pennsylvania into northeastern Ohio. Prosser (1905) introduced the terms Pottsville and Allegheny as formations (now called groups) into Ohio. Lamb (1910) summarized the nomenclature and distribution of Pottsville and Allegheny (lower part) marine limestones (below the Lower Kittanning coal bed) in northeastern Ohio. The distribution and character of Pottsville marine units was summarized by Morningstar (1922), who described and collated all known occurrences of Pottsville marine and brackish water fossil fauna in Ohio that had been noted by the early geologic investigators of the Pennsylvanian. The geologic report of Columbiana County (Stout and Lamborn, 1924), which covers part of the study area, contains only limited descriptions of Pottsville strata because this portion of the Pennsylvanian section is generally buried beneath glacial drift. The geology of the Youngstown area was described by Stevenson (1933), who made extensive use of the stratigraphic nomenclature of Bownocker and Dean (1929).

GEOLOGIC SETTING

In Ohio, the Pennsylvanian Pottsville and Allegheny strata consist of sequences of alternating sandstone, siltstone, and laminated and nonlaminated mudstone that grade vertically and laterally into each other. Minor amounts of marine and nonmarine limestone, clay, coal, flint, and sideritic ironstone, both nodular and banded, also are present. Pennsylvanian strata are slightly thicker in the southeastern part of Ohio (Collins, 1979).

In Ohio, Pennsylvanian rocks everywhere overlie Mississippian strata. The boundary between the two systems is an unconformity that has as much as 400 ft (122 m) of erosional relief (Hyde, 1953). The great variation of relief on the unconformity appears to have strongly influenced the interval thickness and depositional trends of Pottsville rocks older than the informal Lower Mercer limestone (for example, compare isopachs in Figs. 4B and 5G). Strata of Pennsylvanian age in Ohio overlie, in descending stratigraphic order, the Mississippian Maxville Limestone and the Logan and Cuyahoga Formations. The Mississippian units underlying Pennsylvanian rocks vary complexly in the state both because of the irregular sub-Pennsylvanian erosional surface and because there are numerous regional and local disconformities within and below the Maxville Limestone (Morse, 1910; Uttley, 1974).

In northeastern Ohio, the Pennsylvanian strata are underlain by rocks assigned to the Mississippian Cuyahoga Formation. Those rocks are typically greenish gray to gray interbedded very fine grained argillaceous sandstone and shale. In the study area, these strata are locally calcareous and contain thin zones of diminutive marine fossils.

The Pottsville and Allegheny depositional environments in Ohio have been interpreted to range from offshore marine to alluvial plain. Numerous depositional models have been devised to explain the facies relationships for many of the rock types found in Pennsylvanian strata. The models include the following: (1) a repetitive subsidence-controlled shallow-water coastal environment (Stout, 1923), (2) the cyclothem model of Weller (1930) and its modifications (Gray, 1954; Ferm and Williams, 1963), (3) deltaic models based on studies of modern river deltas (Ferm, 1970; Ferm and Horne, 1979), and (4) tectonic models in which deltaic deposition is controlled by subsurface structures (Ferm and Weisenfluh, 1989).

Other models have been invoked to explain the character and distribution of the quartzarenites of the lower part of the Pennsylvanian section such as the informal Sharon conglomerate and the Connoquenessing sandstone (Fig. 2). Some researchers have suggested that the quartzose sandstones were deposited either in beach and barrier-bar environments (Stout, 1916; Gray, 1956; Ferm and Cavaroc, 1969; Donaldson, 1974; Ferm, 1974) or in tidal channels in an epicontinental seaway extending northeast-southwest across the Appalachian basin (Cecil and Englund, 1985). However, detailed studies of the geometry and sedimentology of the quartzarenites in Ohio have concluded that the sandstones were deposited in fluvial environments (e.g., Lamb, 1911; Fuller, 1955; Mrakovich, 1969; Ketring, 1984; Schmidley, 1987; for a regional summary, see also Rice and Schwietering, 1988).

In eastern Ohio, only a few tectonic structures have been identified that may have affected deposition of Pennsylvanian strata (Fig. 1B). The Cambridge arch and the Parkersburg-Lorain syncline (post–Early Permian in age) are broad structures having small amplitudes that fold Pennsylvanian and Permian strata (Collins, 1979). The northern end of the Burning Springs anticline (post-Devonian in age) extends just into Ohio near Marietta and is believed to be related to deep-seated faulting (Shumaker, 1986). Ryder (1991) has suggested the presence of a northeast-southwest–trending normal fault in the subsurface just south of the study area (Fig. 1B), but he assigns movement on the fault to Cambrian time. The Akron, Highlandtown, Middleburg, Smith Township, and Suffield faults (Fig. 1B) are subsurface features mapped by Gray (1982) based on displacement of the Onondaga Limestone (Middle Devonian) and the Berea Sandstone (Early Mississippian age of de Witt, 1970); these are considered to be pre-Pennsylvanian in age. Maximum displacements range from 80 to 180 ft (24 to 55 m); the downthrown side is to the south. The faults represent the westward extension of the Transylvania fracture zone of Root (1992), a major crustal fracture zone extending across Pennsylvania and northeastern Ohio. Root and MacWilliams (1986) suggested that the Akron-Suffield-Smith Township fault zone may have originated as a result of right-lateral wrenching along the Transylvania fault zone. However, in a study of lithofacies patterns in the Allegheny Group, Hook (1985) and Hook and Ferm (1988) developed evidence of possible syndepositional movement along the Highlandtown fault (Fig. 1B).

NOMENCLATURE

Historically, the contact between the Pottsville and Allegheny Formations has been placed at the base of the Brookville coal bed or the top of the Homewood sandstone in Pennsylvania (Wanless, 1975), but neither the Brookville nor the Homewood can be traced into Ohio or even identified with confidence along the Pennsylvania-Ohio border. For example, all of the sandstones or thin coal beds shown in Figure 3 in the interval between the informal Vanport and Lower Mercer limestones are discontinuous. Because those units cannot be identified, the Ohio Geological Survey has elected (in the current bedrock mapping program) to assign the Pennsylvanian strata below the Vanport limestone to Pottsville and Allegheny Groups, undivided. In this report, however, the assignment of Pottsville and Allegheny Groups, undivided, is restricted to those strata between the base of the Vanport limestone and the top of the Upper Mercer limestone or its projection (Fig. 3). Although no specific lithologic distinctions can be made between the Pottsville and Allegheny rocks, (White, 1879; Orton, 1884; Stout, 1918), the subdivision of rocks into Pottsville and Allegheny has persisted primarily because of the association of those names with the economic deposits that they contain (Collins, 1979).

As many as 26 individually named lithologic units have been defined in the Pottsville rocks in Ohio (Collins, 1979).

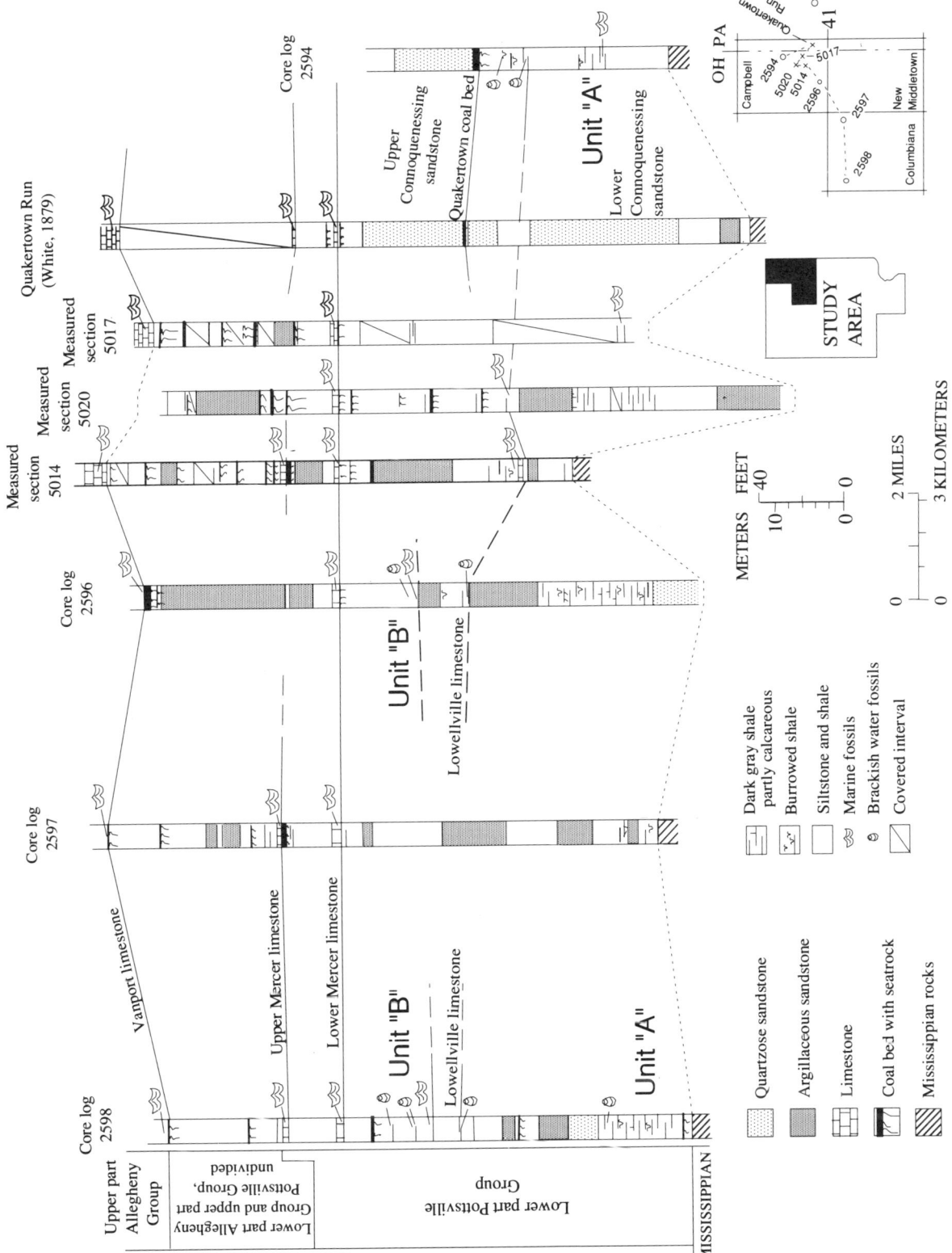

Figure 3. Cross section showing the relation between key beds of the Pottsville Group and the lower part of the Allegheny Group in the northeastern part of the study area. Measured section 5014 contains the type section of the Lowellville limestone, and the measured section in Quakertown Run contains the type section of the Quakertown coal bed.

However, mapping in northeastern Ohio (Slucher and Larsen, 1989; Larsen and Rea, 1990) has demonstrated that many of those lithologic units can be identified only locally and are not persistent, but rather grade laterally into rocks of different lithology. Currently, the Ohio Geological Survey considers all stratigraphic terms below "group" status as informal and is in the process of developing a Lower and Middle Pennsylvanian stratigraphic framework based largely on key beds such as coal beds and marine units that have proven useful in detailed bedrock mapping.

POTTSVILLE MARINE UNITS

Three marine units have been widely recognized in the Pottsville strata of Ohio. In ascending order, they are the Lowellville, Lower Mercer, and Upper Mercer limestones (see Fig. 2). These marine units are best developed and most persistent in the east-central and northeastern part of Ohio. A fourth marine unit, the Boggs limestone, is recognized mainly in central Ohio. Lamb (1910) described and named a limestone, the Howenstein, in the upper part of the Pottsville in southern Stark County; this, however, is probably one of the Mercer limestones. In addition to the limestone units, Morningstar (1922) listed six other units in the Pottsville of southern Ohio that locally contain marine or brackish water fossils. In ascending order, these are the Harrison ore, the Sharon ore or shale above the No. 1 (Sharon) coal, the shales above the Anthony, No. 2 (Quakertown), Bear Run coals, and the Sand Block ore. Of these latter six marine-influenced units, only the Harrison ore has been reported to contain marine fossils in areas other than southern Ohio. However, the Harrison ore occurs at the unconformity at the base of the Pennsylvanian section. As Stout (1918) suggested, the fossils of the Harrison ore may be reworked erosional relics of underlying Mississippian carbonate rocks, or at least relics of older rocks preserved in a regolith. The Harrison ore does not appear to be associated with any widespread marine event (Hansen, 1984); its widely separated deposits may represent several different ages. Morningstar (1922, p. 25) also reported a single locality at which the brackish water brachiopod *Lingula* occurred "in great abundance" in shale above a thin No. 2 (Quakertown) coal in the eastern part of Summit County, about 16 mi (26 km) west of the study area (Fig. 1A). However, the exact identification and correlation of this exposure is uncertain because its location is isolated both stratigraphically and areally. The lack of faunal diversity and abundance in the other marine-influenced units listed by Morningstar suggests deposition in restricted marine to brackish water environments. The units lack continuity; in any case, they are difficult to trace because of their lithology, thinness, and poor exposures.

Morningstar (1922) included two other units in the upper part of the Pottsville, the McArthur shale and the Black flint. However, strata representing their intervals subsequently have been reassigned to the Allegheny Group as the Putnam Hill limestone and the Zaleski flint, respectively (Stout, 1927); neither has been recognized in the study area.

Five marine units have been identified in Pottsville strata in the study area below the Vanport limestone of the Allegheny Group. Three of the units are the widely recognized Lowellville, Lower Mercer, and Upper Mercer limestones. The other two, designated unit "A" and unit "B" in this chapter, were found as a result of the new mapping and core-drilling program of the Ohio Geological Survey.

Figure 3 shows the relationship of the marine units and other named stratigraphic units of the Pottsville Group in a section across the northeastern part of the study area. The study area includes part of White's (1879) area of field investigations, where he traced the Lower and Upper Mercer limestones and the Vanport limestone from western Pennsylvania into Ohio. The Lower Mercer limestone, the most persistent marine unit in the Pottsville section in northeastern Ohio, is used as datum for Figure 3. At the top of the section is the persistent Vanport limestone, which also is a key mapping datum for this area. The base of the section is extended or projected down to underlying Mississippian sandstones and shales of the Cuyahoga Formation. Two of the measured sections contain the type sections of the Lowellville limestone (section 5014) and the Quakertown coal bed (Quakertown Run). Core data in the immediate vicinity of the measured sections allow correlation of the outcrop data into the subsurface. Figure 3 illustrates the difficulty of tracing even named lithologic units in this part of the Pennsylvanian section between relatively closely spaced sections. For example, the Upper and Lower Connoquenessing sandstones cannot be traced far beyond the locality where White (1879) described them in Quakertown Run in northwestern Pennsylvania. Only the Upper Connoquenessing sandstone can be identified with certainty in the immediate vicinity of Quakertown Run and is present only in core 2594 of the holes located in Campbell and Columbiana 7.5-min Quadrangles in Ohio. Both the Lower and Upper Connoquenessing sandstones appear to grade laterally or pinch out into shale and siltstone of the Pottsville Group.

To better understand the distribution of the marine and brackish water units in the study area, distribution and facies maps for each of the five units were constructed from the Ohio Geological Survey core-hole data base and measured stratigraphic sections shown in Figure 4A. The five units are, from oldest to youngest, unit "A", Lowellville limestone, unit "B," and Lower Mercer and Upper Mercer limestones. Facies distribution maps are meant to be generalizations of dominant lithologies inferred during maximum marine conditions occurring within the stratigraphic intervals that contain the subject units.

Unit "A"

Unit "A" is an interval of strata varying from 1 to 40 ft (0.3 to 12 m) in thickness that is identified in core holes by its stratigraphic position and the presence of bioturbation or ma-

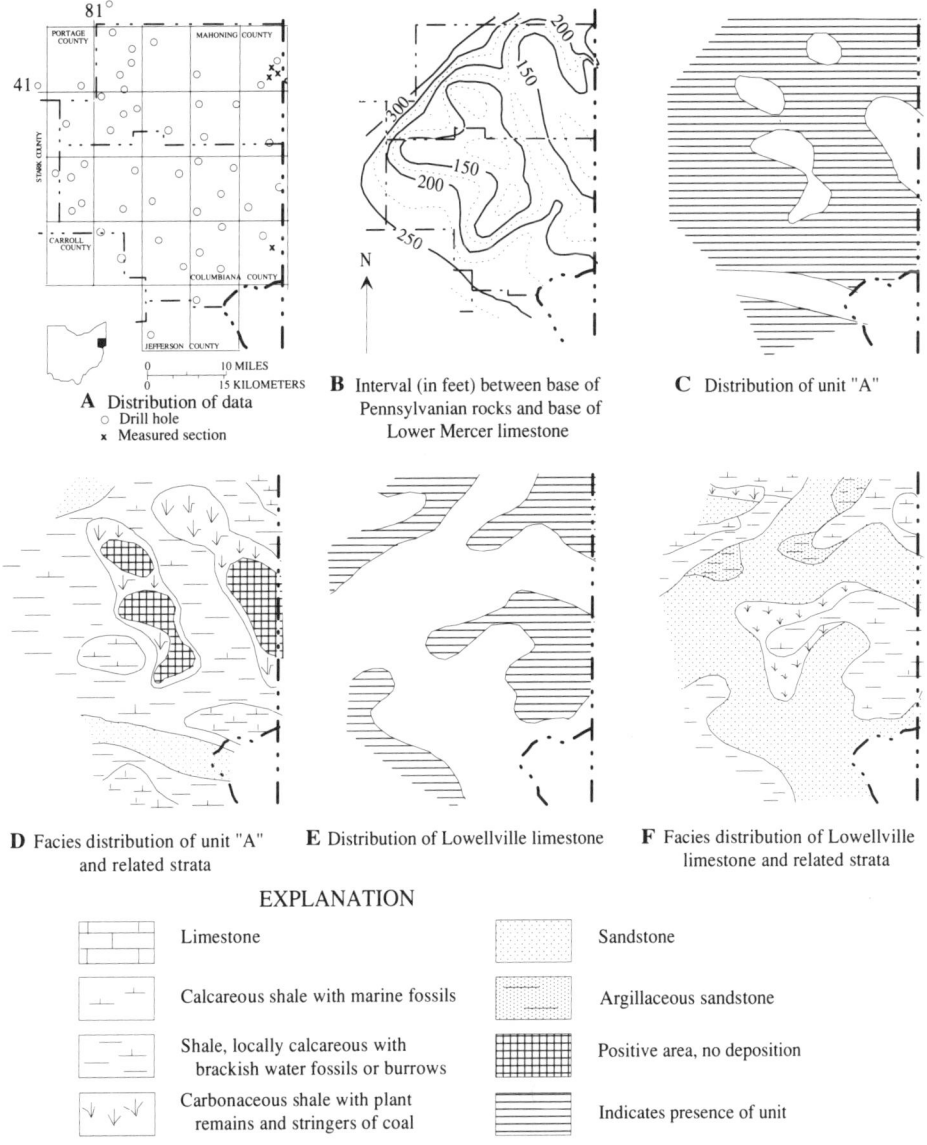

Figure 4. Maps of study area showing location of core data, isopach of basal Pennsylvanian rocks, and syntheses of distribution of facies of marine and brackish water Pottsville units.

rine or brackish water fossils. Contacts, particularly the upper, are commonly gradational. Body fossils found in unit A consist predominantly of *Lingula* and *Orbiculoidea* shells. However, other brachiopods, gastropods, cephalopods, and bivalves are found in calcareous shales in the northeastern corner and southern parts of the area shown in Figure 4D. One core, OGS 2676, in the west-central part of the study area contains a conodont element. Unit A generally is characterized by burrows and other trace fossils, which were used as indicators of at least marginal marine conditions even in the absence of body fossils. Cores of the marine facies of unit A contain dark gray aphanic limestone veinlets and nodules, some of which contain cone-in-cone structures. A facies of thin shale containing stringers of coal onlap Mississippian topographic highs and are laterally equivalent to the marine to brackish water facies of unit A (see Fig. 4D). These carbonaceous-bearing rocks may represent the shoreline facies, and therefore are included with unit A in its distribution map.

Stratigraphically, the base of unit A ranges from approximately 130 to 190 ft (40 to 58 m) below the base of the Lower Mercer limestone. This wide range of interval thickness may be explained by local topographic relief on the Mississippian-Pennsylvanian unconformity of more than 175 ft (53 m) (see Fig. 4B) and by local differences of sediment compaction or by subsequent erosion. The topographic lows (in the northwestern and southwestern parts of Fig. 4B) contain a quartzose sandstone (90 to 95% quartz) that probably represents the locally conglomeratic Sharon sandstone of northeastern Ohio. A coal bed, which locally underlies unit A, is identified by its association with the underlying Sharon sandstone. The coal, commonly called the Sharon coal, generally fills pre-Pennsylvanian valleys (Granchi, 1958). Coal beds at the stratigraphic position of the Sharon coal bed in the study area are thin, multiple beds of limited extent, and thus may be only approximately of the same age as the type Sharon coal in northeastern Ohio and western Pennsylvania. For that reason, the coals are not identified as Sharon on Figure 3. Interpretation of drill-hole data suggests that the marine transgression responsible for deposition of unit A was not a uniform widespread event but infiltrated only those lows where detrital material was not being deposited rapidly. Also, as a result of varying rates of sedimentation, the transgression may have influenced depositional environments in one area for a longer period of time than in other areas, locally appearing as more than a single marine event. As an example of the latter, terrestrial carbonaceous and seat-earth zones locally intertongue with strata containing brackish water fauna in what appears to be a near-shore environment. As a result, unit A ranges greatly in thickness from place to place within the study area.

The variation in thickness and apparent differences of stratigraphic position for unit A are well illustrated in the core logs in Figure 3. In cores 2598 and 2596 (Fig. 3), unit A consists of a relatively thick interval of dark gray shale that is burrowed and locally calcareous; the apparent differences of thickness and stratigraphic position are probably related to three factors: (1) differences of erosion of the upper beds, (2) the compactability of underlying strata, and (3) differences in rates of sedimentation. However, in the intervening core 2597 (Fig. 3), unit A was probably deposited on a slight topographic high of Mississippian rocks and is only represented by a few feet (less than a meter) of burrowed dark gray shale at the base of the Pennsylvanian. In the Quakertown Run section (Fig. 3), the stratigraphic position of unit A is occupied by the fluvial Lower Connoquenessing sandstone, and unit A may either have not been deposited or may have been truncated by erosion. Although other core data clearly suggest intertonguing between marine and continental deposits within a stratigraphic interval that is interpreted to include unit A, there is no evidence in Quakertown Run of that relation between unit A and the Lower Connoquenessing sandstone.

Lowellville limestone

Lamb (1910) named the Lowellville limestone from exposures south of Lowellville, Ohio, first noted by Newberry (1878) along Grindstone Run. Stout (1918) named a marine unit in the east-central outcrop belt the Poverty Run; Morningstar (1922) subsequently correlated the Poverty Run with the Lowellville in northeastern Ohio. In northeastern Ohio, the Lowellville is commonly represented by an upward-coarsening bayfill sequence of shale and siltstone, which is as much as 35 ft (11 m) thick at the type section (measured section 5014, Fig. 3). Although the type section contains 1 to 2 ft (0.3 to 0.6 m) of fossiliferous limestone at its base, the Lowellville is commonly represented by shales that range from very calcareous fossiliferous black shale containing marine fauna to a silty gray shale that is burrowed and only slightly calcareous.

Figure 4E shows that the Lowellville unit in the study area is somewhat more limited in distribution than unit A (see Fig. 4C). Figure 3 uses the base of the Lower Mercer limestone as a datum and illustrates a facies characteristic generally noted in the core-hole data: the thicker and more fossiliferous occurrences of Lowellville appear to have been deposited in topographic lows. As the Lowellville thins, the diversity of marine fauna decreases so that in the thinnest sections only brackish water fauna are found. Because unit A and the Lowellville unit do occur in the same part of the stratigraphic section and are locally less than 30 ft (9 m) apart, their continuity as separate stratigraphic units from drill hole to drill hole can only be demonstrated by such facies maps as Figure 4, D and F. In a few cores the identification of the last occurrence of unit A and the first occurrence of the Lowellville was problematic because that portion of the stratigraphic section consists of intertonguing sequences of carbonaceous claystone and rooted seat earth overlain by thin burrowed zones. It is possible that the marine to brackish water conditions under which unit A was deposited did not entirely withdraw from all parts of the area. Then, a later more extensive transgressive pulse deposited those strata that are now identified as the Lowellville limestone. Cores drilled through the thicker and more marine portions of unit A have lithologies that appear to be identical to those of the Lowellville; the only difference is their stratigraphic positions. Conodonts from the marine unit at the base of section 5017 (Fig. 3) (here identified as unit "A") look similar to those obtained from the Lowellville (G. Merrill, written communication, 1992).

Historically, the Lowellville marine unit (correlated with Poverty Run; see Morningstar, 1922) has been placed stratigraphically above the Massillon sandstone and the Quakertown coal bed (see Fig. 2). The Massillon sandstone of Ohio has been correlated with one or both of the Lower and Upper Connoquenessing sandstones (Orton, 1884), which may coalesce to the west in Ohio to form a single unit. But whatever the correlation of those sandstone units, the stratigraphic position of the type Quakertown coal bed (see the measured section in Quakertown Run, Fig. 3) appears to be above the Lowellville limestone horizon (type section in section 5014) rather than well below the Lowellville, as shown in the standard geologic column for the Pottsville Group in Ohio (see Fig. 2). This physical relationship was noted by Lamb (1910), yet it apparently was never incorporated in the standard geologic column of Ohio. As shown in Figure 3, the Quakertown occurs at the base of, or in, the Upper Connoquenessing sandstone in Quakertown Run. In fact, the geographic occurrence of the Quakertown coal bed is restricted and appears to be limited to a small portion in the extreme northeastern part of the study area. Because the unit otherwise could not be traced in the study area, the name Quakertown should be limited to the thin coal or coaly horizon of the type area. A more suitable name should be found for the economically important coal bed in southern Ohio that presently is improperly designated as "Quakertown" and that is stratigraphically below the Lowellville limestone (Martino et al., 1992a).

Unit "B"

Unit "B" is the thinnest and most restricted of the Pottsville marine zones identified in core holes in the study area (Fig. 5A). Unit B, like unit A, commonly has gradational upper and lower contacts. Where present, it ranges in thickness from a few inches (centimeters) to slightly more than 20 ft (6 m). It consists of a medium gray to dark gray shale that is locally burrowed and that rarely contains a sparse brackish water to marine fauna of *Lingula* and *Orbiculoidea* shells. In two cores, fragments of productid and pelmatozoan shells were found in unit B. Where the unit is thicker, it locally appears to intertongue with nonmarine strata and splits into numerous thin burrowed beds that may contain *Lingula*. Analyses of core data and the facies distribution shown in Figure 5B suggest that the study area was dominated by fluvial clastic deposition during the period of deposition of the unit such that unit B seems to represent only local incursions of marine to brackish water into interdistributary embayments along the margins of the predominant fluvial systems.

Unit B occurs from 30 to 45 ft (9 to 13 m) below the base of the Lower Mercer limestone in the study area. Stratigraphically, it is at about the position of the Boggs limestone (see Fig. 2), as that term was used by Stout (1918, 1927) for an extensive marine limestone bed in Muskingum and Vinton Counties in east-central Ohio. Unfortunately, the name Boggs was used earlier to designate a nonmarine laminated iron ore deposit in Scioto County, southern Ohio (Orton, 1884), and it is not clear whether the marine unit of Stout or unit B of this study area correlates in any way with the Boggs iron ore deposit of Orton (see Hansen, 1986).

Lower Mercer limestone

Of all the marine units in the Pottsville Group of Ohio, the Lower Mercer is the most laterally persistent. It was originally named the Mercer limestone by Rogers (1858) from exposures near Mercer, Mercer County, Pennsylvania. White (1879) later revised it as the Lower Mercer limestone in Pennsylvania and traced it and the Upper Mercer limestone into Mahoning County, Ohio. Orton (1884) adopted White's correlation of the Lower Mercer limestone in Ohio and abandoned the usage of the term "Zoar" for a limestone described by Newberry (1874) in the same apparent stratigraphic position. The Lower Mercer limestone represents the first widespread marine unit of Pennsylvanian age in northeastern Ohio (Fig. 5C). By the time the Lower Mercer was deposited, the sedimentation and compaction effects related to irregular topography at the base of the Pennsylvanian had become almost nonexistent, and the Lower Mercer transgressed across a relatively flat and uniform coastal plain and produced a uniform and easily recognized deposit.

In the study area, the Lower Mercer limestone consists of 1 to 5 ft (0.3 to 1.5 m) of fossiliferous, finely crystalline to micritic, dark gray to black limestone. It commonly has an iron ore or sideritic zone near or at its top. A thin upward-coarsening sequence of shale and siltstone generally overlies the limestone facies, and the lower few inches (centimeters) to few feet (less than a meter) of the shale locally contains marine and brackish water fossils. Although generally persistent, the Lower Mercer limestone is locally truncated by fluvial sandstone in the southern part of the study area. The study area is not large enough to adequately characterize how the sandstone units that appear to trend both northwest and southwest in the southern part of Figure 5D fit into the regional fluvial systems.

Hook (1985) and Hook and Ferm (1988) identified stacked sandstone bodies in Allegheny strata on the south (downthrown) side of the Highlandtown fault (see Fig. 1B); they suggested syndepositional movement on the fault as the probable cause for these and other lithologic trends of Allegheny strata in this area. Most of the facies maps of the Pottsville units in Figures 4 and 5 show similar northwest-trending features paralleling subsurface fault structures shown in Figure 1B. The facies maps of the Lower and Upper Mercer limestones most clearly suggest the occurrence of a persistent fluvial system that generally parallels the Highlandtown fault and overlies what was probably a paleovalley or a fault scarp in the Mississippian rocks prior to the deposition of the Pennsylvanian strata (Fig. 4B). These factors suggest that the position and orientation of the sandstone bodies may be fault controlled.

Upper Mercer limestone

The strata originally assigned to the Mahoning limestone by Rogers (1858) were later reassigned to the newly named Upper Mercer limestone by White (1879). In Ohio, Orton (1878) had designated that limestone as the Gore but later abandoned its usage and adopted White's name of Upper Mercer. The limestone ranges in thickness from 0 to 3.5 ft (1.1 m) and is commonly dark gray to black, aphanic to finely crystalline, and silty to argillaceous and contains abundant marine invertebrate fossils. The top of the limestone can be sideritic or can grade vertically into a zone of siderite nodules, but most commonly the limestone is overlain by black to gray shale that coarsens rapidly upward to siltstone. Generally, only the lower few inches (centimeters) of the overlying shale contain marine fossils. Where the limestone facies is absent, the unit is represented by a variety of lithofacies (Fig. 5, E and F) such as calcareous, burrowed, sandy shale and dark gray clay shale containing a sparse to moderately abundant invertebrate fauna. Locally, zones of coaly shale appear to occupy the position of the Upper Mercer limestone unit.

As previously noted, some depositional trends of the Upper Mercer and related strata appear to parallel subsurface fault traces. In the northwestern part of the study area, the Upper Mercer limestone is locally replaced by either a sandstone unit or a calcareous shale facies near the area of the Smith Township–Suffield faults. Locally in the western part of the study area, other variations of lithofacies associated with the Upper Mercer were found. In OGS core 2616, located in the western part of the study area, three marine units containing limestone and fossiliferous shale occur interbedded with coal, seat earth, and other nonmarine strata in a 20-ft-thick (6-m-thick) section in the stratigraphic position of the Upper Mercer (see Fig. 6). This core does not appear to be located near any known subsurface structure that might have influenced deposition. However, adjacent cores contain only one limestone at the horizon of the Upper Mercer, and the limestone of those can be correlated to any of the three limestones described in the 20-ft-thick (3-m-thick) section of core 2616.

Lamb (1910) noted an "extra" limestone 20 to 30 ft (6 to

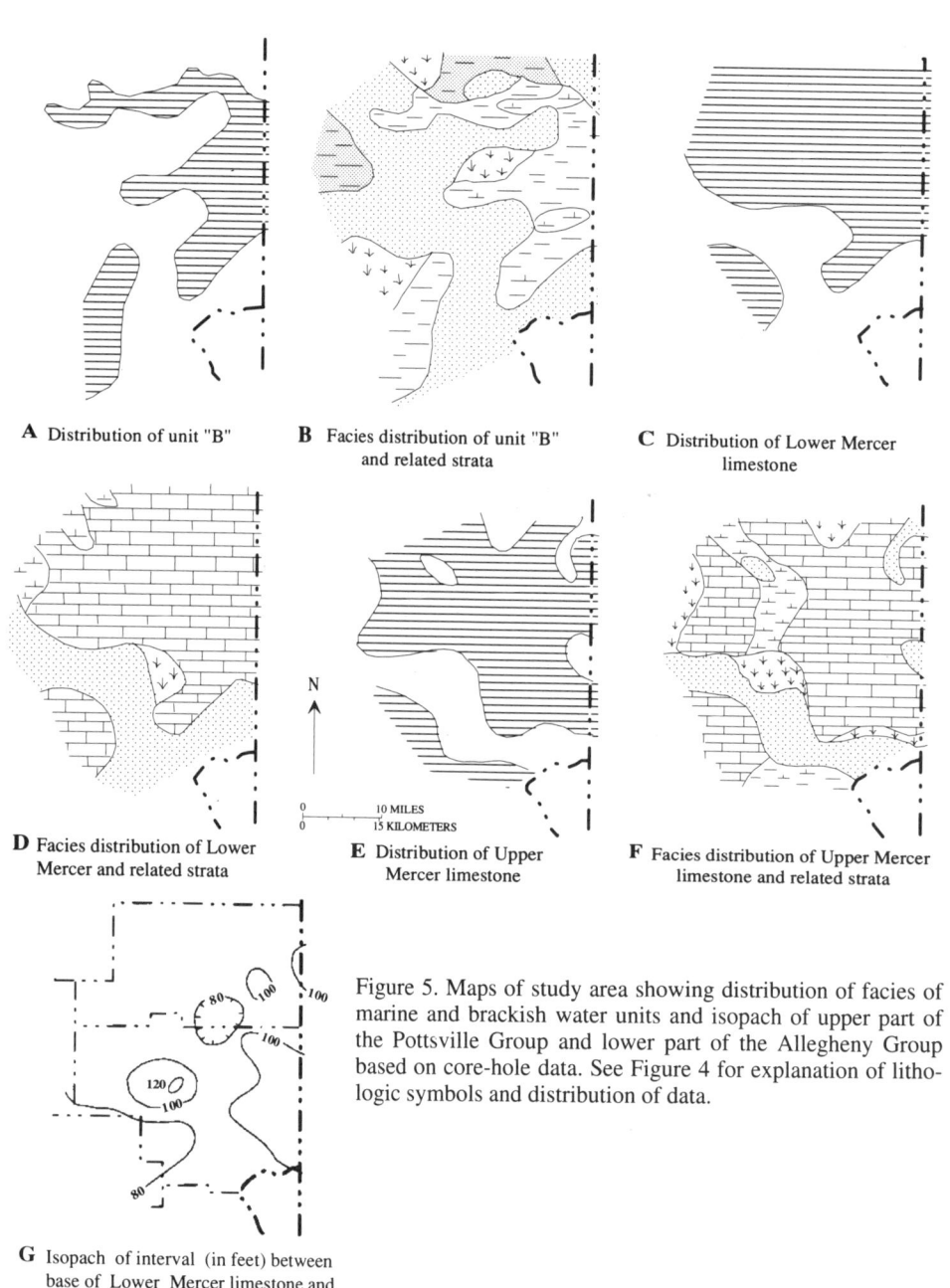

Figure 5. Maps of study area showing distribution of facies of marine and brackish water units and isopach of upper part of the Pottsville Group and lower part of the Allegheny Group based on core-hole data. See Figure 4 for explanation of lithologic symbols and distribution of data.

A Distribution of unit "B"

B Facies distribution of unit "B" and related strata

C Distribution of Lower Mercer limestone

D Facies distribution of Lower Mercer and related strata

E Distribution of Upper Mercer limestone

F Facies distribution of Upper Mercer limestone and related strata

G Isopach of interval (in feet) between base of Lower Mercer limestone and base of Vanport limestone. Hachures indicate direction of thinning

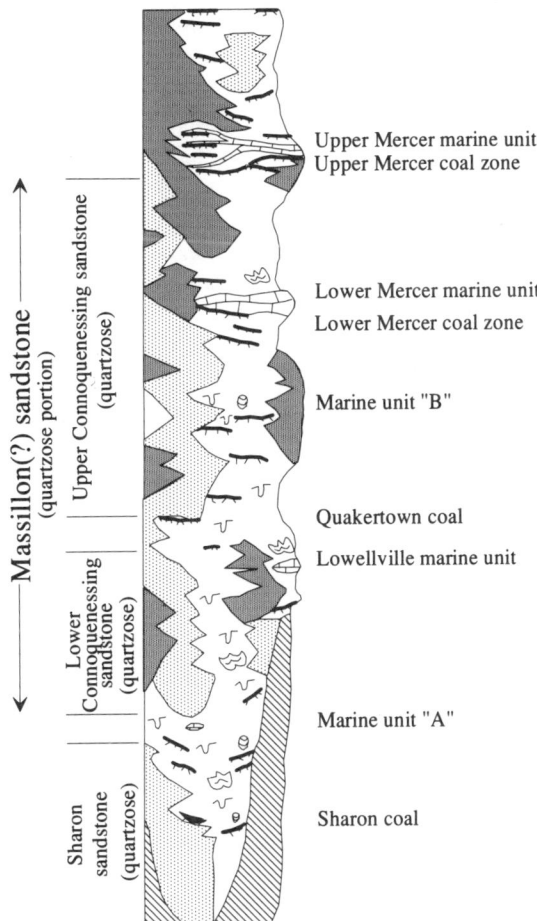

Figure 6. Schematic diagram of the stratigraphic column for the Pottsville Group and lower part of the Allegheny Group, undivided, northeastern Ohio, showing the discontinuity of named units based on data from the new geologic mapping program. Interval ranges from 187 to 404 ft (57 to 123 m) in thickness. See Figure 3 for lithologic descriptions. All units are considered informal.

9 m) above what he identified as the Upper Mercer in the Nimishillen Valley of southern Stark County, west of the study area, which he named the Howenstein limestone. Although Lamb attempted to trace the Howenstein eastward, Sturgeon (1943) was able to show that Lamb had mistakenly correlated the Vanport limestone in central Mahoning County with the Howenstein. Subsequent studies by the Ohio Geological Survey support Sturgeon's conclusions. However, the occurrence of the Howenstein limestone as described by Lamb in the Nimishillen Valley remains a problem. DeLong and White (1963) noted that, except for the section described by Lamb, no stratigraphic sections are known that contain both the Mercer limestones and the Howenstein limestone. Therefore, they interpreted the Howenstein to be the Upper Mercer, and made the Upper Mercer and Lower Mercer limestones of Lamb (1910) their Lower Mercer and Boggs limestones, re-

spectively. DeLong and White pointed out that isolated and limited exposures of the Lower and Upper Mercer limestones in Stark County make physical identification and correlation of these units (and the Boggs limestone) uncertain at best. However, the occurrence of three limestone beds in the Upper Mercer marine zone in OGS core 2616 suggests that the Howenstein probably represents a local upper split of the Upper Mercer limestone. Detailed mapping will likely identify other cases of limestone units splitting into several beds in Pennsylvanian strata of the Appalachian basin (see, for example, the Putnam Hill limestone as described by DeLong and White, 1963).

DISCUSSION AND CONCLUSIONS

Figure 6 is a generalization of the stratigraphic column for the Pottsville Group and is constructed mostly from data gathered by the Ohio Geological Survey core-drilling program in northeastern Ohio (1985–1989). Although the column contains many of the same key beds that are in the standard geologic column for the Pottsville of Ohio (Fig. 2), some named units, such as most of the coal beds and sandstone beds, cannot be correlated from the northeastern to the east-central and southern parts of the state. Analysis of the cores shows that many named units are absent or cannot be readily identified in large parts of the study area. For example, the Lower and Upper Connoquenessing sandstones of western Pennsylvania cannot be readily correlated with the Massillon sandstone of northeastern Ohio. These units are generally characterized as massively bedded quartzose sandstones. Newberry (1874) named and placed the Massillon above the No. 1 (Sharon) and below the No. 2 (Quakertown) coal beds, thus making it correlative with the Lower Connoquenessing sandstone, as later defined by White (1879). The Ohio Geological Survey, however, in what seems to be a redefinition of the name, has traditionally identified Massillon as the sandstone that overlies the No. 2 (Quakertown) coal bed (Stout, 1918, 1927; Bownocker and Dean, 1929; Collins, 1979). The difficulty of identifying individual sandstones within this interval led Orton (1884) to suggest that the name Massillon be applied to both Lower and Upper Connoquenessing sandstones in Ohio; Chance (1879) and Winslow and White (1966) also suggested that the Lower and Upper Connoquenessing sandstones be treated as a single unit. Cores from the study area, as shown in Figure 3 and schematically illustrated for the entire study area in Figure 6, indicate that a variety of lithologies occur in the interval, and field estimates of the mineral content of the sandstones range from 60 to 90% quartz. Thus, the sandstones of this part of the Pottsville section cannot be traced with confidence even within the study area. The usefulness of the names Massillon and Connoquenessing (or any other named sandstones) as key regional stratigraphic units for Ohio, then, is highly questionable, and they are used in Figure 6 only to illustrate the stratigraphic diversity of what would be considered the type Massillon in northeastern Ohio.

Of particular interest is the occurrence of marine-influenced strata in the lower parts of the Pennsylvanian section in northeastern Ohio at the same apparent stratigraphic position as that in southern Ohio. The southern transgressions appear to be more readily defined because they are commonly associated with coal beds. Martino et al. (1992a, b) identified at least four marine or brackish water transgressions in a 140-ft (43-m) interval between the Sharon coal at the base of the section and the Poverty Run (approximately equivalent to Lowellville of this report) limestone in roadcuts east of Jackson (Fig. 1A) in southern Ohio. Those marine units may all be equivalent to the sparsely fossiliferous and burrowed shales of unit A, which is found below the Lowellville limestone in the cores of the study area in northeastern Ohio (see Fig. 3, core logs 2598 and 2596). However, additional stratigraphic and paleontologic studies are needed to establish that relationship. The marine units of this interval in both northeastern and southeastern Ohio appear to have been deposited in approximately the same kind of restricted environments associated with alluviated valleys and estuaries of a broad coastal plain, far from any area of significant subsidence and subject mostly to subtle eustatic changes or compaction-induced subsidence.

It is evident that the earliest marine transgressions flooded into topographic lows of the eroded Mississippian landscape. Possibly as a result of differential compaction, later transgressions tended to reoccupy the position of earlier lows, but increased input of clastic sediments and rising sea level produced complicated patterns of distribution and facies in the lower part of the Pottsville section. Numerous cores from the study area contain paleoslump features about 1 to 10 ft (0.3 to 3 m) thick in the interval between the base of the Lower Mercer marine unit and the base of the Pennsylvanian section. The slumping of these strata may have been triggered by tectonism, channel cutting, or rapid deposition of fine-grained sediments (Allen, 1982) that was amplified by the irregular topography of the sub-Pennsylvanian surface and differential compaction in that part of the section. However, slump features have not been reported previously from this part of the stratigraphic section in other parts of Ohio, and the abundance of the features in northeastern Ohio therefore may be characteristic of only that area. Not until the time of the deposition of the Lower Mercer did marine transgressions extend across the entire study area and produce a relatively uniform widespread marine unit. Structural movements may have played an important part in the deposition and distribution of some marine and continental units, but the study area is too small to verify that hypothesis.

The discontinuity and the lack of biostratigraphic control of many of the stratigraphic units that make up the Pottsville Group in this study area, particularly those below the Lower Mercer limestone, show that they generally cannot be correlated regionally. Thus, local names should be applied to these stratigraphic units rather than names projected from distant parts of the basin; the use of stratigraphic terms from southern Ohio in northeastern Ohio, and the converse, as shown in Figure 2, is untenable. The difficulty in distinguishing between the Pottsville and Allegheny Groups is another case that suggests a need to redefine much of the Pennsylvanian stratigraphic section for Ohio on the basis of terms developed in Ohio. Doubtless, any new stratigraphic framework for this part of the Pennsylvanian section must depend primarily on the recognition and identification of key marine marker beds. The new detailed stratigraphic information developed in this study has important consequences concerning our ideas about the depositional framework of Pennsylvanian marine and continental units in this part of the Appalachian basin, and it illustrates the broad utility of the results that have been achieved by the core-drilling and new geologic mapping program of the Ohio Geological Survey.

REFERENCES CITED

Allen, J.R.L., 1982, Sedimentary structures, their character and physical basis: Developments in sedimentology 30B, v. 2: New York, Elsevier, 663 p.

Bownocker, J. A., and Dean, E. S., 1929, Analysis of the coals of Ohio: Ohio Division of Geological Survey Bulletin 34, 360 p.

Cecil, C. B., and Englund, K. J., 1985, Origin of quartzarenites in Upper Mississippian and Lower Pennsylvanian of Appalachian Basin [abs.]: American Association of Petroleum Geologists Bulletin, v. 69, no. 9, p. 1434.

Chance, H. M., 1879, A special survey made in 1875 along the Beaver and Shenango Rivers in Beaver, Lawrence, and Mercer Counties: 2nd Geological Survey of Pennsylvania, Report of Progress 5, pt. 2, p. 185–228.

Collins, H. R., 1979, The Mississippian and Pennsylvanian (Carboniferous) Systems in the United States—Ohio: U.S. Geological Survey Professional Paper 1110-E, 26 p.

DeLong, R. M., and White, G. W., 1963, Geology of Stark County: Ohio Division of Geological Survey Bulletin 61, 209 p.

de Witt, W., Jr., 1970, Age of the Bedford Shale, Berea Sandstone, and Sunbury Shale in the Appalachian and Michigan basins, Pennsylvania, Ohio, and Michigan: U.S. Geological Survey Bulletin 1294-G, 11 p.

Donaldson, A. C., 1974, Pennsylvanian sedimentation of central Appalachians, in Briggs, G., ed., Carboniferous of the southeastern United States: Geological Society of America Special Paper 148, p. 47-78.

Douglass, R. C., 1979, The distribution of fusulinids and their correlation between the Illinois basin and the Appalachian basin, in Palmer, J. E., and Dutcher, R. R., eds., Depositional and structural history of the Pennsylvanian System of the Illinois basin, Part 2, Invited papers—Field trip 9, Ninth International Congress of Carboniferous Stratigraphy and Geology: Illinois State Geological Survey Guidebook Series, no. 15, p. 15–30.

Douglass, R. C., 1987, Fusulinid biostratigraphy and correlations between the Appalachian and Eastern Interior basins: U.S. Geological Survey Professional Paper 1451, 95 p.

Ferm, J. C., 1970, Allegheny deltaic deposits, in Morgan, J. P., ed., Deltaic sedimentation, modern and ancient: Society of Economic Paleontologists and Mineralogists Special Publication 15, p. 246–255.

Ferm, J. C., 1974, Carboniferous environmental models in eastern United States and their significance, in Briggs, G., ed., Carboniferous of the southeastern United States: Geological Society of America Special Paper 148, p. 79–95.

Ferm, J. C., and Cavaroc, V. V., Jr., 1969, A field guide to Allegheny deltaic deposits in the upper Ohio Valley, with a commentary on deltaic aspects of Carboniferous rocks in the northern Appalachian Plateau: Pittsburgh and Ohio Geological Societies, Guidebook for Annual Field Trip, 21 p.

Ferm, J. C., and Horne, J. C., eds., 1979, Carboniferous depositional environments in the Appalachian region: Columbia, University of South Carolina, Carolina Coal Group, 760 p.

Ferm, J. C., and Weisenfluh, G. A., 1989, Evolution of some depositional models in Late Carboniferous rocks of the Appalachian coal fields, in Lyons, P. C., and Alpein, B., eds., Peat and coal; Origin, facies, and depositional models: International Journal of Coal Geology, v. 12, p. 259–292.

Ferm, J. C., and Williams, E. G., 1963, A model for cyclic sedimentation in the Appalachian Pennsylvanian: American Association of Petroleum Geologists Bulletin, v. 47, no. 2, p. 356–357.

Ferm, J. C., and 6 others, 1979, Allegheny correlations, in Ferm, J. C., and Horne, J. C., eds., Carboniferous depositional environments in the Appalachian region: Columbia, University of South Carolina, Carolina Coal Group, p. 13–17.

Fuller, J. O., 1955, Source of Sharon conglomerate of northeastern Ohio: Geological Society of America Bulletin, v. 66, p. 159–176.

Granchi, J. A., 1958, Coal resources of the Pottsville Formation: Ohio Division of Geological Survey Report of Investigations, no. 36, 53 p.

Gray, H. H., 1954, Stratigraphy and sedimentation of Pottsville rocks near Beach City, Ohio [Ph.D. dissertation]: Columbus, Ohio State University, 123 p.

Gray, H. H., 1956, Petrology of the Massillon sandstone at the type locality: Ohio Journal of Science, v. 56, no. 3, p. 138–146.

Gray, J. D., 1982, Subsurface structure mapping in eastern Ohio, in Gray, J. D., et al., eds., An integrated study of the Devonian-age black shale in eastern Ohio: U.S. Department of Energy, Report DOE/ET/12131-1399, p. 3.1–3.13.

Hansen, M. C., 1984, Middle Carboniferous depositional events; The Sharon Sandstone, a basal Pennsylvanian quartz arenite in the north-central Appalachian basin, in Shumaker, R. C., comp., Proceedings, Appalachian Basin Industrial Associates, Columbus, Ohio, Spring Meeting, 1983: Columbus, Ohio State University, p. 148–173.

Hansen, M. C., 1986, Microscopic Chondrichthyan remains from Pennsylvanian marine rocks of Ohio and adjacent areas [Ph.D. dissertation]: Columbus, Ohio State University, 536 p.

Hildreth, S. P., 1828, Miscellaneous observations of the coal, diluvial, and other strata of certain portions of the state of Ohio: American Journal of Science and Arts, v. 13, no. 1, p. 38–40.

Hildreth, S. P., 1833, Observations on the saliferous rock formation, in the valley of the Ohio: American Journal of Science and Arts, v. 24, no. 1, p. 46–68.

Hildreth, S. P., 1836, Observations of the bituminous coal deposits of the valley of the Ohio, and the accompanying rock strata; With notices of the fossil organic remains and the relics of vegetable and animal bodies, illustrated by a geological map, by numerous drawings of plants and shells, and by views of interesting scenery: American Journal of Science, Arts, v. 29, no. 1, p. 1–154.

Hook, R. W., 1985, A paleoenvironmental model for the occurrence of vertebrate fossils in Carboniferous coal-bearing strata [Ph.D. dissertation]: Lexington, University of Kentucky, 72 p.

Hook, R. W., and Ferm, J. C., 1988, Paleoenvironmental controls on vertebrate-bearing abandoned channels of the Upper Carboniferous: Paleogeography, Paleoclimatology, and Paleoecology, v. 63, p. 159–181.

Hyde, J. E. (Marple, M. F., ed.), 1953, Mississippian formations of central and southern Ohio: Ohio Division of Geological Survey Bulletin 51, 355 p.

Ketring, C. L., Jr., 1984, Paleogeography and subsurface geometry of the "Sharon" Conglomerate (Pennsylvanian) in Jackson and Gallia Counties, Ohio [M.S. thesis]: Columbus, Ohio State University, 103 p.

Lamb, G. F., 1910, Pennsylvanian limestones of northeastern Ohio below the Lower Kittanning coal: Ohio Naturalist, v. 10, no. 5, p. 89–135.

Lamb, G. F., 1911, The Mississippian-Pennsylvanian unconformity and the Sharon conglomerate: Journal of Geology, v. 19, no. 2, p. 104–109.

Larsen, G. E., 1991, Historical development and problems within the Pennsylvanian nomenclature of Ohio: Ohio Journal of Science, v. 91, no. 1, p. 69–76.

Larsen, G. E., and Rea, R. G., 1990, Bedrock geology of the Alliance quadrangle, Ohio: Ohio Division of Geological Survey Open-File Map BG-C2H1, scale 1:24,000.

Martino, R. L., Rice, C. L., and Slucher, E. R., 1992a, Stop 3: section above No. 2 coal bed near Jackson, Ohio, in measured sections M3, M4, M5, and M6, in Rice, C. L., Martino, R. L., and Slucher, E. R., Regional aspects of Pottsville and Allegheny stratigraphy and depositional environments of Ohio and Kentucky (Geological Society of America, 1992 annual meeting, Cincinnati, field trip 4): U.S. Geological Survey Open-File Report 92-558, p. 21–23.

Martino, R. L., Rice, C. L., and Slucher, E. R., 1992b, Stop 2: Basal Pennsylvanian strata near Jackson, Ohio, in Rice, C. L., Martino, R. L., and Slucher, E.R., Regional aspects of Pottsville and Allegheny stratigraphy and depositional environments of Ohio and Kentucky (Geological Society of America, 1992 annual meeting, Cincinnati, field trip 4): U.S. Geological Survey Open-File Report 92-558, p. 13–16.

Merrill, G. K., 1974, Pennsylvanian conodont localities in northeastern Ohio: Ohio Division of Geological Survey Guidebook 3, 25 p.

Merrill, G. K., 1991, Advances in Appalachian Pennsylvanian conodont biostratigraphy: Northeastern-Southeastern Sections Meeting, Geological Society of America Abstracts with Programs, v. 23, no. 1, p. 103.

Moore, R. C., chairman, et al., 1944, Correlation of Pennsylvanian formations of North America: Geological Society of America Bulletin, v. 55, no. 6, p. 657–706.

Morningstar, H., 1922, Pottsville fauna of Ohio: Ohio Division of Geological Survey Bulletin 25, 312 p.

Morse, W. C., 1910, The Maxville limestone: Ohio Division of Geological Survey Bulletin 13, 128 p.

Mrakovich, J. V., 1969, Sedimentary structures and depositional environment of the Sharon Conglomerate in western Summit, eastern Medina, and northeastern Wayne Counties [M.S. thesis]: Kent, Ohio, Kent State University, 92 p.

Murphy, J. L., 1973, The Noble Limestone Member (Conemaugh Group, Pennsylvanian)—New occurrence in Noble and Guernsey Counties, Ohio: Ohio Journal of Science, v. 73, no. 1, p. 42–46.

Newberry, J. S., 1874, The Carboniferous System: Ohio Division of Geological Survey, v. 2, pt. I, Geology, p. 81–180.

Newberry, J. S., 1878, Report on the geology of Jefferson County; Mahoning County: Ohio Division of Geological Survey, v. 3, p. 716–814.

Orton, E., 1878, Supplemental report on the geology of the Hanging Rock district: Ohio Division of Geological Survey, v. 3, p. 883–941.

Orton, E., 1883, The lower coal measures of Ohio: Ohio Mining Journal, v. 1, no. 3, p. 97–109.

Orton, E., 1884, The stratigraphical order of the lower coal measures of Ohio: Ohio Division of Geological Survey, v. 5, Economic Geology, p. 1–128.

Prosser, C. S., 1905, Revised nomenclature of the Ohio geological formations: Ohio Division of Geological Survey Bulletin 7, 36 p.

Rice, C. L., and Schwietering, J. F., 1988, Fluvial deposition in the central Appalachians during the Early Pennsylvanian: U.S. Geological Survey Bulletin 1839-B, 10 p.

Rogers, H D., 1858, The geology of Pennsylvania: Philadelphia, Lippincott, v. 1, 586 p.; v. 2, 1046 p.

Root, S. I., 1992, Effect of the Transylvania fracture zone on evolution of the western margin of the central Appalachian basin, in Bartholomew, M. J., Hyndman, D. W., Mogk, D. W., and Mason, R., eds., Basement tectonics 8—Characterization and comparison of ancient and Mesozoic continental margins, Proceedings, International Conference of Basement Tectonics, 8th, Butte, Montana, 1988: Dordrecht, The Netherlands, Kluwer, p. 469–480.

Root, S. I., and MacWilliams, R. H., 1986, The Suffield fault, Stark County, Ohio: Ohio Journal of Science, v. 86, no. 4, p. 161–163.

Ryder, R. T., 1991, Stratigraphic framework of Cambrian and Ordovician rocks in the central Appalachian basin from Richland County, Ohio, to Rockingham County, Virginia: U.S. Geological Survey Miscellaneous Investigation Series Map I-2264.

Schmidley, E. B., 1987, The sedimentology, paleogeography and tectonic setting of the Pennsylvanian Massillon sandstone in east-central Ohio [M.S. thesis]: Akron, Ohio, University of Akron, 193 p.

Shumaker, R. C., 1986, The effect of basement structure on sedimentation and detached structural trends within the Appalachian basin, in McDowell, R. C., and Glover, L., III, eds., The Lowry volume—Studies in Appalachian geology: Blacksburg, Virginia Polytechnic Institute Department of Geological Sciences, Memoir 3, p. 67–81.

Slucher, E. R., and Larsen, G. E., 1989, Bedrock geology of the East Palestine quadrangle, Ohio and Pennsylvania: Ohio Division of Geological Survey Open-File Map BG-C1G5, scale 1:24,000.

Stevenson, E. L., 1933, The geology of the Youngstown region [M.S. thesis]: Columbus, Ohio State University, 129 p.

Stout, W., 1916, Geology of southern Ohio including Jackson and Lawrence Counties and parts of Pike, Scioto, and Gallia: Ohio Division of Geological Survey Bulletin 20, 723 p.

Stout, W., 1918, Geology of Muskingum County: Ohio Division of Geological Survey Bulletin 21, 351 p.

Stout, W., 1923, Origin of coal formation clays, in Stout, W., Stull, R. T., McCaughey, W. J., and Demorest, D. J., Coal formation clays of Ohio: Ohio Division of Geological Survey Bulletin 26, p. 533–568.

Stout, W., 1927, Geology of Vinton County: Ohio Division of Geological Survey Bulletin 31, 402 p.

Stout, W., and Lamborn, R. E., 1924, Geology of Columbiana County: Ohio Division of Geological Survey Bulletin 28, 408 p.

Sturgeon, M. T., 1943, Contributions to the stratigraphy of the Allegheny Series in Columbiana and Mahoning Counties, Ohio—Part I, Stratigraphy and correlation of the coals and limestone below the Lower Kittanning coal: Ohio Journal of Science, v. 43, no. 6, p. 235–249.

Sturgeon, M. T., and DeLong, R. M., 1964, Revision of some stratigraphic names between the Lower and Middle Kittanning coals in eastern Ohio: Ohio Journal of Science, v. 64, no. 1, p. 41–43.

Sturgeon, M. T., and Merrill, W. M., 1949, An additional fossiliferous member in the Allegheny Formation (Pennsylvanian) of Ohio: Ohio Journal of Science, v. 49, no. 1, p. 1–11.

Uttley, J. S., 1974, The stratigraphy of the Maxville Group of Ohio and correlative strata in adjacent areas [Ph.D. dissertation]: Columbus, Ohio State University, 252 p.

Wanless, H. R., 1939, Pennsylvanian correlations in the Eastern Interior and Appalachian coal fields: Geological Society of America Special Paper 17, 130 p.

Wanless, H. R., 1975, Appalachian region, in McKee, E. D., and Crosby, E. J., comps., Paleotectonic investigations of the Pennsylvanian System in the United States—Part 1, Introduction and regional analysis of the Pennsylvanian System: U.S. Geological Survey Professional Paper 853, pt. 1, p. 17–62.

Weller, J. M., 1930, Cyclic sedimentation of the Pennsylvanian period and its significance: Journal of Geology, v. 38, no. 2, p. 97–135.

White, G. W., and Totten, S. M., 1985, Glacial geology of Columbiana County, Ohio: Ohio Division of Geological Survey Report of Investigations 129, 25 p.

White, I. C., 1879, Special report on the correlation of the coal measures in western Pennsylvania and eastern Ohio: Pennsylvania Geological Survey Report of Progress QQ, p. 215–303.

Winslow, J. D., and White, G. W., 1966, Geology and groundwater resources of Portage County, Ohio: U.S. Geological Survey Professional Paper 511, 80 p.

MANUSCRIPT ACCEPTED BY THE SOCIETY FEBRUARY 1, 1994

Geological Society of America
Special Paper 294
1994

Revised stratigraphy and nomenclature for the Middle Pennsylvanian Kanawha Formation in southwestern West Virginia

Bascombe M. Blake, Jr., and Alan F. Keiser
West Virginia Geological and Economic Survey, Morgantown, West Virginia 26505-0879
Charles L. Rice
U.S. Geological Survey, National Center, MS 926, Reston, Virginia 22092

ABSTRACT

The stratigraphy of the Kanawha Formation in West Virginia has been confused by regional miscorrelations of many units. To resolve these inconsistencies, this report has: (1) revised and defined three widely distributed marine units as the Betsie, Dingess, and Winifrede Shale Members of the Kanawha Formation (Middle Pennsylvanian); (2) extended the name "Fire Clay" into West Virginia from Kentucky for a coal bed regionally identified by its flint clay (tonstein) parting and miscorrelated in different areas of West Virginia as the older Hernshaw coal bed or the younger Chilton coal bed; and (3) reestablished the stratigraphic positions of several key coal beds that have been regionally miscorrelated from their type areas. A stratigraphic section parallel to depositional strike, from the Kanawha River Valley in central West Virginia to the Tug Fork of the Big Sandy River in southwestern West Virginia, shows the correlation and continuity of marine members and coal beds of the middle part of the Kanawha Formation.

INTRODUCTION

The Kanawha Formation of West Virginia is composed largely of coal-bearing sequences of sandstones, siltstones, and shales. These strata were deposited in a moderately subsiding foreland basin as part of a west-northwestward–prograding clastic wedge shed from eastern and southeastern highlands during the Middle Pennsylvanian. Lateral variations in lithology and thickness of the strata are generally so great that correlative units are at times difficult to recognize, even in closely spaced sections (Wanless, 1946). This lithologic variability has led to the development of numerous depositional models for the Kanawha Formation including coastal plain (Tankard, 1986; Martino, this volume), beach/barrier-back barrier (Hobday and Horne, 1977; Ferm and Weisenfluh, 1989), deltaic-alluvial plain (Horne and Ferm, 1978; Flores and Arndt, 1979; Donaldson and Shumaker, 1981; Tankard, 1986), and shallow marine. Historically, the Kanawha Formation has been subdivided by key coal beds of regional extent, as well as by locally persistent sandstone beds and a few areally extensive, fossiliferous shales and limestones of marine origin. The strata of the Kanawha Formation are gently folded and commonly poorly exposed. Additionally, as the Kanawha thickens from about 700 ft (210 m) in the Kanawha River Valley near Charleston to more than 2,000 ft (600 m) near Bolt in Raleigh County to the southeast (Fig. 1), the number of coal beds and other stratigraphic units also increases (Arkle, 1974). These stratigraphic complexities have presented formidable obstacles to the development of a regionally consistent nomenclature for the Kanawha Formation.

Central to the development of geologic nomenclature in West Virginia are the county geologic reports and accompanying geologic maps (scale, 1:62,500) published between 1906 and 1939. The county reports by the West Virginia Geological and Economic Survey that were most influential in the development of the nomenclature presently used for the Kanawha Formation in West Virginia were published prior to 1920: Boone County (Krebs and Teets, 1915), Fayette County (Hennen and Teets, 1919), Kanawha County (Krebs and Teets,

---

Blake, B. M., Jr., Keiser, A. F., and Rice, C. L., 1994, Revised stratigraphy and nomenclature for the Middle Pennsylvanian Kanawha Formation in southwestern West Virginia, *in* Rice, C. L., ed., Elements of Pennsylvanian Stratigraphy, Central Appalachian Basin: Boulder, Colorado, Geological Society of America Special Paper 294.

41

Figure 1. Index map of part of West Virginia showing location of stratigraphic section A-A' (Fig. 9), location of boreholes, and type localities for the Dingess and Winifrede Shale Members of the Kanawha Formation. Labelled counties were the subject of geologic reports (see text) that influenced the nomenclature of the Kanawha Formation.

1914), Logan and Mingo Counties (Hennen and Reger, 1914), Raleigh County (Krebs and Teets, 1916), and Wyoming and McDowell Counties (Hennen and Gawthrop, 1915). Figure 1 shows the counties covered by these geologic reports.

Many of the key coal beds and marine units of the Kanawha Formation were initially identified and described in the Kanawha County report by Krebs and Teets (1914). Their data were based largely on earlier stratigraphic studies of the coal measures in the Kanawha River Valley by White (1891, 1903, 1908). The authors of the other county reports adapted and modified the stratigraphic nomenclature of the Kanawha County report to suit their individual stratigraphic needs. Because the subdivision of the Kanawha Formation in most of the county reports depended on the identification of the thickest and most persistent coal beds that were of commercial interest at that time, stratigraphy in individual county geologic reports is generally internally consistent. However, individual coal beds were miscorrelated between mining districts, which were separated by areas of undeveloped and rugged terrain. Thus, the stratigraphic nomenclature between the Kanawha River area and the area of the Tug Fork of the Big Sandy River has proven to be inconsistent (see Fig. 2).

A detailed stratigraphic column intended for use as a standard reference section in all county geologic reports was prepared by R. V. Hennen at the direction of White (1914). That stratigraphic section did not fully take into account contemporaneous stratigraphic work being done in other parts of West Virginia. Consequently, some key coal beds and marine units of the Kanawha Formation were not recognized in the Kanawha River Valley. Later attempts to insert these coal beds and marine units into this standard section for the Kanawha Formation introduced miscorrelations, and, in some cases, resulted in the misplacement of sequences of beds. For example, the Seth Limestone of Krebs and Teets (1915) and the Alma coal bed of the Tug Fork area were incorrectly projected into the Kanawha River Valley section because of their relative positions, above and below, respectively, a coal bed mistakenly identified as "Cedar Grove" in the Tug Fork area (see Fig. 2).

Recent stratigraphic studies by the West Virginia Geological and Economic Survey Coal Section (Blake et al., 1981) and geologic studies in correlative strata of the Breathitt Formation in eastern Kentucky (e.g., Huddle et al., 1963) have clearly demonstrated the utility of key beds and sequences of key beds for analyzing the complex Middle Pennsylvanian stratigraphy. Marine shale and limestone units in West Virginia and Kentucky are more extensive and identifiable throughout the basin than the more discontinuous coal beds. Earlier workers recognized the marine fossil-bearing shales and limestones as important stratigraphic and economic units. Unfortunately, several marine units were locally misidentified and given more than one name. For synopses of the geologic reports on West Virginia counties and subsequent geologic investigations of the Kanawha Formation, see Arkle (1974), Arkle et al. (1979), and Arndt (1979).

This report revises the nomenclature of the Kanawha Formation in West Virginia, removes long-standing miscorrelations, and provides a common stratigraphic section based on accurate correlation of regionally extensive marine shale and limestone units and key coal beds. A secondary purpose of this report is to define three key marine members in the Kanawha Formation. Unless otherwise noted, the stratigraphic names used in this chapter are those established in the Kanawha County report of Krebs and Teets (1914).

KEY MARINE UNITS

Dark shales and siltstones containing marine, brackish, and freshwater invertebrate fauna are found in many parts of the Kanawha Formation in West Virginia and correlative strata in eastern Kentucky and southwestern Virginia. Invertebrate fossils commonly occur in the shales directly overlying coal beds. Several of the reported marine units are thick and areally extensive enough to be identified and mapped widely in the coal field. Those units, which may be more than 100 ft (30 m) thick, mostly consist of coarsening-upward sequences of shale, siltstone, and sandstone deposited in shallow epicontinental seas. In general, the basal contacts of the marine units are sharp and commonly occur at the base of fossiliferous shales or fine-grained sandstones. The upper contacts of the marine zones are commonly disconformable with overlying strata, commonly a fluvial sandstone, indicating a period of nondeposition or truncation by an overriding fluvial system.

Series	Formation	KENTUCKY: Big Sandy District (Modified from Rice & Smith, 1980; Rice et al., 1987)	KENTUCKY: Williamson 7.5' quadrangle (Modified from Alvord & Trent, 1962)	WEST VIRGINIA: Tug Fork Area (Hennen & Reger, 1914)	WEST VIRGINIA: Kanawha River Valley (Modified from Hennen & Teets, 1914; Arndt et al., 1979b)	WEST VIRGINIA: **This Report**	Formation	Series
Middle Pennsylvanian (part)	Breathitt (part)	Hazard coal zone	Winifrede coal bed	Winifrede/Buffalo Creek coal bed	Winifrede coal bed	Winifrede coal zone	Kanawha (part)	Middle Pennsylvanian (part)
		Magoffin Member	Magoffin Beds of Morse (1931)	*Buffalo Creek Limestone of Hennen & Reger (1914)*	Winifrede Limestone of White (1908)	**Winifrede Shale Member**		
		Taylor coal bed	Taylor coal bed	*Chilton "A" coal bed*	Chilton coal bed	Chilton coal zone		
		Fire Clay coal bed	Fire Clay coal bed	*Chilton coal bed*	*Hernshaw (?) coal bed*	Fire Clay coal zone		
		Unnamed marine zone	Unnamed marine zone	(Not reported)	*Seth Limestone of Krebs (1915)* ④	Unnamed marine unit		
		Whitesburg coal bed	Whitesburg coal bed	Hernshaw coal bed	Cedar Grove coal bed	Cedar Grove coal bed		
		Kendrick Shale Member	Kendrick Shale of Jillson (1919)	Dingess Limestone of Hennen & Reger (1914)	(Position commonly described as occupied by the Peerless sandstone)	**Dingess Shale Member**		
		Williamson coal bed	Williamson coal bed	Williamson coal bed	*Alma coal bed*	Williamson coal zone		
		Elkins Fork Shale of Morse (1931)	Elkins Fork Shale of Morse (1931)	*Seth Limestone of Krebs (1915)* ③	Campbell Creek Limestone of White (1885)	Campbell Creek Limestone of White (1885)		
		Upper Elkhorn coal zone: Upper Elkhorn No. 3 coal zone	Nosben coal bed	*Cedar Grove coal bed*	Campbell Creek coal zone: Peerless coal bed	**Peerless coal zone**		
			Sidney coal bed	*Lower Cedar Grove coal bed*	No. 2 Gas coal bed	**No. 2 Gas coal zone**		
		Upper Elkhorn coal zone: Upper Elkhorn Nos. 1 & 2 coal zone	Alma coal bed	Alma coal bed	Powellton coal bed	Powellton coal bed		
		Crummies Member	*Campbell Creek Limestone of White (1885)* ①	*Campbell Creek Limestone of White (1885)* ③	Cannelton Limestone of White (1885)	Cannelton Limestone of White (1885)		
		Lower Elkhorn coal zone	Pond Creek coal bed	*Campbell Creek coal bed*	Eagle coal bed	Eagle coal bed		
		Betsie Shale Member	*Cannelton Limestone of White (1885)* ②	*Cannelton Limestone of White (1885)* ③	Eagle Limestone and Shale of White (1891)	**Betsie Shale Member**		
		Matewan coal bed	(Not discussed)	Matewan coal bed	Unnamed coal bed	**Matewan coal bed**		

① as used by Alvord, 1971
② as used by Trent, 1965
③ as used by Hennen & Reger, 1914
④ as used by Hennen & Teets, 1919

Bold Usage this report
Italic Erroneous nomenclature-usage in previous reports

Figure 2. Chart showing the revision of names used in this report and the correlation of key marine units, coal beds, and coal zones of the middle part of the Kanawha Formation in West Virginia, with equivalent units in eastern Kentucky.

Invertebrate fossils most commonly occur in the basal transgressive part of the marine shales or in strata deposited in the offshore prodelta environments. The middle and upper parts of the marine units commonly are barren of invertebrate fossils, possibly because of rapid sedimentation rates during the regressive part of the marine sequence. Locally, invertebrate fossils occur throughout a marine unit or are abundant in thin but discontinuous beds within the middle and upper parts of a marine unit. These latter occurrences are possibly due to slow sedimentation rates or to stream channel migration or avulsion, which provided relatively sediment-free environments for the establishment of faunal communities in the rapidly deposited regressive part of the marine sequence.

The marine units are characterized by large ellipsoidal, arenaceous limestone concretions (Fig. 3), which are locally fossiliferous and which may contain cone-in-cone structure. Limestone concretions commonly occur at distinct horizons and may coalesce into irregular, discontinuous beds of fossiliferous limestone, particularly near the bases of the units. Some of the original names for the marine units, such as the Campbell Creek Limestone of White (1885) and the Cannelton Limestone of White (1885), apparently refer only to the limestone beds and concretions and not to the marine units as a whole (see Krebs and Teets, 1914). Our studies show that the marine units of the Kanawha Formation contain only minor amounts of limestone and are best characterized as interbedded to interlaminated sequences of shale, siltstone, and very fine to fine-grained sandstone.

A common feature of the Middle Pennsylvanian marine shales and siltstones are thin, diffuse bands and discontinuous nodular beds of siderite (Fig. 3) (Weller, 1930). Locally, transgressive lag deposits, which are generally extensively burrowed, consisting of argillaceous, calcareous, very fine to fine-grained sandstone with abundant siderite pebbles, mud clasts (rip-ups), and invertebrate shell material, are present at the bases of the marine units. These deposits generally mark the presence of a ravinement surface eroded during marine transgression (Swift, 1968; Liu and Gastaldo, 1989; Liu, 1990) but could also represent tempestites. Brachiopods, pelecypods, cephalopods, crinoids, and solitary corals are locally abundant in several of the Middle Pennsylvanian marine units (Henry and Gordon, 1979); unfortunately, most are long-ranging forms and only a few have ranges short enough to be useful for stratigraphic studies.

Some workers have emphasized the deltaic and diachronous nature of the stratigraphic units that make up the Kanawha Formation (see, for example, Flores and Arndt, 1979; Ferm and Weisenfluh, 1989) and have described the marine sequences generally as interdistributary bayfills. Although some thin marginal-marine, brackish or freshwater deposits may fall into that category, such analyses fail to recognize the regional continuity of the major marine units, which resulted from rapid transgressions of the sea across very extensive, flat coastal plains. Nor do such analyses account for the uniform nature of the prograding detrital sediments, which are punctuated by numerous, widespread, thick coal beds that filled the receiving basin. Conversely, many marine units have been widely recognized and their utility for stratigraphic subdivision in Middle Pennsylvanian rocks of the Appalachian basin was recognized by White (1914) and was demonstrated by the mapping done by the U.S. Geological Survey and the Kentucky Geological Survey during 1960 to 1978 (see Rice and Smith, 1980).

Many of the marine units have recognizable gamma-ray log signatures caused by the affinity between uranium and organic material in the marine shales (Potter et al., 1984). Thus, the organic material–rich shales at the bases of the marine units produce a greater gamma-ray deflection than the coarser grained siltstones and sandstones in the upper parts of the units that contain less organic matter (Fig. 4). These marine units may be difficult to identify by gamma-ray logs when the

Figure 3. Type locality of the Cannelton ("Stockton") Limestone of White (1885) in U.S. Highway 60 roadcut, north side of Kanawha River opposite Montgomery, West Virginia. The Cannelton extends from about road level to base of channel-fill sandstone. Discoidal limestone concretions near base, for which the unit was named, are as much as 5 ft (1.5 m) in diameter. Well-developed laminations in shale and siltstone are due largely to thin beds and diffuse bands of siderite.

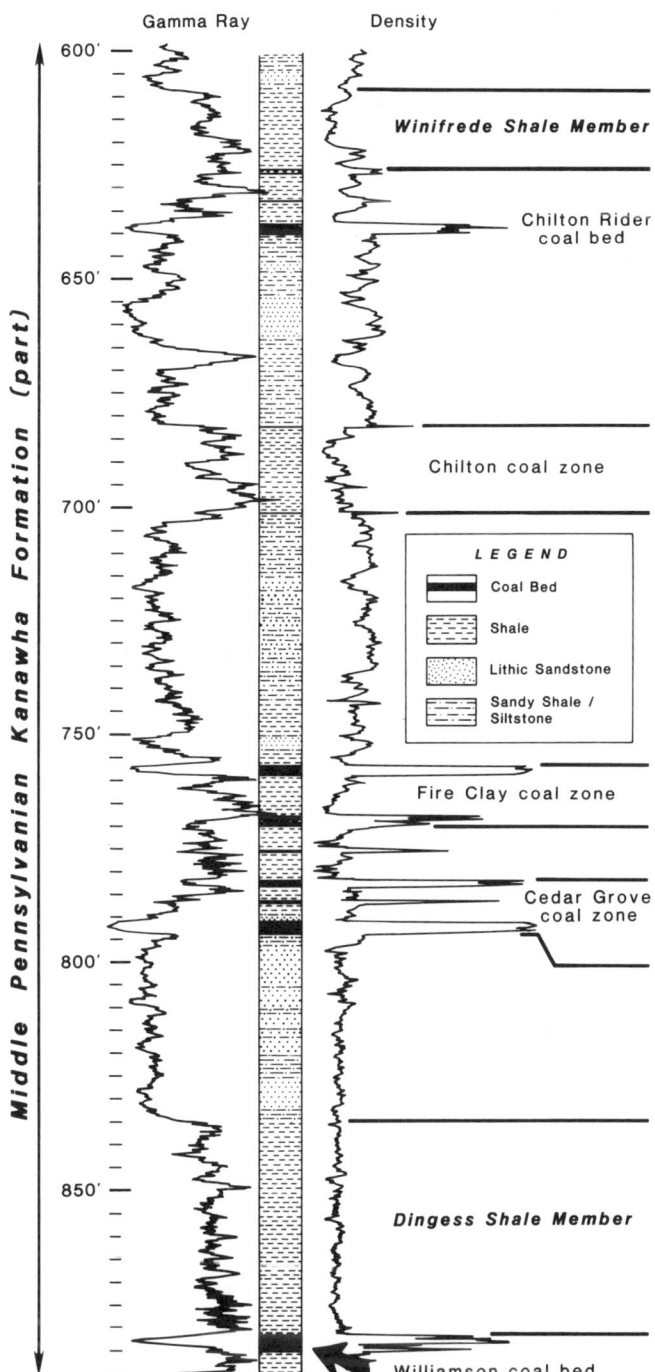

Figure 4. Lithic log of diamond drillhole core of middle part of the Kanawha Formation and corresponding gamma-ray and density (gamma-gamma) logs. Location is near section 4, Figure 1. A zone of several gamma-ray peaks occurs in the lower 10 ft (3 m) of the Dingess Shale Member. The reduced gamma-ray response in the lower 5 ft (1.5 m) of the Winifrede Shale Member may be due to the presence of a calcareous transgressive sandstone at the base of the unit. Gamma-ray and density log curves are slightly misaligned vertically, possibly due to tool spacing.

coarsening-upward sequences are thin or incomplete because of truncation by overriding fluvial systems.

On the basis of their regional distribution, three of the marine units in the Kanawha Formation are recognized as key elements in the Middle Pennsylvanian stratigraphic framework proposed in this chapter. They are, from oldest to youngest, the Betsie, Dingess, and Winifrede Shale Members.

Betsie Shale Member

The Betsie Shale Member was named by Rice et al. (1987) to replace the miscorrelated Eagle Limestone of White (1891). The Eagle was named for exposures of fossiliferous limestone and shale near Eagle about 1.5 mi (2.5 km) southeast of Montgomery, Fayette County, West Virginia. The unit was described about 2 mi (3.2 km) southeast of Eagle, on the north side of the Kanawha River at Alloy, West Virginia, by Englund et al. (Fig. 25, 1979) as a 2-ft-thick (0.6 m) calcareous, fossiliferous siltstone at the top of a 40-ft-thick (12 m) dark shale. Hennen and Reger (1914), in an attempt to extend the Eagle from the Kanawha River Valley to the Tug Fork Valley in Mingo County, incorrectly assigned the name Eagle to a stratigraphically lower marine unit in Mingo County (not shown in Fig. 2 but below the Matewan coal bed) that is probably the Dorothy Limestone and Shale of Krebs and Teets (1916). Hennen and Reger also miscorrelated the actual stratigraphic position of the Eagle in the Tug Fork Valley; in its place they showed the younger Cannelton ("Stockton") Limestone of White (1885). Their incorrect usage of the name "Cannelton" was extended into Kentucky by Huddle and Englund (1966) and by several U.S. Geological Survey geologic quadrangle maps in Kentucky (for example, Trent, 1965; Alvord and Miller, 1972). Thus, the Cannelton generally was incorrectly positioned in correlation charts of Pennsylvanian strata for eastern Kentucky (see, for example, Rice and Smith, 1980). Subsequently, following Hennen and Reger's usage of the name Cannelton, the name Eagle Limestone was misapplied to older strata in many of these same reports in Kentucky, resulting in an unfortunate stratigraphic confusion for one of the most extensive and recognizable stratigraphic units in the central Appalachian basin.

Because the name Eagle was preoccupied by several other formations in North America, Rice et al. (1987) formally renamed the marine unit the Betsie Shale Member of the Breathitt Formation in Kentucky and Tennessee, of the Wise Formation in Virginia, and of the Kanawha Formation in West Virginia; they designated a distinctive 135-ft-thick (41 m) shale and siltstone section near Betsie Gap, Bell County, Kentucky, as the stratotype. Rice et al. (1987) demonstrated the continuity of the Betsie Shale Member with the Eagle Limestone of White (1891) in West Virginia, and extended the Betsie into West Virginia on that basis. An easily accessible reference section for the Betsie Shale Member in West Virginia is near the town of Chapmanville in Logan County where it is well exposed at road level for more than 0.25 mi (0.40 km) in the first roadcut south of Chapmanville along U.S. Highway 119 on the west side of the Guyandotte River. The Betsie Shale Member is more than 100 ft (30 m) thick in southern West Virginia and is as much as 150 ft (46 m) thick in some areas of Kentucky and Virginia. It commonly contains marine invertebrate fossils and may contain several horizons of limestone concretions (Fig. 5A) in addition to that locally found at the base of the member (Fig. 5B). The basal shale beds of the Betsie characteristically produce a large curve deflection on gamma-ray logs (Fig. 6) that locally may be more than 400 American Petroleum Institute (API) units.

In many parts of West Virginia, the Betsie Shale Member is below the lowest mined coal bed in the Kanawha Formation. Thus, the Betsie occurs in a little known part of the section, and its persistence and distinctive character provide a useful datum for analyses of strata of both the lower part of the Kanawha Formation and upper part of the underlying New River Formation of Early Pennsylvanian age.

Dingess Shale Member

The Dingess Limestone was originally described as part of the former Kanawha series by Hennen and Reger (1914, p.165) in Logan and Mingo Counties as a gray to brown, commonly hard and siliceous, lenticular, locally ferriferous, and fossiliferous limestone ranging in thickness from 0 to 5 ft (0 to 1.5 m). The Dingess is here redefined as a thick sequence of marine shale, siltstone, and sandstone that commonly contains a discontinuous, nodular limestone bed or limestone concretions near its base; it is here revised as a member of the Kanawha Formation and is called the Dingess Shale Member.

Figure 5. Betsie Shale Member of the Kanawha Formation in roadcut of U.S. Highway 119 on west side of Guyandotte River near Chapmanville, West Virginia. This is the designated reference section for the Betsie in West Virginia. A, Several horizons of limestone concretions indicated by arrows. Member extends from base of limestone concretion just above car hood to base of sandstone. B, Detail of undulating basal ravinement surface just below discoidal limestone concretion. (Hammer at contact gives scale.)

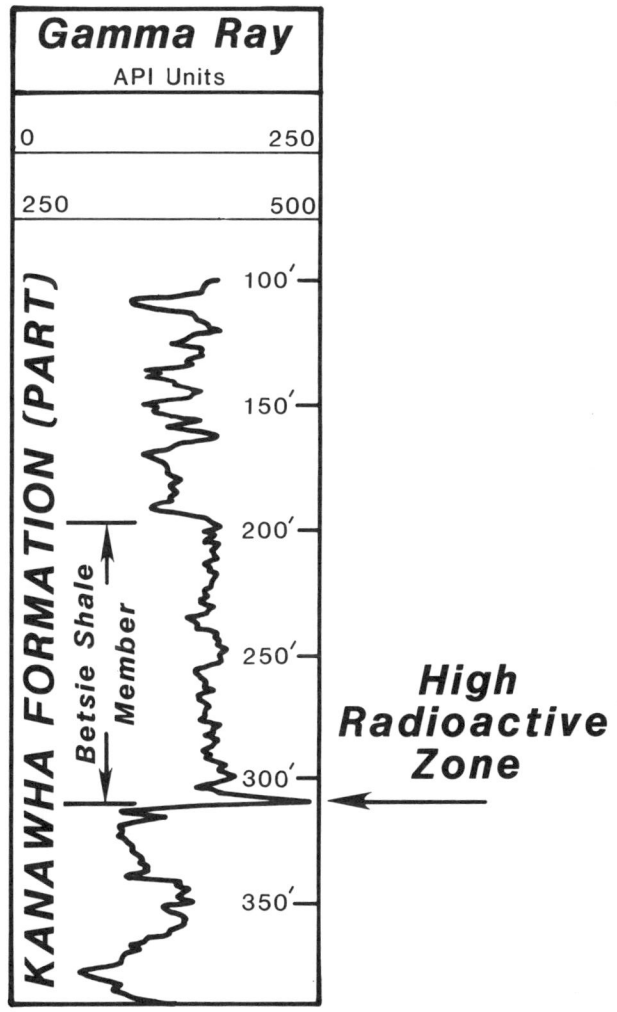

Figure 6. Portion of gamma-ray log (well number 059-801, see Fig. 1) of the Betsie Shale Member of the Kanawha Formation showing its characteristic signature and the highly radioactive zone at the base of the unit.

The unit is a useful stratigraphic marker bed that Hennen and Reger mapped throughout the two counties and identified in many other parts of southern West Virginia. They initially identified the unit, which contains fossils "in profusion," in a faulted area now known to be associated with the Warfield fault, about 1 mi (1.6 km) southeast of Dingess, Mingo County (Fig. 1). Later, this member was traced from Winifrede Junction in Kanawha County, southeastward to the Tug Fork (see Hennen and Reger, 1914). Drillhole data, reconnaissance mapping, and faunal analyses (T. W. Henry, oral communication, 1988) now suggest that the section near Dingess described by Hennen and Reger is the younger Winifrede Shale Member (discussed below). In order to avoid any possible confusion as to its identity, we designate an area about 16 mi (26 km) south of Dingess (Fig. 1), between Williamson and Matewan (7 mi or 11 km southeast of Williamson), West Virginia, as the type area of the Dingess Shale Member. This is an area of uncomplicated structure where the member is generally fossiliferous and as much as 80 ft (24 m) thick (Alvord, 1971). In this area, the Dingess is easily identified as the medium to dark gray, silty to sandy clay shale that directly overlies the extensively mined Williamson coal bed. The Dingess has been recognized as a useful marker bed by West Virginia Geological and Economic Survey geologists for coal resource studies and regional correlation in all parts of southwestern West Virginia (Blake et al., 1981).

The Dingess Shale Member is as much as 40 ft (12 m) thick in the Kanawha River Valley. Locally, it has been removed by channelling and replaced by fluvial channel-fill sandstone in the Kanawha River Valley (Henry, 1984). It has not been identified in measured sections of the proposed Pennsylvanian System stratotype in West Virginia (Englund et al., 1979) and probably was removed by erosion in those sections. Where the Dingess is relatively thick in southern West Virginia, its distinctive gamma-ray signature (Fig. 4) makes it easy to identify in the subsurface.

The Dingess Shale Member of the Kanawha Formation and the Kendrick Shale Member of the Breathitt Formation have been correlated based on their stratigraphic position above the Williamson coal bed in both West Virginia and Kentucky (Huddle and Englund, 1966; Rice, 1980; Blake et al., 1989). The Kendrick is widespread in eastern Kentucky and has been mapped into southwestern Virginia (Miller, 1969) and northeastern Tennessee (Rice and Newell, 1990).

Winifrede Shale Member

A 1-ft-thick (0.3 m) fossiliferous limestone bed discovered by White (1908, p. 431) was shown as the Winifrede Limestone of the former Kanawha series by White (*in* Krebs and Teets, 1914, p. 641) in a general section for the Kanawha series; it occurs about 66 ft (20 m) below the Winifrede coal bed at its type locality. The Winifrede Limestone was described by Price (*in* Krebs and Teets, 1914) on South Hollow of Fields Creek near Winifrede Junction, Kanawha County, West Virginia (Fig. 1). This marine unit, here redefined as the Winifrede Shale Member of the Kanawha Formation, generally consists of 10-15 ft (3 to 4.5 m) of dark gray, fossiliferous, commonly sandy shale in the Kanawha River valley, which may locally contain a thin discontinuous, nodular limestone at its base. It thickens southwestward to more than 50 ft (15 m) in west-central Logan County and is probably more that 90 ft (27 m) thick in southwesternmost West Virginia (Alvord, 1971). On the basis of their common stratigraphic position with respect to the Dingess Shale Member and certain extensively mined coal beds, the Winifrede has been correlated with the Magoffin Member of the Breathitt Formation in Kentucky by Wanless (1939, p. 53), Outerbridge (1976), and Blake et al. (1989). The Winifrede Member and its correlatives are probably among the most widely recognized and mapped Middle Pennsylvanian units in the central part of the Appalachian basin, extending from east-central West Virginia to Scott County, Tennessee (Englund, 1968). In Kentucky, this horizon was earlier referred to as the "Fossil Limestone" by Wanless (1946). Hennen and Teets (1919, p. 243) suggested that the Buffalo Creek Limestone of Logan and Mingo Counties (Hennen and Reger, 1914) is also equivalent to the Winifrede. Our reconnaissance mapping confirms their conclusion, and we therefore assign the marine strata previously referred to as Buffalo Creek Limestone (here abandoned) to the Winifrede Shale Member.

OTHER REGIONAL MARINE UNITS

Three other marine units have played important roles in the development of a stratigraphic framework for the Middle Pennsylvanian rocks in West Virginia. These are, from bottom to top, the Dorothy Limestone and Shale of Krebs and Teets (1916), the Cannelton ("Stockton") Limestone of White (1885), and the Campbell Creek Limestone of White (1885).

Dorothy Limestone and Shale of Krebs and Teets (1916)

The Dorothy Limestone and Shale of Krebs and Teets (1916) is a little known marine unit below the Betsie Shale Member. In the Tug Fork area of West Virginia, as well as in Kentucky and Virginia, strata equivalent to the Dorothy were incorrectly identified as the Eagle Limestone of White (1891) by Alvord and Miller (1972). This unit, which may be correlative with the Oceana Limestone of Hennen and Gawthrop (1915), occurs above the Lower War Eagle(?) coal bed (see Fig. 8). The Dorothy has not been mapped but is probably extensive in the southern part of West Virginia and the central part of the Appalachian basin in Virginia and southeastern Kentucky. The Dorothy Limestone and Shale of Krebs and Teets (1916) is here retained as an informal member because of lack of data concerning its mappability, character, thickness, and distribution.

Cannelton Limestone of White (1885)

The Cannelton Limestone of White (1885), at its type locality, is about 50 ft (15 m) above the Eagle coal bed across the Kanawha River from Montgomery, Fayette County, West Virginia (Hennen and Teets, 1919). This marine–to–brackish water unit generally consists of coarsening-upward shale and siltstone beds that contain concretions and lenses of arenaceous limestone as much as 5 ft (1.5 m) thick (Fig. 3). In the Tug Fork drainage, both in West Virginia (Hennen and Reger, 1914) and in Kentucky (see, for example, Huddle and Englund, 1966, Rice and Smith, 1980), the Cannelton Limestone of White (1885) was identified as the stratigraphically higher Campbell Creek Limestone of White (1885). Although many U.S. Geological Survey reports for Kentucky indicate the presence of a coarsening-upward unit containing large limestone concretions above the Pond Creek (Lower Elkhorn) coal zone equivalent to the Eagle coal zone in West Virginia (Fig. 2), the Cannelton was rarely identified because it contained few marine fossils.

In southeastern Kentucky, a marine shale and siltstone unit as much as 160 ft (49 m) thick has been identified above the Path Fork coal bed (Tazelaar and Newell, 1974) and is probably equivalent to the Cannelton as shown in Figure 2. The basal beds of the unit are abundantly fossiliferous and contain a diverse mollusk assemblage in the area of Cranks Ridge, Harlan County, Kentucky (T. W. Henry, oral communication, 1989). We have noted a restricted brachiopod/bivalve fauna from scattered locations in southwestern West Virginia. This unit is well exposed near Crummies, Kentucky, where the underlying Path Fork coal bed is extensively mined. As Figure 2 shows, geologic reports in Kentucky previously miscorrelated the marine shale unit at Crummies as the Campbell Creek Limestone of White (1885) (see Huddle and Englund, 1966). This unit has been formally named the Crummies Member of the Breathitt Formation by Rice et al. (1994).

Campbell Creek Limestone of White (1885)

The Campbell Creek Limestone of White (1885) is named for exposures along Campbell Creek near its confluence with the Kanawha River in Kanawha County about 4 mi (6.5 km) southeast of Charleston, West Virginia. Reconnaissance mapping indicates that the unit underlies the Williamson coal bed (zone) and is equivalent to the fossiliferous beds named the Seth Limestone by Krebs and Teets (1915) for exposures near Seth, Boone County, West Virginia. Sporadic occurrences of marine or brackish invertebrate fossils have been reported by coal company geologists; we have identified this unit from scattered localities in southern West Virginia. U.S. Geological Survey reports for Kentucky indicate that shales containing limestone concretions and, locally, marine or brackish-water fossils occur at the same horizon as the Campbell Creek Limestone of White (1885) (see Fig. 2). This unit is identified in Kentucky as the Elkins Fork Shale of Morse (1931). Although the shales and limestones of the Cannelton and Campbell Creek locally contain invertebrate fossils, the units are not considered regionally mappable because they are discontinuous and their boundaries are difficult to define. Therefore, in West Virginia, the Cannelton Limestone and Campbell Creek Limestone are retained as informal members as originally defined by White (1885).

COAL-BED NOMENCLATURE

Coal beds are considered to be informal stratigraphic units because of their discontinuous nature and because they commonly cannot be distinguished by physical appearance. Also, because of their tendency to split into two or more beds, coal beds commonly occur in coal zones that may be a few feet to tens of feet (meters) thick. However, coal beds and coal zones are very important in stratigraphic correlation of the Pennsylvanian coal measures, and coal names tend to take on a formal character to which they are not entitled. It is not surprising that coal beds (zones) may have different names in widely separated areas of the coal basin. But the names of coal beds should identify discrete beds in their type areas; the use of those names should not be extended beyond those areas except as part of related stratigraphic sequences that include regionally recognized marker beds such as the marine shale and limestone beds discussed above.

A discussion of several coal beds (zones) of local and regional importance in West Virginia will help to eliminate the confusion in stratigraphic correlation between the Kanawha River Valley and Tug Fork drainage in Mingo County. In order of importance, these are the Fire Clay, Chilton, Hernshaw, Williamson, Cedar Grove, and Matewan coal beds (zones).

Fire Clay, Chilton, and Hernshaw coal beds

One of the most useful stratigraphic marker beds in the central Appalachian basin is a persistent and distinctive flint clay parting or tonstein, 2 to 12 in. (5 to 30 cm) thick, that occurs in a coal bed about midway between the bases of the Winifrede and Dingess Shale Members of the Kanawha Formation in West Virginia (see Fig. 9) and equivalent strata in adjacent states (Wanless, 1946). Hennen and Reger (1914) identified the coal bed and tonstein parting along Huff Creek and Dingess Run, near Logan, West Virginia (section 7, Fig. 9), and correlated the coal bed with the Chilton coal bed of Kanawha County. However, the tonstein has not been found in the Chilton coal at its type locality (Wanless, 1939, p. 55), and later reconnaissance and mapping in the Kanawha Valley (Arndt et al., 1979a, b; Henry, 1984) indicate that the type Chilton is stratigraphically higher than the "Chilton" coal bed of Logan County. The uncertainty of the position of the coal beds in this part of the stratigraphic section led Arndt et al. (1979b) to identify the coal bed containing the flint clay in the Harewood section (near Boomer, West Virginia) of the proposed Pennsylvanian System stratotype as the "Hernshaw(?)" coal bed. Following this usage, Keiser et al. (1987) correlated the Hernshaw coal bed with the Fire Clay (Hazard No. 4) coal bed of eastern Kentucky on the basis of the occurrence, petrography, and areal extent of the associated tonstein. The correlation of the tonstein-bearing coal bed of the Kanawha Formation with the Fire Clay coal bed of eastern Kentucky was also suggested by Wanless (1939) and was confirmed by Eble (1988) and Eble and Grady (1990) on the basis of palynologic data. Recent mapping we have done, as yet unpublished, has also demonstrated that the tonstein parting occurs in a coal bed 30 to 50 ft (9 to 15 m) above the Hernshaw coal bed in its designated type area along Lens Creek near Hernshaw, Kanawha County, West Virginia. Because of the confusion surrounding the names used to identify the coal bed that contains the tonstein in West Virginia, and because of its regional stratigraphic importance, the name Fire Clay is extended from Kentucky into West Virginia to designate this coal bed; it replaces the names Hernshaw and Chilton for the coal bed at this horizon.

Petrographic, x-ray, and electron-probe analyses by various workers suggest that the tonstein and its accessory minerals represent an altered volcanic ashfall; this subject is discussed in detail by Rice, Belkin, et al. (this volume). Some of the accessory minerals, such as apatite, zircon, and sanidine, contain radioactive elements that produce a distinctive gamma-ray signature (Fig. 7) that aids in identifying the coal bed and tonstein in geophysical logs of boreholes.

Williamson and Cedar Grove coal beds

The Williamson coal bed, which occurs at or near the base of the Dingess Shale Member, was named for William-

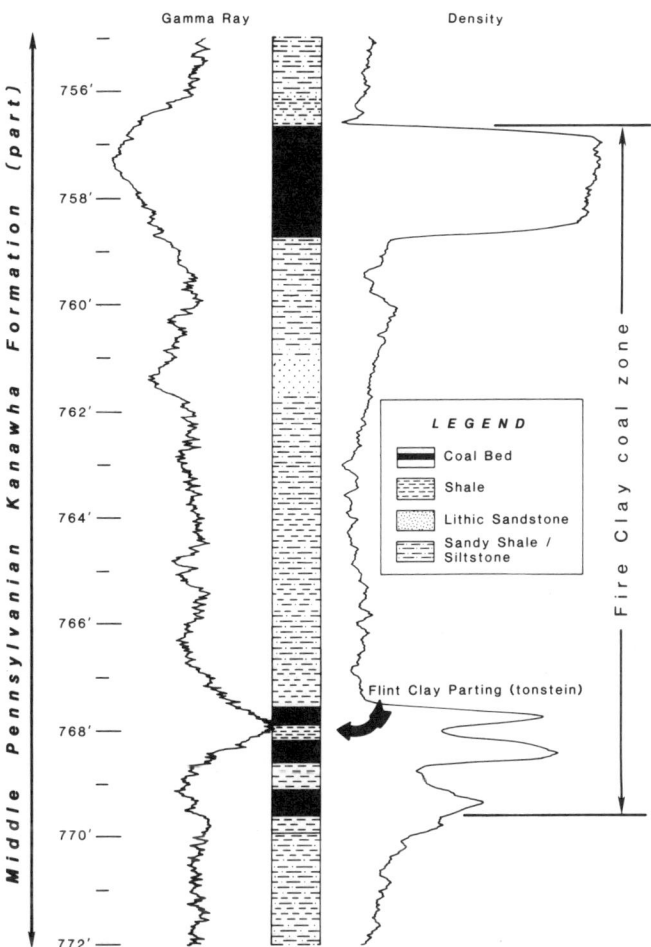

Figure 7. Expanded section of gamma-ray and density (gamma-gamma) logs of the Fire Clay coal zone (see Fig. 4) run to estimate coal-bed thickness and quality. Gamma-ray log shows the high radioactivity (deflection to the right on the gamma-ray curve) of the flint clay (tonstein) parting. Differences in the appearance in the log curves in this and Figure 4 result from changes in tool speed, sampling interval, and sensitivity.

son Creek near Williamson, Mingo County, West Virginia (Fig. 1) (Hennen and Reger, 1914). The Dingess marine unit was only briefly mentioned in the Kanawha County Report and the Williamson coal bed was not mentioned at all (Krebs and Teets, 1914). In the absence of a complete and well-defined stratigraphic section, Hennen and Reger (1914) miscorrelated the Cedar Grove coal bed of Kanawha County with a coal bed in the Campbell Creek coal zone below the Williamson coal bed in Logan and Mingo Counties. This miscorrelation of coal beds was carried back to the Kanawha River Valley in the Fayette County Report of Hennen and Teets (1919), where they incorrectly placed the Cedar Grove coal bed stratigraphically below both the Seth and Dingess limestones (see Fig. 2).

Our reconnaissance mapping in the Kanawha River valley indicates that the Cedar Grove coal bed occupies a stratigraphic position between the Dingess Shale Member (below) and the Fire Clay coal zone (above) (Fig. 8). In thicker parts of the Kanawha Formation, this position may be occupied by as many as five coal beds, here designated the Cedar Grove coal zone. In Logan and Mingo Counties (Hennen and Reger, 1914) and adjacent areas of Kentucky, we have assembled coal beds formerly known as the "Cedar Grove Rider," "Upper Cedar Grove," "and "Lower Cedar Grove" to the older Campbell Creek coal zone, which occurs between the older Cannelton and the younger Campbell Creek Limestones of White (1885) (Fig. 8). Because many coal beds in the coal fields in this part of the stratigraphic section have been identified as Campbell Creek coal beds, the name Campbell Creek is retained for coal beds equivalent to the coal beds in the Peerless coal zone and the No. 2 Gas coal zone (equivalent to the Upper Elkhorn coal zone of eastern Kentucky).

Matewan coal bed

The Matewan coal bed was named for exposures near the community of the same name about 7 mi (11 km) southeast of Williamson, West Virginia, on the Tug Fork of the Big Sandy River by Hennen and Reger (1914). The Matewan is a multiple bedded coal or coal zone that occurs at or near the base of the Betsie Shale Member. Locally, in the Kanawha River valley (Englund et al., 1979), a thin, unnamed coal bed occupies this stratigraphic position. However, the Matewan coal bed was miscorrelated in the Kanawha River Valley by Hennen and Teets (1919) with what is probably the uppermost coal bed of the Eagle coal zone, the Eagle "A" coal bed (Fig. 8). As with other units discussed herein, our stratigraphic framework retains the original definition of the Matewan coal bed as established in the type area.

DISCUSSION AND CONCLUSIONS

Figure 8 shows a generalized stratigraphic column for the Kanawha Formation of central and southwestern West Virginia, including all of the marine units and coal beds and zones described in this chapter. Because of the rapid southeastward thickening of the section, there will undoubtedly be a need in the future to recognize and name other stratigraphic units, particularly as the understanding of the biostratigraphy of Pennsylvanian coal beds and marine units increases. Most of the stratigraphic confusion of the past has resulted from miscorrelations of coal beds in the middle part of the formation, which were the result of a lack of an understanding of the number, character, or stratigraphic value of the extensive marine units. The Fire Clay coal bed and its flint clay parting were not recognized as a regional stratigraphic marker bed in West Virginia.

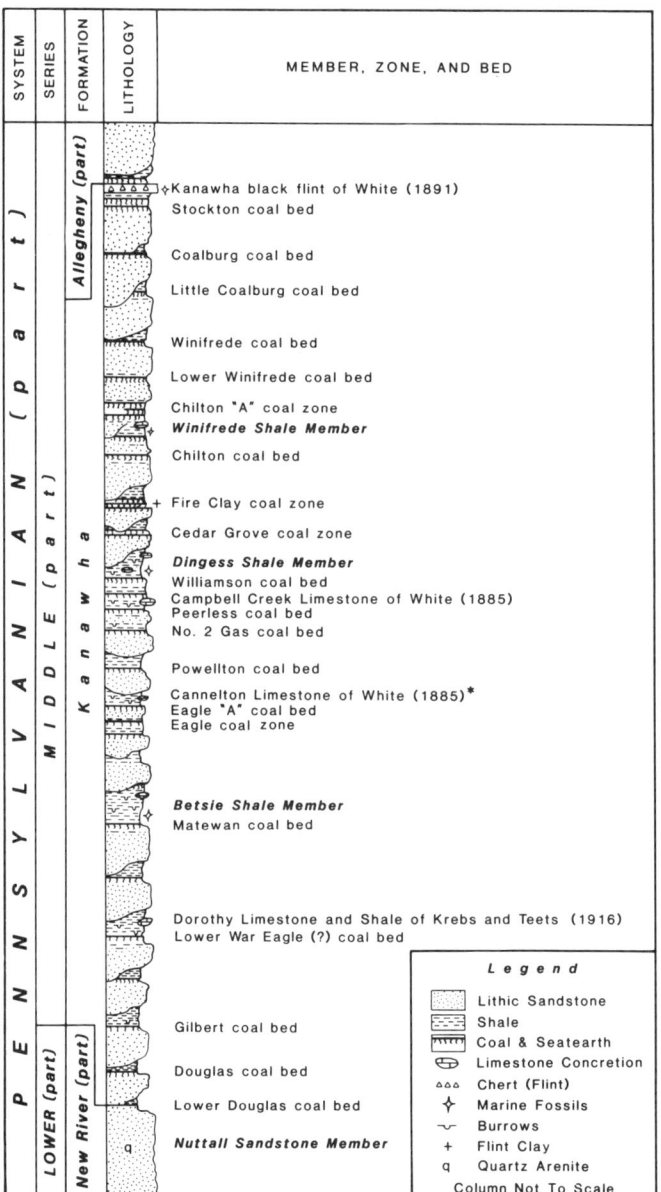

Figure 8. Generalized stratigraphic column of the Kanawha Formation of central and southwestern West Virginia. Formally named units of the New River and Kanawha Formations are shown in italics.

However, the position of the Fire Clay halfway between the Winifrede and Dingess Shale Members is invaluable for distinguishing and identifying these similar marine units.

The stratigraphic section (Fig. 9) illustrates the continuity of the units discussed in this report. It was constructed from logs of drillholes extending approximately along depositional strike from the area of Montgomery on the Kanawha River to the area of Williamson on the Tug Fork (see Fig. 1). This study was concerned only with the middle part of the formation as shown in the cross section. The lower part of the Kanawha Formation is thickest and contains many more minable coal beds south-southeast of the study area. Coal beds and marker beds in the upper part of the formation are more useful in the northern portion of the study area, having been eroded to the south. In the area of the cross section, the continuity of most coal beds also has been determined by coal resource studies that used fossil-bearing marine units for stratigraphic control (see West Virginia Geological and Economic Survey coal resource open file reports). Previously, miscorrelations were generally made between sections 4 and 6 in Figure 9, an area where the Betsie Shale is below drainage and its position must be determined from oil and gas well data. Although the Winifrede Shale Member is relatively thin in the northeastern half of the area depicted by the cross section, it is generally very fossiliferous and can be readily identified by its stratigraphic position. The Winifrede is further distinguished by a distinctive Atokan brachiopod fauna that has been recognized in many outcrops and coal exploration drillholes across southwestern West Virginia (Henry and Gordon, 1979).

West Virginia Geological and Economic Survey personnel and coal company geologists have long recognized the utility of stratigraphic studies of the three extensive and mappable marine units, the Betsie, Dingess, and Winifrede Shale Members. The revision of these three units as members of the Kanawha Formation simplifies stratigraphic description and provides an improved framework for coal exploration and resource calculation in West Virginia. Additionally, recognition of these well-known stratigraphic units provides a basin-wide continuum with formal equivalents in adjacent states and promotes development of a better regional Pennsylvanian synthesis.

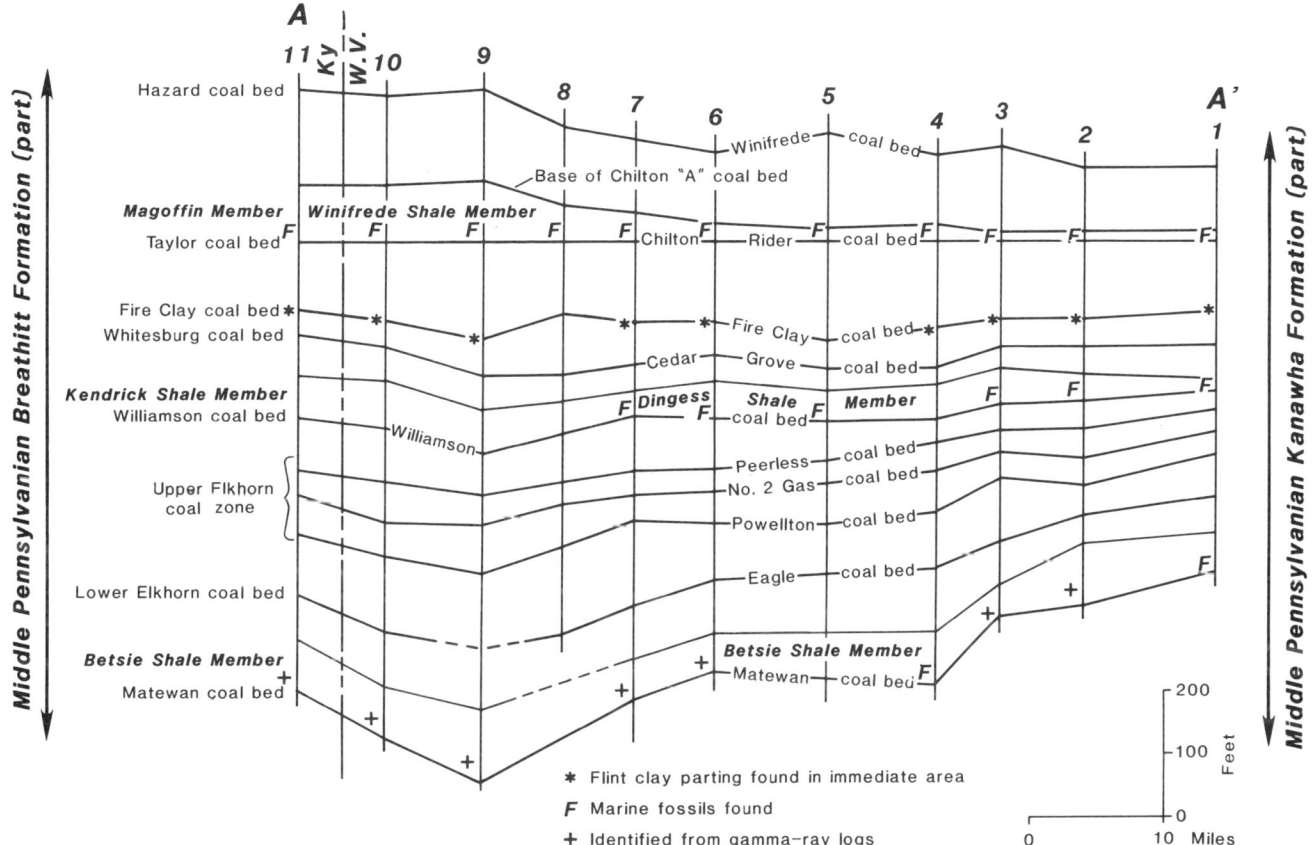

Figure 9. Stratigraphic section of the middle part of the Kanawha Formation showing continuity of marine units and coal beds between the Tug Fork of the Big Sandy River and the Kanawha River (see Fig. 1). Sections are composites of two or three diamond drillholes identified in Table 1.

TABLE 1. LOCATIONS OF DIAMOND DRILLHOLES AND WIRE-LINE LOGS FROM OIL AND GAS WELLS USED TO COMPILE COMPOSITE STRATIGRAPHIC SECTIONS IN FIGURE 9*

Section	Identification†	Latitude	Longitude
1	WVGES #150-71 (ddh)	38°12'25"	81°20'30"
	WVGES #186-1 (ddh)	38°11'22"	81°20'28"
	WVGES #188-78 (ddh)	38°12'15"	81°20'08"
2	WVGES #162-80 (ddh)	38°06'16"	81°29'10"
	WVGES #150-96 (ddh)	38°05'36"	81°28'14"
	WVGES #037-1949 (OG)	38°07'55"	81°25'59"
3	WVGES #162-85 (ddh)	38°02'28"	81°32'38"
	WVGES #005-1397 (OG)	38°01'53"	81°32'38"
4	WVGES #40-89 (ddh)	37°58'58"	81°36'52"
	WVGES #117-28 (OG)	37°59'04"	81°37'47"
5	WVGES #59-54 (ddh)	37°54'34"	81°41'54"
	WVGES #197-78 (ddh)	37°54'46"	81°41'17"
6	WVGES #60-19 (ddh)	37°52'45"	81°50'15"
	WVGES #103-19 (ddh)	37°52'04"	81°51'03"
	WVGES #100-01 (OG)	37°50'54"	81°52'00"
7	WVGES #56-67 (ddh)	37°48'28"	81°57'57"
	WVGES #56-79 (ddh)	37°47'28"	81°57'32"
	WVGES #100-05 (OG)	37°48'36"	81°58'29"
8	WVGES #56-83 (ddh)	37°44'59"	82°01'01"
	WVGES #197-7 (ddh)	37°44'18"	82°01'17"
	WVGES #56-85 (ddh)	37°44'17"	82°02'32"
9	WVGES #190-01 (ddh)	37°41'02"	82°06'50"
	WVGES #111-12 (OG)	37°41'23"	82°06'17"
10	WVGES #58-17 (ddh)	37°41'14"	82°13'07"
	WVGES #059-989 (OG)	37°41'46"	82°12'53"
11	Island Creek Coal Co. #S-502 (ddh)	37°40'50"	82°19'52"
	Kentucky #31-718 (OG)	37°38'27"	82°20'47"

*WVGES #150-71 signifies the West Virginia Geological and Economic Survey control point number.
†ddh = Diamond drillholes; OG = Oil and gas wells.

ACKNOWLEDGMENTS

This report is the product of the Cooperative Geologic Mapping (COGEOMAP) project begun in 1986 between the West Virginia Geological and Economic Survey and the U.S. Geological Survey to map the geology of a series of quadrangles and to establish stratigraphic continuity between the area of the proposed Pennsylvanian System stratotype in the Kanawha River Valley and the drainage area of the Tug Fork of the Big Sandy River. It combines COGEOMAP mapping data with previous studies by the Coal Section of the West Virginia Geological and Economic Survey, and voluminous coal exploration and oil and gas logs contributed by many coal mining and oil and gas companies and geologic consultants of southern West Virginia and adjacent parts of Kentucky.

REFERENCES CITED

Alvord, D. C., 1971, Geologic map of parts of the Naugatuck and Delbarton quadrangles, eastern Kentucky: U.S. Geological Survey Geologic Quadrangle Map GQ-879, scale 1:24,000.

Alvord, D. C., and Miller, R. L., 1972, Geologic map of the Elkhorn City quadrangle, Kentucky-Virginia and part of the Harman quadrangle, Pike County, Kentucky: U.S. Geological Survey Geologic Quadrangle Map GQ-951, scale 1:24,000.

Alvord, D. C., and Trent, V. A., 1962, Geology of the Williamson quadrangle in Kentucky: U.S. Geological Survey Geologic Map GQ-187, scale 1:24,000.

Arkle, T., Jr., 1974, Stratigraphy of the Pennsylvanian and Permian Systems of the central Appalachians, in Briggs, G, ed., Carboniferous of the southeastern United States: Geological Society of America Special Paper 148, p. 5–29.

Arkle, T., Jr., and 9 others, 1979, The Mississippian and Pennsylvanian Systems in the United States—West Virginia and Maryland: U.S. Geological Survey Professional Paper 1110-D, 35 p.

Arndt, H. H., 1979, Middle Pennsylvanian Series, in Englund, K. J., Arndt, H. H., and Henry, T. W., eds., Proposed Pennsylvanian System stratotype, Virginia–West Virginia: American Geological Institute Guidebook Series no. 1, p. 73–80.

Arndt, H. H., Englund, K. J., and Gillespie, W. H., 1979a, Road log—Third day, in Englund, K. J., Arndt, H. H., and Henry, T. W., eds., Proposed Pennsylvanian System stratotype, Virginia–West Virginia: American Geological Institute Guidebook Series No. 1, p. 15–25.

Arndt, H. H., Henry, T. W., Pfefferkorn, H. W., and Windolph, J. F., Jr., 1979b, Road log—Fourth day, in Englund, K. J., Arndt, H. H., and Henry, T. W., eds., Proposed Pennsylvanian System stratotype, Virginia–West Virginia: American Geological Institute Guidebook Series no. 1, p. 29–39.

Blake, B. M., Jr., Keiser, A. F., and Reppert, R. S., 1981, Geophysical logs and marine zones as useful coal exploration tools in southern West Virginia [abs.]: American Association of Petroleum Geologists Bulletin v. 65, p. 901–902.

Blake, B. M., Jr., Keiser, A. F., and Eble, C. F, 1989, Stop 32—Bolt Mountain section, in Cecil, C. B., and Eble, C. F., eds., Carboniferous geology of the Eastern United States St. Louis, Missouri, to Washington, D.C., June 28–July 8, 1989, Field trip guidebook T143 for the 28th International Geological Congress: Washington D.C.: American Geophysical Union, p. 95–97.

Donaldson, A. C., and Shumaker, R. C., 1981, Late Paleozoic molasse of central Appalachians, in Miall, A. D., ed., Sedimentation and tectonics in alluvial basins: Geological Association of Canada Special Paper 23, p. 99–124.

Eble, C. F., 1988, Palynology and paleoecology of a Middle Pennsylvanian coal bed from the central Appalachian basin [Ph.D. thesis]: Morgantown, West Virginia University, 495 p.

Eble, C. F., and Grady, W. C., 1990, Paleoecological interpretation of a Middle Pennsylvanian coal bed in the central Appalachian basin, U.S.A.: International Journal of Coal Geology, v. 16, p. 255–286.

Englund, K. J., 1968, Geology and coal resources of the Elk Valley area, Tennessee and Kentucky: U.S. Geological Survey Professional Paper 572, 59 p.

Englund, K. J., Arndt, H. H., and Henry, T. W., eds., 1979, Proposed Pennsylvanian System stratotype, Virginia and West Virginia: American Geological Institute Guidebook Series no. 1, 138 p.

Ferm, J. C., and Weisenfluh, G. A., 1989, Evolution of some depositional models in Upper Carboniferous rocks of the Appalachian coal fields, in Lyons, P. C., and Alpern, B., eds., Peat and coal—Origin, facies, and depositional models: International Journal of Coal Geology, v. 12, p. 259–292.

Flores, R. M., and Arndt, H. H., 1979, Depositional environments of Middle Pennsylvanian Series in proposed Pennsylvanian System stratotype, *in* Englund, K. J., Arndt, H. H., and Henry, T. W., eds., Proposed Pennsylvanian System stratotype, Virginia–West Virginia: American Geological Institute Guidebook Series no. 1, p. 115–121.

Hennen, R. V., and Gawthrop, R. M., 1915, Wyoming and McDowell Counties: West Virginia Geological and Economic Survey [County Report], 783 p.

Hennen, R. V., and Reger, P. B., 1914, Logan and Mingo Counties: West Virginia Geological and Economic Survey [County Report], 776 p.

Hennen, R. V., and Teets, D. D., Jr., 1919, Fayette County: West Virginia Geological and Economic Survey [County Report], 1,002 p.

Henry, T. W., 1984, Geology of the Mammoth quadrangle, Kanawha and Clay Counties, West Virginia: U.S. Geological Survey Geologic Quadrangle Map GQ-1576, scale 1:24,000.

Henry, T. W., and Gordon, M., Jr. 1979, Late Devonian through early Permian(?) invertebrate faunas in proposed Pennsylvanian System stratotype area, *in* Englund, K. J., Arndt, H. H., and Henry, T. W., eds., Proposed Pennsylvanian System stratotype, Virginia-West Virginia: American Geological Institute Guidebook Series no. 1, p. 97–103.

Hobday D. K., and Horne, J. C., 1977, Tidally influenced barrier island and estuarine sedimentation in the Upper Carboniferous of southern West Virginia: Sedimentary Geology, v. 18, p. 97–122.

Horne, J. C., and Ferm, J. C., 1978, Carboniferous depositional environments: Eastern Kentucky and West Virginia: University of South Carolina, Field Guide, 151 p.

Huddle, J. W., and Englund, K. J., 1966, Geology and coal reserves of the Kermit and Varney area, Kentucky: U.S. Geological Survey Professional Paper 507, 83 p.

Huddle, J. W., Lyons, E. J., Smith, H. L., and Ferm, J. C., 1963, Coal reserves of eastern Kentucky: U.S. Geological Survey Bulletin 1120, 247 p.

Jillson, W. R., 1919, The Kendrick Shale—A new calcareous fossil horizon in the coal measures of eastern Kentucky: Kentucky Department of Geology and Forestry, ser. 5 [of Kentucky Geological Survey], Mineral and Forest Resources of Kentucky, v. 1, no. 2, p. 96–104.

Keiser, A. F., Grady, W. C., and Blake, B. M., Jr., 1987, Flint clay parting in the Hernshaw coal of West Virginia [abs.]: Eleventh International Congress of Carboniferous Stratigraphy and Geology, Beijing, China, Abstracts of Papers, v. 2, p. 447.

Krebs, C. E., and Teets, D. D., Jr., 1914, Kanawha County: West Virginia Geological and Economic Survey [County Report], 776 p.

Krebs, C. E., and Teets, D. D., Jr., 1915, Boone County: West Virginia Geological and Economic Survey [County Report], 648 p.

Krebs, C. E., and Teets, D. D., Jr., 1916, Raleigh County and western part of Mercer and Summers Counties: West Virginia Geological and Economic Survey [County Report], 648 p.

Liu, Y., 1990, Depositional environments of the upper Mary Lee coal zone, Lower Pennsylvanian "Pottsville" Formation, northwestern Alabama, *in* Gastaldo, R.A., Demko, T.M., and Liu, Yuejin, eds., Carboniferous coastal environments and paleocommunities of the Mary Lee coal zone, Mariona and Walker Counties, Alabama: Geological Society of America Guidebook for Fieldtrip 6, Southeastern Section Meeting, April 7-8, 1990, p. 21–39.

Liu, Y., and Gastaldo, R.A., 1989, Characteristics and implications of a ravinement surface in the upper Pottsville Formation, northwestern Alabama [abs.]: Geological Society of America Abstracts with Programs, v. 21, p. 48.

Miller, R. L., 1969, Pennsylvanian Formations of southwest Virginia: U.S. Geological Survey Bulletin 1280, 62 p.

Morse, W .C., 1931, Pennsylvanian invertebrate faunas: Kentucky Geological Survey, ser. 6, v. 36, p. 293–348.

Outerbridge, W. F., 1976, The Magoffin Member of the Breathitt Formation, *in* Cohee, G. V., and Wright, W. B., eds., Changes in stratigraphic nomenclature in the U.S. Geological Survey, 1975: U.S. Geological Survey Bulletin 1422-A, p. A64–A65.

Potter, P. E., Maynard, J. B., and Pryor, W. A., 1984, Sedimentology of shale: New York, Springer-Verlag, 303 p.

Rice, C. L., 1980, Kendrick Shale Member of the Breathitt Formation of eastern Kentucky, *in* Sohl, N. F., and Wright, W. B., eds., Changes in stratigraphic nomenclature by the U.S. Geological Survey, 1979: U.S. Geological Survey Bulletin 1502-A, p. A117–A122.

Rice, C. L. and Newell, W. L., 1990, Geology of part of the Jellico East quadrangle, Campbell and Claiborne Counties, Tennessee: U.S. Geological Survey Geologic Quadrangle Map GQ-1674, scale 1:24,000.

Rice, C. L., and Smith, H. J., 1980, Correlation of coal beds, coal zones, and key stratigraphic units in the Pennsylvanian rocks of eastern Kentucky: U.S. Geological Survey Miscellaneous Field Studies Map MF-1188.

Rice, C. L., Currens, J. C., Henderson, J. A., and Nolde, J. B., Jr., 1987, The Betsie Shale Member—A datum for exploration and stratigraphic analysis of the lower part of the Pennsylvanian in the central Appalachian basin: U.S. Geological Survey Bulletin 1834, 17 p.

Rice, C. L., Henry, T. W., and Chesnut, D. R., Jr., 1994, The distribution and biostratigraphy of the Crummies Member (new name) of the Breathitt Formation in Pennsylvanian rocks of eastern Kentucky, *in* Sando, W. J., ed., Shorter contributions to paleontology and stratigraphy: U.S. Geological Survey Bulletin 2073-A, A1–A9.

Swift, D.J.P., 1968, Coastal erosion and transgressive stratigraphy: Journal of Geology, v. 76, p. 444–456.

Tankard, A. J., 1986, Depositional response to foreland deformation in the Carboniferous of eastern Kentucky: American Association of Petroleum Geologists Bulletin, v. 70, p. 853–868.

Tazelaar, J. F., and Newell, W. L., 1974, Geologic map of the Evarts quadrangle and part of the Hubbard Springs quadrangle, southeastern Kentucky and Virginia: U.S. Geological Survey Geologic Quadrangle Map GQ-914, scale 1:24,000.

Trent, V. A., 1965, Geology of the Matewan quadrangle in Kentucky: U.S. Geological Survey Geologic Quadrangle Map GQ-373, scale 1:24,000.

Wanless, H. R., 1939, Pennsylvanian correlations in the Eastern Interior and Appalachian coal fields: Geological Society of America Special Paper 17, 130 p.

Wanless, H. R., 1946, Pennsylvanian geology of a part of the southern Appalachian coal field: Geological Society of America Memoir 13, 162 p.

Weller, J. W., 1930, Cyclic sedimentation of the Pennsylvanian Period and its significance: Journal of Geology, v. 38, no. 2, p. 7–16.

White, I. C., 1885, Resume of the work of the U.S. Geological Survey in the Great Kanawha Valley during the summer of 1884: The Virginia's, v. 6, p. 7–16.

White, I. C., 1891, Stratigraphy of the bituminous coal field of Pennsylvania, Ohio, and West Virginia: U.S. Geological Survey Bulletin 65, 212 p.

White, I. C., 1903, Levels above tides, True meridians; Report on coal: West Virginia Geological and Economic Survey, v. II, 725 p.

White, I. C., 1908, Supplementary coal report: West Virginia Geological and Economic Survey, v. II(A), 720 p.

White, I. C., 1914, Introduction, in Krebs, C.E., and Teets, D.D., Jr., Kanawha County: West Virginia Geological and Economic Survey [County Report], p. xvii–xxviii.

MANUSCRIPT ACCEPTED BY THE SOCIETY FEBRUARY 1, 1994

Geological Society of America
Special Paper 294
1994

Palynostratigraphy of selected Middle Pennsylvanian coal beds in the Appalachian basin

Cortland F. Eble
Kentucky Geological Survey, 228 MMRB, University of Kentucky, Lexington, Kentucky 40506

ABSTRACT

Selected Middle Pennsylvanian coals and one Upper Pennsylvanian coal from three outcrop sections and one exploratory drillcore in the central and northern Appalachian basin were analyzed for their spore content. In the studied assemblages, the most frequently encountered taxa were species of *Lycospora, Laevigatosporites, Punctatisporites, Densosporites* (and related crassicinulate genera), *Granulatisporites* (and related sphaerotriangular genera), and *Florinites*. Mid–Middle Pennsylvanian assemblages are generally dominated by *Lycospora*, with *L. pellucida, L. pusilla, L. granulata, L. orbicula,* and *L. micropapillata* being the most common species. A few beds in this interval, however, show more even distributions of *Lycospora* (produced by arboreous lycopsids) and forms related to tree ferns, calamites, and cordaites. In general, a stratigraphic upward increase in tree fern taxa is observed, with upper Middle Pennsylvanian coals being codominated or dominated by tree fern spores.

In addition, the range zones of the following taxa appear to be useful for palynologic delineation of Middle Pennsylvanian strata on both intra- and interbasinal scales: *Microreticulatisporites sulcatus, Triquitrites sculptilis, Laevigatosporites globosus, Radiizonates difformis-rotatus, Torispora securis, Thymospora pseudothiessenii, Murospora kosankei, Mooreisporites inusitatus, Granasporites medius, Schulzospora, Densosporites, Schopfites, Lycospora, Cirratriradites,* and *Vestispora.* The recognition of these taxa in Appalachian spore assemblages allows for comparison and correlation of Pennsylvanian strata in the Eastern and Western Interior basins of North America and the Upper Carboniferous of Western Europe.

INTRODUCTION

Fossil spores and pollen have long been of value in the correlation and relative age dating of Pennsylvanian coal beds in North America. The utility of palynomorphs as a biostratigraphic tool is of particular importance in basins dominated by terrestrial sequences, as other biostratigraphic methods are often of limited value in these areas. The Appalachian basin of the eastern United States is an example of such an area where palynology is very useful in the identification and correlation of individual coal beds and coal zones.

This study focuses on the palynology of Middle Pennsylvanian coal beds from the Appalachian basin, with special emphasis being placed on the stratigraphic utility of selected spore taxa. The section of strata, included in three of the four sample locations, extends from the No. 2 Gas coal bed, which occurs in the middle of the Kanawha Formation, to the Stockton coal bed, which occurs at the top of the Kanawha Formation (Fig. 1b). The fourth section, the Chestnut Ridge–Interstate 68 section, was incorporated into the study to emphasize the rapid thinning of Middle Pennsylvanian strata in a southwest to northeast direction, and also to demonstrate the usefulness of spores in correlating over large areas. Because this section contains both Middle and lower Upper Pennsylvanian strata (Fig. 1a), spore taxa in coal beds of the Allegheny Formation and lower Conemaugh Group will also be reported on.

Eble, C. F., 1994, Palynostratigraphy of selected Middle Pennsylvanian coal beds in the Appalachian basin, *in* Rice, C. L., ed., Elements of Pennsylvanian Stratigraphy, Central Appalachian Basin: Boulder, Colorado, Geological Society of America Special Paper 294.

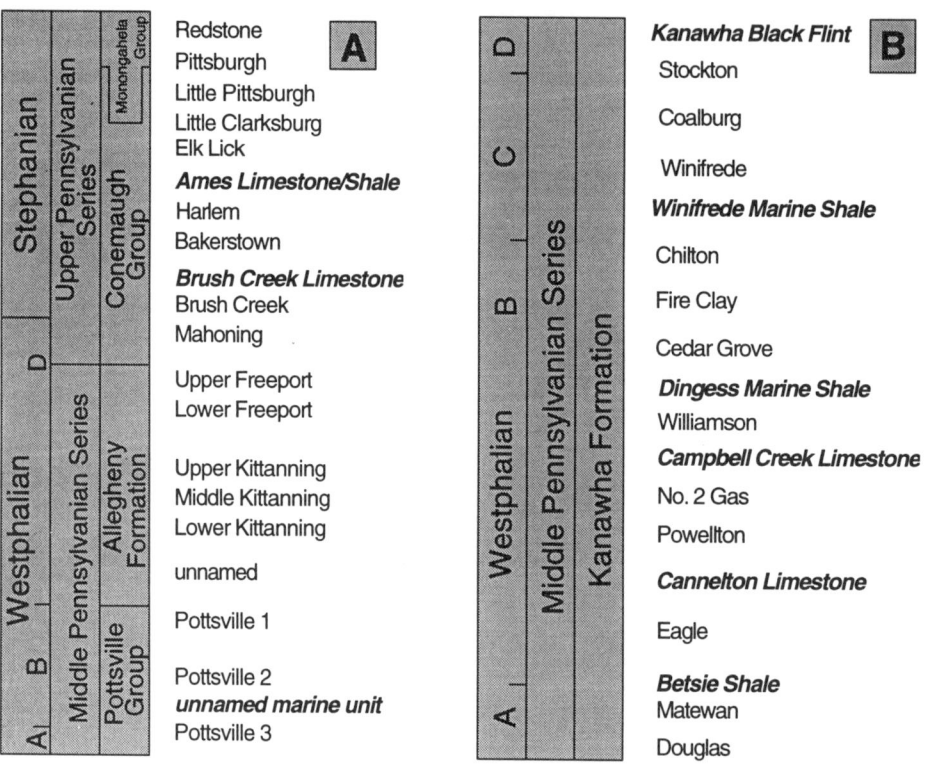

Figure 1. Generalized stratigraphic columns. A, Middle and Upper Pennsylvanian coal beds exposed along Interstate 68 on the west flank of the Chestnut Ridge anticline in northern West Virginia. B, Middle Pennsylvanian coal beds in southern West Virginia; correlative beds in neighboring eastern Kentucky are listed in Blake et al. (this volume). Important marine horizons are shown in italics and boldface type. Coal bed and marine zone nomenclature are based on Blake et al. (this volume).

PREVIOUS WORK

Correlation of Upper Carboniferous strata in Western Europe and the Soviet Union has been the subject of numerous studies (Smith and Butterworth, 1967; Butterworth, 1969; Butterworth and Smith, 1976; Owens et al., 1978). Peppers (1984) has provided a comprehensive review of this subject. The investigations of Clayton et al. (1977) and Owens (1984) are of particular value because they synthesized available data. In North America, studies by Thiessen and Staud (1923), Thiessen and Wilson (1924), Cross (1947), Kosanke (1950, 1973, 1982, 1984, 1988a,–c), Schemel (1957), Peppers (1964, 1970, 1979), Cropp (1960, 1963), Upshaw (1967), Clendening (1974), Ravn (1979), Ravn and Fitzgerald (1982), Phillips and Peppers (1984), and Eble and Gillespie (1984, 1986, 1989) have all reported on the application of palynostratigraphy to Pennsylvanian coal-bearing strata.

The reports by Peppers (1984) and Ravn (1986) are especially important. Peppers (1984) divided Pennsylvanian strata of the Illinois basin (Eastern Interior basin) into 11 spore assemblage zones spanning Morrowan- to Virgilian-age strata, and also proposed a correlation of Pennsylvanian strata in the Illinois basin with the Upper Carboniferous of Western Europe. Ravn (1986) identified the range zones of stratigraphically useful taxa in the Western Interior basin of Iowa, correlated Iowa coal beds with those of the Eastern Interior basin (Illinois), and proposed four spore assemblage zones for both intra- and interbasinal characterization and correlation. For the present work, these two investigations are of particular value because they allow for detailed comparison and correlation of Middle and Upper Pennsylvanian strata among the Appalachian, Eastern Interior, and Western Interior basins.

MATERIALS AND METHODS

Mid-Middle Pennsylvanian coal beds were sampled from two outcrop sections and one exploratory drillcore in the central Appalachian basin, and one outcrop section in the northern Appalachian (Dunkard) basin (Fig. 2). Table 1 lists the coal beds sampled, their thicknesses, and corresponding maceration numbers. Coal samples were first mechanically stage crushed to -20 mesh (850 mµ), and then riffled to obtain a representative sample of approximately 30 to 40 g (ASTM, 1989). From this subsplit, five g of well-mixed coal were subsequently removed for maceration. Maceration procedures followed those outlined by Barss and Williams (1973) and Doher (1980), with

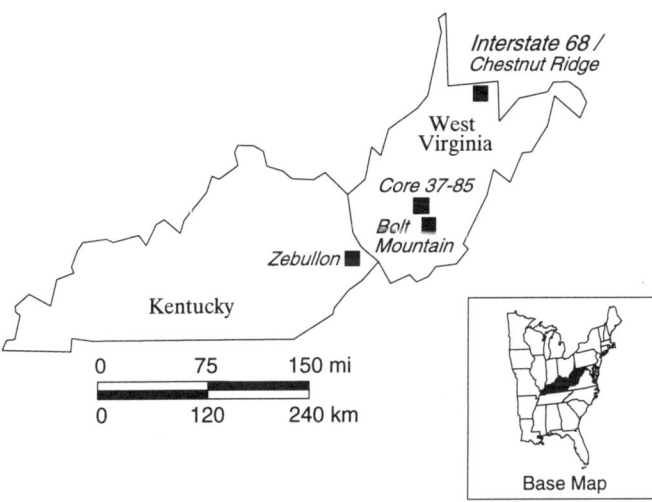

Figure 2. Sample location map.

minor modifications to achieve the most satisfactory results. Canada balsam and Permount, a synthetic piccolyte resin, were used as mounting media.

To obtain the relative distribution of palynomorph taxa, 250 spores were counted from each maceration. A 250-spore count/sample has been adopted by other workers (Kosanke, 1973, 1984; Ravn, 1979, 1986), and is based on providing a minimum practical figure for statistical consideration (Patterson and Fishbein, 1989). In addition, the remainder of each slide was scanned in order to record any forms not identified during the statistical count. Typically, two slides per maceration were counted and scanned.

RESULTS

The most commonly encountered genera in the studied assemblages were *Lycospora, Laevigatosporites, Punctatisporites, Densosporites, Granulatisporites* (and related sphaerotriangular genera), and *Florinites*. Figures 3 through 8 illustrate the distributions of the more abundant forms for each location.

Zebullon section

A large road cut section along U.S. Highway 119 in Pike County, Kentucky, informally referred to as the Zebullon section, exposes approximately 200 m of Middle Pennsylvanian Breathitt Formation strata. The coal beds/zones that were sampled at this location are, in ascending order, the Upper Elkhorn No. 3, Upper Elkhorn No. 3 Rider?, Williamson, Whitesburg, and Fire Clay?. Two laterally persistent marine shale units that assist correlation efforts, the Elkins Fork and Kendrick, are also present in the Zebullon section (Fig. 3).

All of the coal beds in this section are dominated by *Lycospora,* except for the upper split of the Whitesburg coal zone, which contains high percentages of fern (*Punctatisporites minutus, Punctatisporites minutus* and *Apiculatasporites saetiger*) and calamite taxa (*Calamospora* and larger species of *Laevigatosporites*). The most common species of *Lycospora* are *L. pellucida, L. pusilla, L. granulata, L. orbicula,* and *L. micropapillata. Lycospora torquifer* and *L. rotunda* are also recorded consistently, but in minor amounts (generally less than 2 to 3%) (Fig. 4). Except for occasional increases in abundance of other taxa (e.g., *Florinites* in the Whitesburg coal and *Granasporites medius* in the Williamson coal bed), the overall homogeneity of spore floras in this section is striking.

WVGS Core 37-85

Nine coal beds were sampled from a core whose location was designated 37-85 by the West Virginia Geological and Economic Survey. The drillhole is located on the Belle 7.5-min Quadrangle in Kanawha County, West Virginia (Fig. 2). The coals that occur in this core are, in ascending order, the Lower No. 2 Gas, Upper No. 2 Gas, Lower Cedar Grove, Upper Cedar Grove, Little Chilton, Chilton, Chilton Rider, and Chilton "A." The Winifrede marine shale, a distinctive unit that is equivalent to the Magoffin Member of eastern Kentucky, occurs in core 37-85 between the Chilton Rider and Chilton "A" coal beds (Fig. 5).

The palynofloras from this core are, like those obtained from Zebullon section coal beds, very similar overall, with *Lycospora* being the most abundant genus. However, the introduction of three stratigraphically important taxa in coal beds toward the top of the core may assist correlation efforts. These taxa are *Triquitrites sculptilis* and *Microreticulatisporites sulcatus,* both of which are first observed in the Little Chilton coal, and *Laevigatosporites globosus,* which first occurs in the Chilton coal bed. In the proposed Pennsylvanian stratotype of Englund et al. (1979), Kosanke (1988a) reported the first consistent occurrence of *L. globosus* in the Chilton(?) coal bed, and *T. sculptilis* in the underlying Hernshaw coal bed. (Kosanke also reported a somewhat anomalous occurrence of this taxon in the No. 2 Gas coal bed.) In addition, Eble (1988) reported scattered occurrences of *T. sculptilis* and *M. sulcatus* from the Fire Clay (reported as Hernshaw) coal bed. In the Illinois Basin, Peppers (1984) reported that these taxa have their origins in the upper part of the Abbott Formation (Atokan Series; NG and SF palynomorph zones) in the Illinois basin. In the Western Interior basin of Iowa, Ravn (1986) recorded *T. sculptilis* from thin, discontinuous coals of the Kilbourn Formation (VA zone). The overlying Blackoak and Cliffland coals of the Kalo Formation mark the first occurrences of *L. globosus* and *M. sulcatus* (SG zone) (Fig. 6).

Bolt Mountain section

The Bolt Mountain section occurs along the westernmost margin of Raleigh County, West Virginia, and represents the thickest sequence of mid-Middle Pennsylvanian strata sampled

TABLE 1. LIST OF SAMPLES USED IN THIS CHAPTER

Coal Bed/Zone	Description	Maceration	Coal Bed/Zone	Description	Maceration
Zebullon Section, Pike County, Kentucky, Meta Quadrangle			**Bold Mountain Section** (continued)		
?Fire Clay	0.15 m coal bed	337	Chilton "A"	0.06 m rider coal	136
?Fire Clay	0.18 m coal bench	338	Chilton "A"	0.14 m rider coal	135
?Fire Clay	0.31 m coal bench	339	Chilton "A"	0.18 m coal bench	134
Whitesburg	0.15 m coal bench	297	Coal Bed/Zone	Description	Maceration
Whitesburg	0.15 m coal bench	298			
Whitesburg	0.15 m coal bench	303	Chilton "A"	0.10 m coal bench	133
Whitesburg	0.15 m coal bench	305	Chilton "A"	0.10 m coal bench	132
Whitesburg	0.21 m coal bench	340	Little Chilton	0.19 m coal/shale bench	130
Whitesburg	0.21 m coal bench	341	Little Chilton	0.25 m coal bench	129
Whitesburg	0.27 m coal bed	342	Cedar Grove	0.55 m coal bed	127
Whitesburg	0.04 m coal bed	407	Cedar Grove	0.25 m coal bench	124
Whitesburg	0.20 m coal bench	343	Cedar Grove	0.13 m coal bench	122
Whitesburg	0.18 m coal bench	344	Cedar Grove	0.24 m coal bench	120
Williamson	0.21 m coal bench	345	Cedar Grove	0.19 m coal/shale bench	119
Williamson	0.18 m coal bench	346	Williamson	0.20 m coal bench	118
Williamson	0.31 m underclay	347	Williamson	0.37 m coal bench	117
Williamson	0.08 m coal bench	348	Upper No. 2 Gas	0.20 m coal/shale bench	114
?Up. Elkhorn #3 Rider	0.21 m coal bed	349	Upper No. 2 Gas	0.20 m coal bench	113
Up. Elkhorn #3	0.24 m coal bench	350	Lower No. 2 Gas	1.07 m coal bench	111
Up. Elkhorn #3	0.09 m coal/shale bench	351	Lower No. 2 Gas	0.46 m coal bench	109
Up. Elkhorn #3	0.03 m shale parting	352	Lower No. 2 Gas	0.12 m coal bench	107
Up. Elkhorn #3	0.05 m coal/shale bench	353	**Route 48-Chestnut Ridge Section, Monongalia-Preston Counties, West Virginia, Lake Lynn Quadrangle**		
Up. Elkhorn #3	0.24 m shale parting	354			
Up. Elkhorn #3	0.24 m coal bench	355	Pottsville 1	0.40 m coal bed	195
WVGS Core 37-85, Kanawha County, West Virginia, Cedar Grove Quadrangle			Pottsville 2	0.31 m coal bed	196
			Pottsville 3	0.31 m coal bed	197
Chilton "A"	0.76 m coal bed	663	Unnamed bed	0.33 m coal bed	675
Chilton Rider	0.46 m coal bed	655	Lower Kittanning	0.31 m coal bench	198
Chilton	0.31 m coal bed	656	Lower Kittanning	0.21 m coal bench	200
Little Chilton	0.12 m coal bed	657	Middle Kittanning	0.31 m coal bed	201
Upper Cedar Grove	0.52 m coal bed	658	Upper Kittanning	0.31 m coal bench	202
Lower Cedar Grove	0.31 m coal bed	659	Upper Kittanning	0.15 m coal bench	204
Upper No. 2 Gas	0.52 m coal bed	660	Lower Freeport	0.18 m coal bench	205
Lower No. 2 Gas	0.52 m coal bench	661	Lower Freeport	0.27 m coal bench	207
Lower No. 2 Gas	0.27 m coal bench	662	Upper Freeport	0.31 m coal bench	208
Bold Mountain Section, Raleigh County, West Virginia, Arnett-Pilot Knob Quadrangles			Upper Freeport	0.27 m coal bench	209
			Upper Freeport	0.43 m coal bench	210
Stockton	0.49 m coal bench	149	Upper Freeport	0.18 m coal bench	211
Stockton	0.59 m coal bench	148	Upper Freeport	0.23 m coal bench	212
Stockton	0.43 m coal bench	147	Upper Freeport	0.38 m coal bench	213
Coalburg	0.88 m coal bed	145	Upper Freeport	0.31 m coal bench	214
?Upper Winifrede	0.23 m coal bench	143	Mahoning	0.15 m coal bench	215
?Upper Winifrede	0.40 m coal bench	141	Mahoning	0.31 m coal bench	216
?Lower Winifrede	0.13 m coal/shale bench	139	Mahoning	0.31 m coal bench	217
?Lower Winifrede	0.40 m coal bench	138	Mahoning	0.31 m coal bench	218
			Mahoning	0.31 m coal bench	219

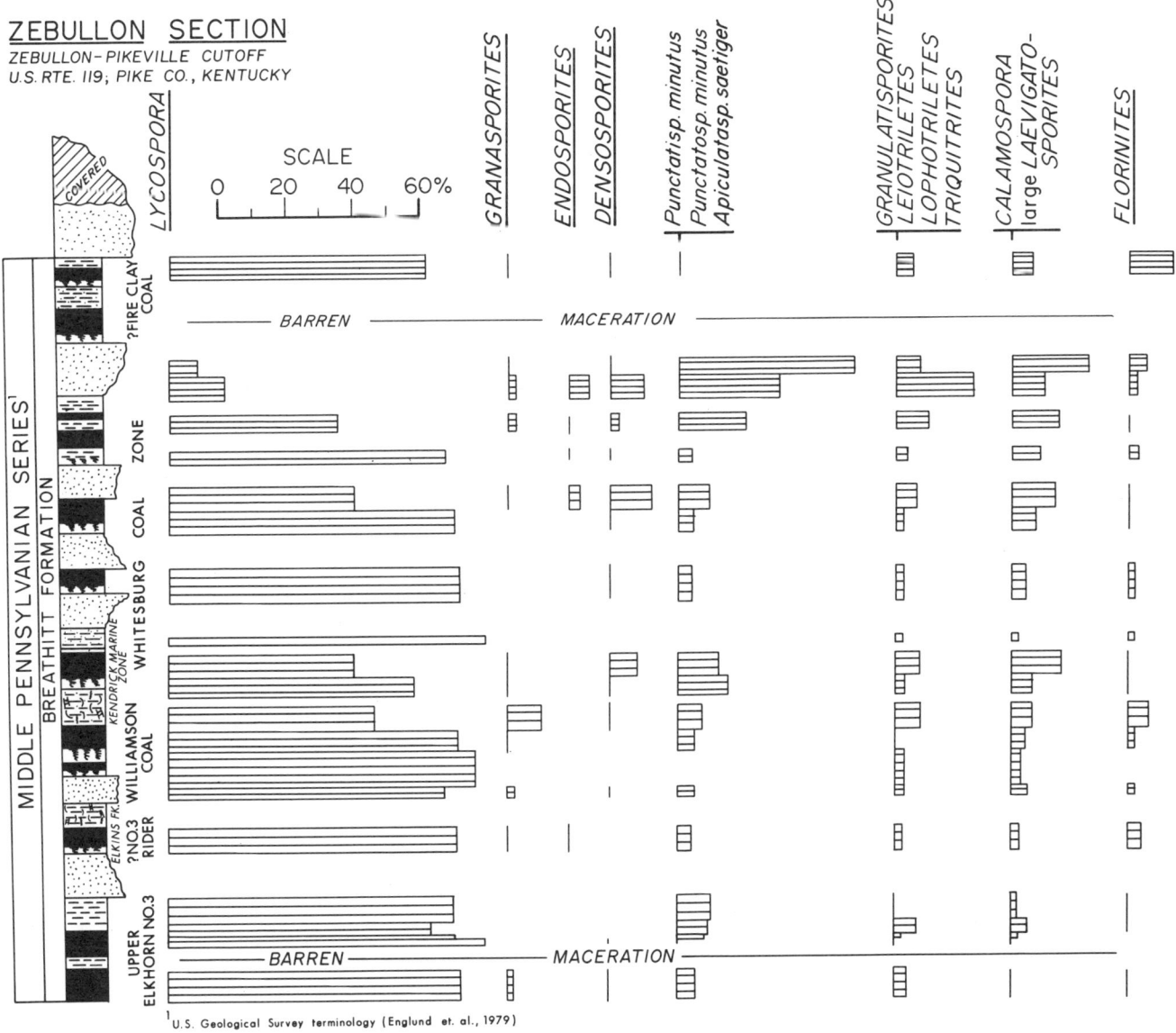

Figure 3. Distribution of selected palynomorph taxa in the Zebullon section.

for this study. The Bolt Mountain section exposes nearly 610 m of the Kanawha Formation, from just below the Gilbert coal bed at the base of the section to an unnamed sandstone just above the Stockton coal bed at the top. All the coals in the section were sampled; however, because of the high rank of the coals, and subsequent poor recovery of palynomorphs in the lower part of the section, only the following coal beds/zones are discussed in this study. In ascending order, they are the Lower No. 2 Gas, Upper No. 2 Gas, Williamson, Cedar Grove, Little Chilton, Chilton "A," Lower Winifrede, Upper Winifrede, Coalburg, and Stockton. Two laterally continuous marine zones, the Dingess (equivalent with the Kendrick marine shale of eastern Kentucky; see Blake and Rice, this volume) and Winifrede (Magoffin of eastern Kentucky), were also identified in this section (T. W. Henry, oral communication, 1989).

These units greatly assist correlation efforts throughout the basin (Fig. 7).

The No. 2 Gas through Williamson coal beds in the Bolt Mountain section are dominated by *Lycospora,* with other taxa represented in relatively minor amounts. From the Cedar Grove coal zone upward, however, the palynomorph distribution becomes more equally distributed, with fern, calamite, and cordaite taxa all occurring more frequently. The top two coals in the section, the Coalburg and Stockton, in particular, are dominated or codominated by fern spores. This type of palynomorph distribution is similar to the distribution that was observed in core 37-85; the Chilton "Group" coals (Chilton, Chilton Rider, and Chilton "A") in 37-85 contained more fern and calamite spores, in contrast to the coals beneath, which were all *Lycospora*-dominant. In the previously discussed Ze-

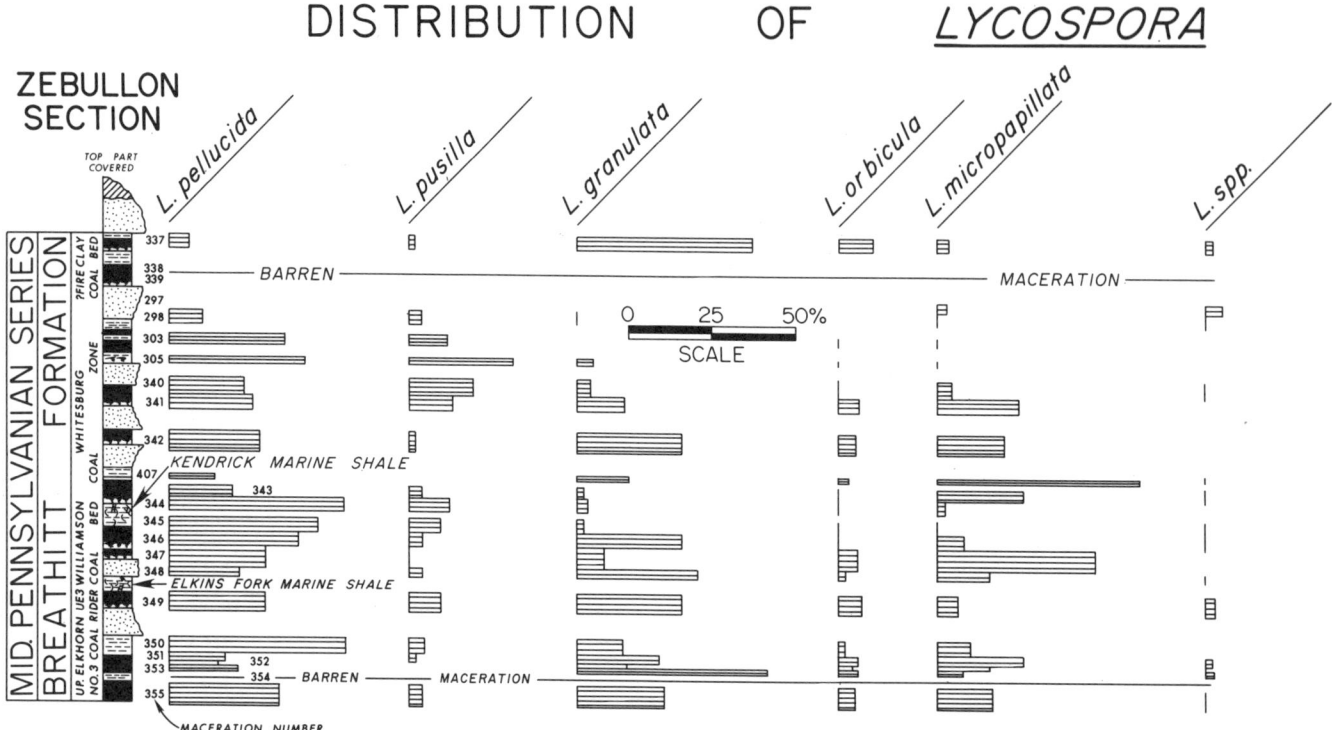

Figure 4. *Lycospora* species distribution in the Zebullon section.

bullon section, a majority of the coal bed palynofloras are *Lycospora*-dominant, with most resembling the lower portions of both Bolt Mountain and core 37-85. The occurrence of more fern-lycopod codominant assemblages in the Whitesburg coal zone is consistent with the Cedar Grove assemblages from Bolt Mountain.

Stratigraphically useful spore taxa that occur in the Bolt Mountain section include *Triquitrites sculptilis*, *Microreticulatisporites sulcatus*, *Laevigatosporites globosus*, *Radiizonates difformis-rotatus*, and *Torispora securis*. *Triquitrites sculptilis* and *M. sulcatus* were first recorded at the level of the Little Chilton coal bed, with *Laevigatosporites globosus* beginning its range in the overlying Chilton "A." First occurrences of these two taxa agree well with their ranges in core 37-85. In the upper part of the Bolt Mountain section, *Radiizonates difformis* and the closely related species *R. rotatus*, were first observed in the Winifrede coal bed, although Kosanke has reported these species from the subjacent Chilton "A" coal bed in the Pennsylvanian stratotype area (Kosanke, 1988a). In any case, strata immediately above the Winifrede marine shale appears to mark the base of the range of *Radiizonates difformis* and *R. rotatus*.

Torispora securis was observed only in the Stockton coal in the Bolt Mountain section, reinforcing the earlier range zone reports by Kosanke (1984) and Eble and Gillespie (1986). *Torispora securis* begins its range zone lower in the Illinois basin than it does in the Appalachian basin. If the thickened exine of *Torispora* represents a desiccation prevention mechanism, its earlier occurrence in the Illinois basin may indicate that conditions in the Illinois basin were somewhat drier than those in the Appalachians during late Kanawha (Abbott) time. Based on overall spore assemblages, Eble and Gillespie (1986) considered the "Kanawha splints," the Winifrede, Coalburg, and Stockton coal beds, to correlate with the *Radiizonates difformis* (RD) palynomorph assemblage of Peppers (1984), which is Upper Atokan to Lower Desmoinesian in age, and the Westphalian C of Western Europe. In the Western Interior basin, *Radiizonates difformis* occurs throughout the Kilbourn Formation, but terminates in the overlying Blackoak coal of the Kalo Formation. The Blackoak coal also marks the basal range zone for *Torispora securis* (Fig. 6).

Chestnut Ridge–U.S. Route 48 section

In contrast to the nearly 610 m of lower and mid-Middle Kanawha Formation strata exposed in the Bolt Mountain section, less than 60 m of Pottsville strata are preserved in northern West Virginia where U.S. Route 48 transects the Chestnut Ridge Anticline (see Fig. 1). Three thin (less than 0.41 m), discontinuous Pottsville coal beds designated, in descending order, the Tionesta, Quakertown, and Sharon by Fonner et al. (1981), were studied. Other coal beds in this section that were studied include, in ascending order, an unnamed lower Allegheny Formation coal bed (originally mapped as a Pottsville Group coal), the Lower, Middle, and Upper Kittanning, Lower and Upper Freeport, and Mahoning coal beds.

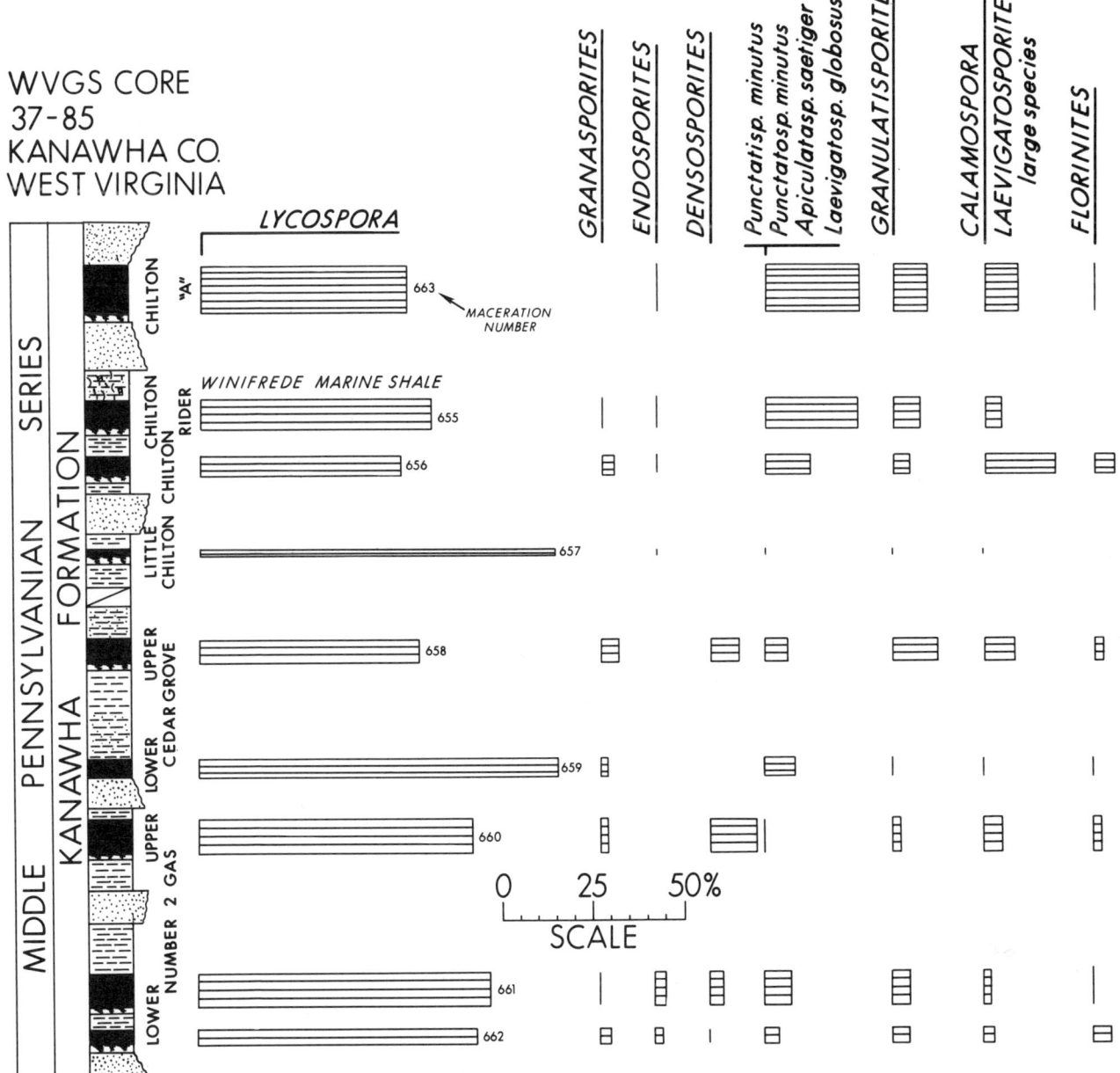

Figure 5. Distribution of selected palynomorph taxa in core 37-85.

Pottsville Group coal beds

The coal beds in the Pottsville Group, designated, in descending order, 1, 2, and 3, contain palynofloras dominated by *Lycospora,* or codominated by *Lycospora, Laevigatosporites,* and *Florinites* (e.g., Pottsville coal 1) (Fig. 8). Kosanke (1984) reported on the spore assemblages from the Sharon and Quakertown coals from southeastern Ohio. The Sharon is codominated by *Lycospora* and *Densosporites,* and contains *Schulzospora,* but no *Laevigatosporites.* The Quakertown is dominated by *Lycospora* and contains both *Schulzospora* and *Laevigatosporites,* which are stratigraphically important spore taxa in Lower and lower Middle Pennsylvanian rocks. In the Appalachian Basin *Schulzospora* occurs throughout the Lower Pennsylvanian and ends its range zone in the lower part of the Kanawha Formation just below the Betsie Shale. In contrast, *Laevigatosporites* begins its range zone in the upper part of the Lower Pennsylvanian (Kosanke, 1982, 1984; Eble and Gillespie, 1989), expands rapidly in the Middle Pennsylvanian, and extends through the Upper Pennsylvanian (Kosanke, 1988a–c).

Based on the range zones of these two taxa, the Pottsville 2 and 3 coals in the Chestnut Ridge section are probably not the Quakertown and Sharon coal beds. The Pottsville 3 coal contains *Schulzospora* sp., and relatively abundant *Laevigato-*

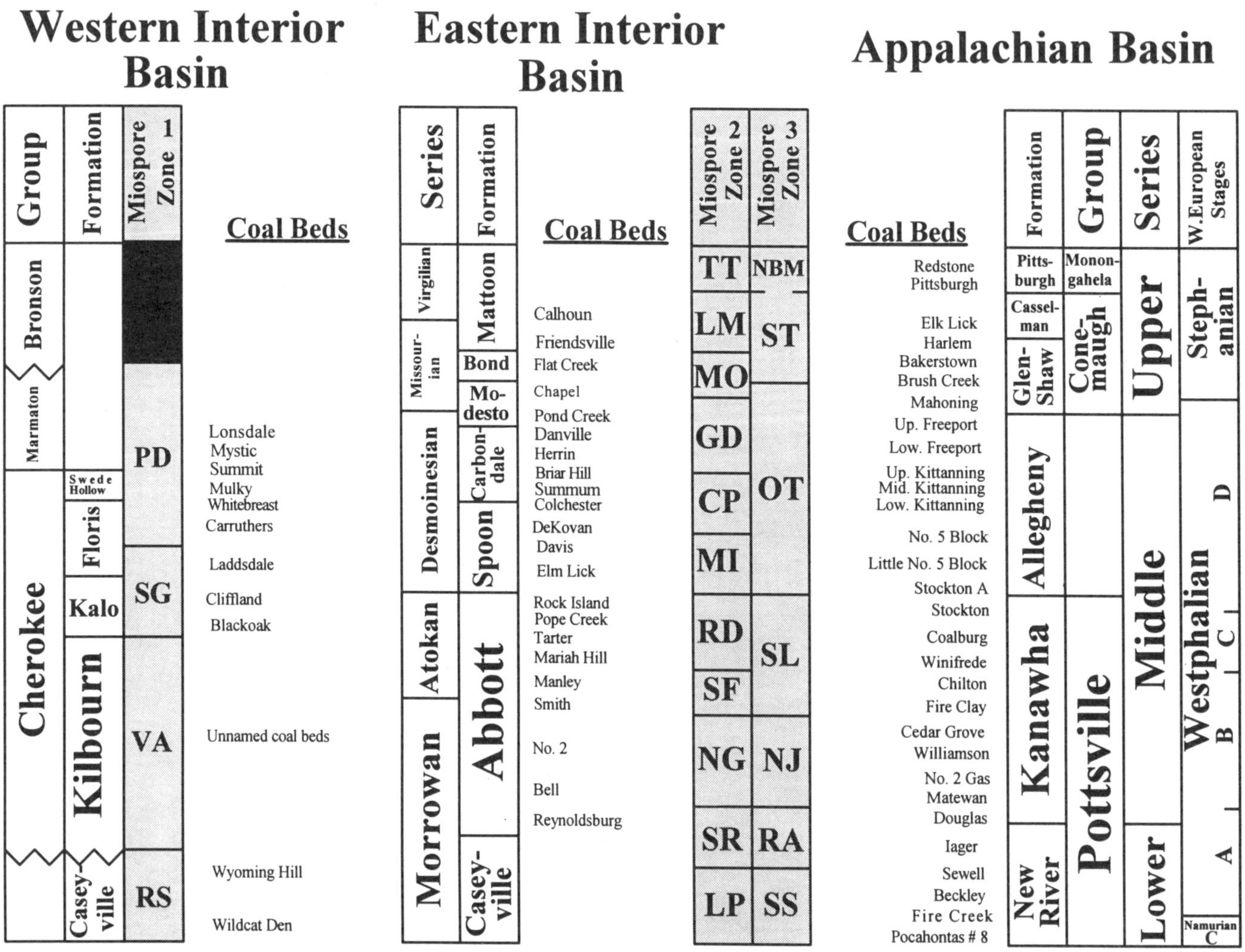

Figure 6. Comparison of Appalachian coal-bed miospore assemblages with palynofloral zonations for the Eastern Interior basin (Peppers, 1984) and Western Interior basin (Ravn, 1986). Coal-bed correlations between the Western and Eastern Interior basins are from Ravn (1986). Unless noted otherwise in the text, coal-bed correlations between the Appalachian and Interior basins are approximate. Placement of Western European stage boundaries applies to the Appalachian basin only.

sporites. The Pottsville 2 coal contains an abundance of *Laevigatosporites*, but no *Schulzospora*. The Pottsville 1 coal contains an assemblage similar to that of the Pottsville 2 coal, but has more *Florinites*. Since the author is not aware of any palynologic data on the Tionesta coal, the proposed correlation by Fonner et al. (1981) cannot be verified at this time.

The overall palynofloras of the Pottsville 1, 2, and 3 coals suggest the following correlations with the central Appalachian basin (Fig. 9). *Triquitrites sculptilis* and *Microreticulatisporites sulcatus*, but not *Laevigatosporites globosus*, were recorded from the Pottsville 1 coal. Assemblages of similar composition were recorded from the Little Chilton coal in core 37-85 and the Bolt Mountain section. The Pottsville 2 coal contains a *Lycospora*-dominant palynoflora, but has none of the above three stratigraphically diagnostic taxa. This sequence resembles assemblages recorded from the No. 2 Gas through Alma "A" and Williamson coals in core 37-85 and the Bolt Mountain section, and the Upper Elkhorn No. 3 through Whitesburg coal zone in the Zebulon section. Unfortunately, this is a thick sequence of strata to suggest equivalence with, but further work in this interval should allow for more precise correlation. Based on the presence of *Schulzospora* sp. and abundant *Laevigatosporites*, the Pottsville 3 coal correlates with the lower part of the Kanawha Formation, more specifically, the interval from just below the Eagle coal bed to the base of the formation.

Using these correlations, it appears that all of the Pottsville in the Chestnut Ridge section is of mid–Middle Pennsylvanian age, thus emphasizing the rapid thinning of Pennsylvanian strata in a northeastern direction (Donaldson and Eble, 1989). The Mississippian-Pennsylvanian unconformity in the Chestnut Ridge area, located approximately 3 to 4 m below the Pottsville

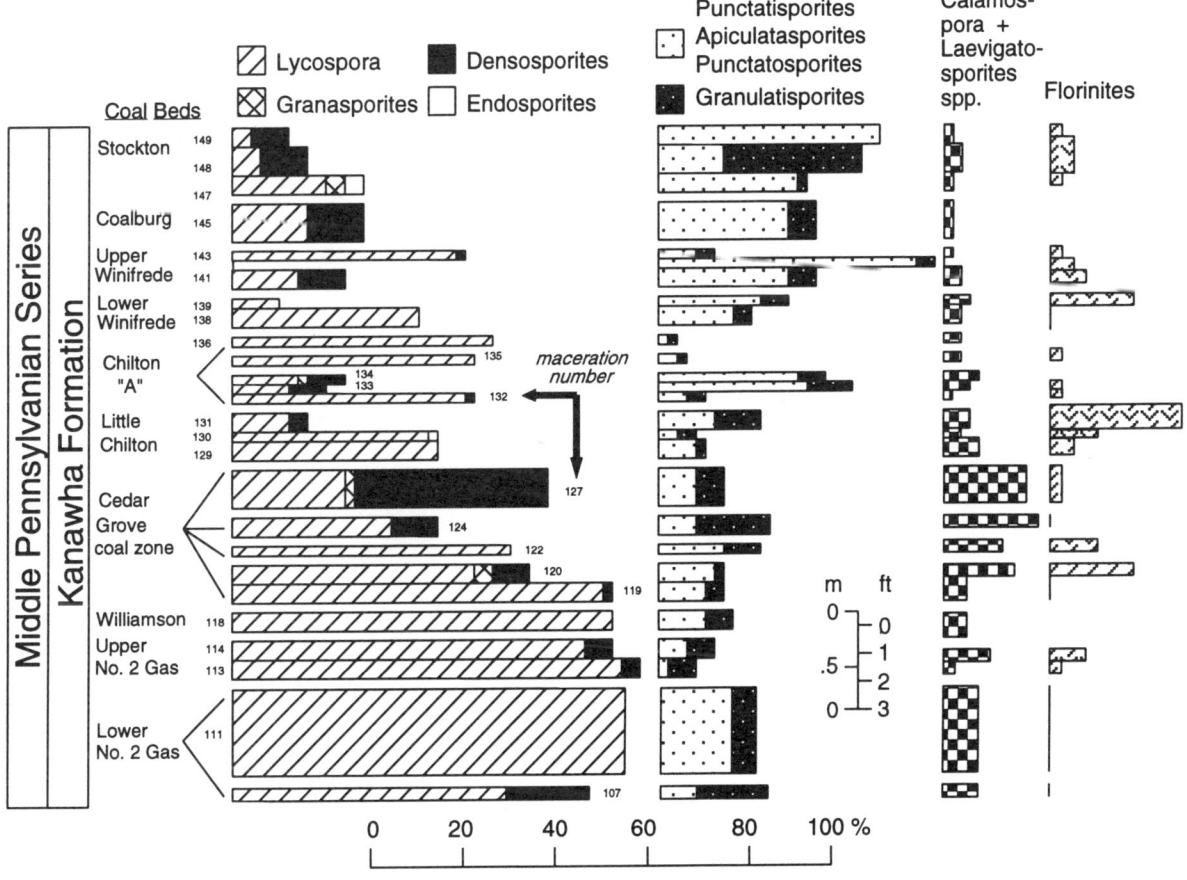

Figure 7. Distribution of miospores, grouped together according to affinity, in coal beds from the Bolt Mountain section in southern West Virginia.

3 coal, was apparently an erosional surface during all of the Early Pennsylvanian (Pocahontas and New River Formations of southern West Virginia), and probably part of the early Middle Pennsylvanian (Kanawha Formation) as well.

Allegheny Formation coal beds

An 0.3-m-thick coal bed originally mapped as the Tionesta coal (Pottsville Group) was sampled and analyzed for palynomorphs. In addition to yielding a diverse assemblage codominated by lycopod and fern taxa (Fig. 8), the stratigraphically diagnostic forms *Torispora securis, Triquitrites sculptilis, Thymospora pseudothiessenii, Microreticulatisporites sulcatus,* and *Murospora kosankei* were also recorded from this coal bed. This type of assemblage correlates with the lower part of the Charleston Sandstone of southern West Virginia, and is perhaps equivalent to the Little or Lower No. 5 Block coal (Kosanke, 1984, 1988a) (Fig. 9). The assemblage from this coal bed also correlates with the Upper Atokan/Lower Desmoinesian of the Illinois basin (*Cadiospora magna-Mooreisporites inusitatus,* MI, palynomorph assemblage of Peppers, 1984).

The Lower, Middle, and Upper Kittanning coal beds contain very similar spore assemblages, all codominated by lycopod and fern taxa (Fig. 8). Perhaps the most significant forms that occur in the Kittannings are *Schopfites dimorphus* (and the closely related species *S. colchesterensis*), and *Densosporites. Schopfites dimorphus* is primarily restricted to the Kittanning coals in the northern Appalachian Basin, although Denton (1957) has reported *Schopfites* from the stratigraphically lower Brookville coal bed in eastern Ohio. However, Frederiksen (1961) did not record *Schopfites dimorphus* from either the Brookville or overlying Lower Clarion coal in western Pennsylvania. Based on the range of *S. dimorphus,* Kosanke (1973, 1984) proposed that the Lower Kittanning correlates with the No. 6 Block coal in southern West Virginia, the Princess No. 6 coal in eastern Kentucky, and the Colchester (No. 2) coal in Illinois. Similarly, Peppers (1970) and Kosanke (1973) correlated the Middle Kittanning coal bed with the Princess No. 7 coal, and Springfield (No. 5) coal (Peppers, 1970). However, the results of this study indicate that the spore floras of these latter two coal beds also compare favorably with the Upper Kittanning coal.

The range zone terminus of *Densosporites* in the Appalachian Basin has been reported as occurring in the Upper

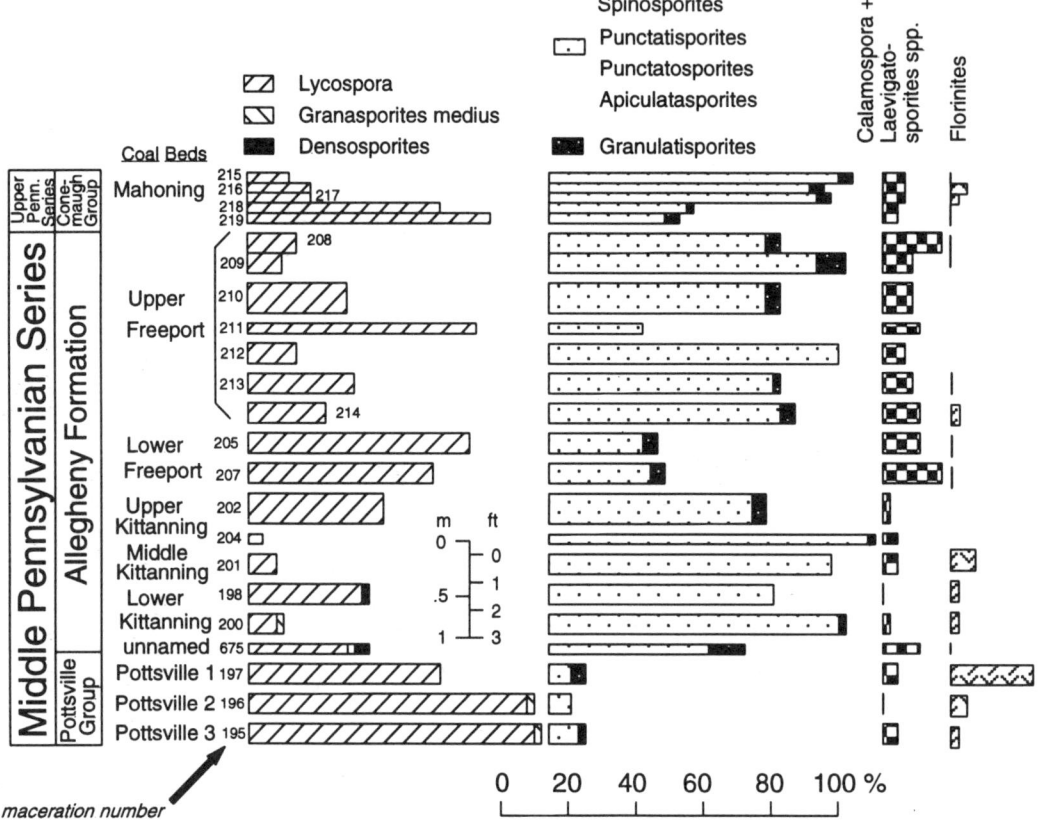

Figure 8. Distribution of miospores, grouped together according to affinity, in Middle and Upper Pennsylvanian coal beds exposed along Interstate 68 on the west flank of the Chestnut Ridge anticline in northern West Virginia.

Kittanning Rider (Lower Freeport?) (Schemel, 1957), Upper Kittanning (Eble, 1986), and Middle Kittanning coal beds (Denton, 1957; Kosanke, 1973). In my opinion, this minor difference is probably one of semantics (i.e., inconsistent nomenclature usage in different areas of the basin), or the result of coal bed misidentification in the field. In any case, knowing that *Densosporites* terminates in the upper part of the Kittanning coal group is probably sufficient for most correlation studies.

The palynofloras of the Lower and Upper Freeport and Mahoning coal beds are very similar in overall composition, being codominated by lycopod and fern spores. This interval does, however, contain the range zone termini of several stratigraphically important taxa that appear to be associated with the Westphalian-Stephanian transition (Phillips et al., 1974; Phillips and Peppers, 1984; Phillips et al., 1985).

Schopfites dimorphus and *S.* sp. were recorded from the Lower Freeport coal, but not higher in the Chestnut Ridge section. Schemel (1957) reported a similar relationship. Preliminary results (C. F. Eble, unpublished data, 1985) also indicate *Schopfites dimorphus* to be present in the Clarion coal bed of north-central West Virginia and eastern Ohio. These findings slightly extend the known range of *Schopfites* in the Appalachian Basin. With further documentation, the range zone termini of *Schopfites* and *Densosporites* may assist in the differentiation between the Lower and Upper Freeport coals, and also between the Upper Kittanning and Lower Freeport coal beds.

Granasporites medius (=*Cappasporites distortus,* Ravn et al., 1986) is found no higher than the Upper Freeport coal in the Chestnut Ridge section. Likewise, the following taxa have their range zone tops in the overlying Mahoning coal bed: *Lycospora* (sporadic occurrences have been noted higher), *Cirratriradites, Vestispora, Torispora securis, Thymospora pseudothiessenii, Triquitrites sculptilis, Murospora kosankei,* and *Mooreisporites inusitatus* (Fig. 10). These extinctions were verified through examination of stratigraphically higher coal beds (Eble, unpublished data), and through the works of Schemel (1957) and Kosanke (1988a).

When compared with the Illinois basin spore zonation of Peppers (1984), the interval from the Lower Kittanning through Mahoning coal correlates with the *Schopfites colchesterensis–Thymospora pseudothiessenii* (CP) and *Lycospora granulata–Cappasporites distortus* (GD) spore assemblage zones (Upper Desmoinesian); the boundary between the two zones is drawn at the Upper Kittanning coal, or perhaps the Lower Freeport, should the occurrence of *Schopfites* in the latter bed be substantiated by further work. In Iowa, Ravn (1986) reported that

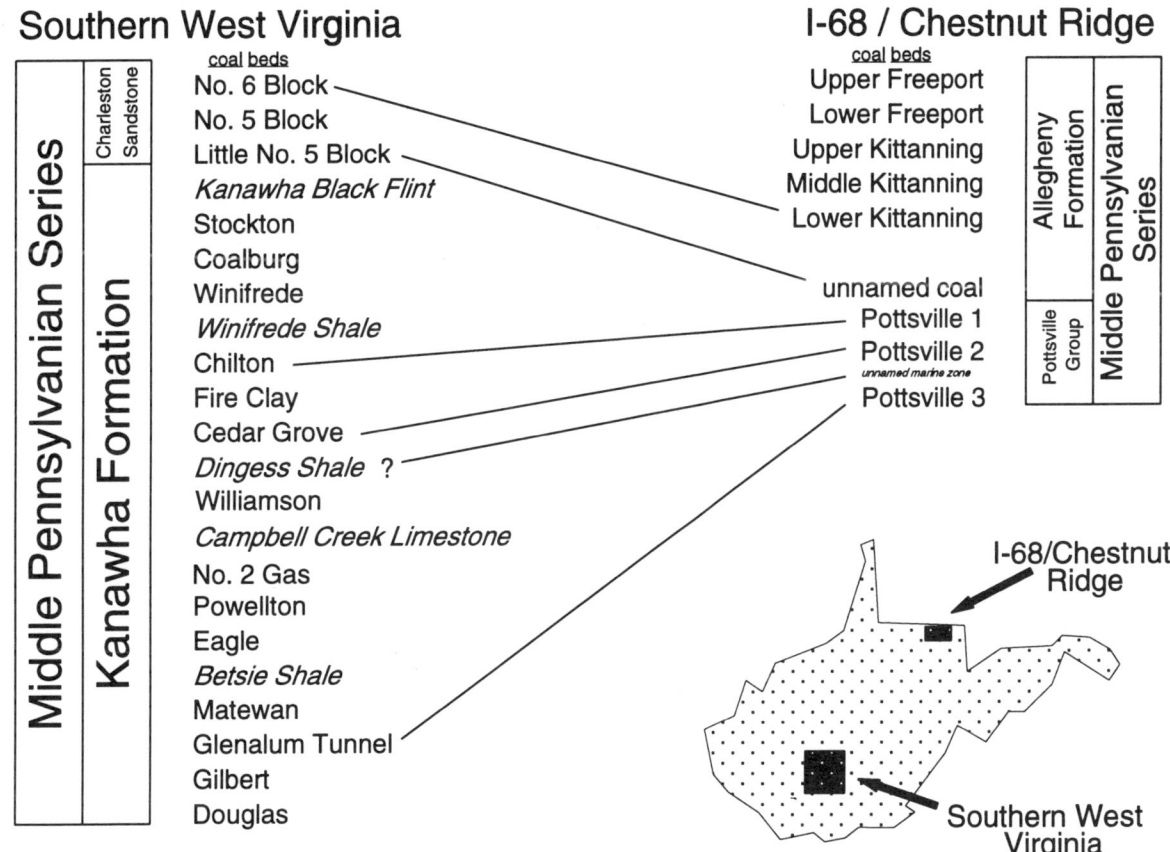

Figure 9. Approximate palynostratigraphic correlation of Pottsville Group and Allegheny Formation coal beds from the Interstate 68–Chestnut Ridge section in northern West Virginia with coals from southern West Virginia. Major marine units that assist correlation in the southern part of the state, and across the central Appalachian basin, are shown in italics.

Schopfites dimorphus occurs from the base of the Swede Hollow Formation to the base of the Marmaton Group. This, coupled with the extinction of *Densosporites* at the base of the Swede Hollow Formation, suggests that the Lower Kittanning–Mahoning interval is at least partially correlative with the Swede Hollow Formation–Marmaton Group? interval (Fig. 6).

Also of interest is the species distribution of *Lycospora* in the Allegheny Formation coals of this section (Fig. 11). *Lycospora granulata* and *L. pusilla* are the dominant forms in the Lower, Middle, and Upper Kittanning coals, Lower and Upper Freeport coals, and the Mahoning coal bed, whereas *Lycospora pellucida* is a more common component of the Pottsville coal palynofloras. This stratigraphic change in species dominance of *Lycospora* has been noted in the Illinois basin (Peppers, 1979, 1984; Phillips et al., 1985), and apparently reflects a shift in major lycopod tree species from *Lepidophloios harcourtii* to *L. halii*. Also, the persistence of *Lycospora pusilla* in the Allegheny coal spore floras suggests that *Lepidodendron hickii* (the plant which bore *L. pusilla*) was a common component of paleomires during late Middle Pennsylvanian time in the Appalachian Basin.

SUMMARY

The use of spores and pollen as a biostratigraphic tool in Pennsylvanian strata is well known. Their application is particularly valuable in dominantly terrestrial sequences where other paleontologic methods, notably marine invertebrates, are of limited value. The Middle Pennsylvanian of the Appalachian basin is a good example of such a sequence. Through the use of outcrop section and drillcore samples, the present study has documented the palynology of selected Middle Pennsylvanian coal beds, and noted the range zones of several stratigraphically useful spore taxa. These taxa, which are of value in both intra- and interbasinal correlation efforts, are: *Microreticulatisporites sulcatus, Triquitrites sculptilis, Laevigatosporites globosus, Radiizonates difformis-rotatus, Torispora securis, Thymospora pseudothiessenii, Murospora kosankei, Mooreisporites inusitatus, Granasporites medius, Schulzospora, Densosporites, Schopfites, Lycospora, Cirratriradites,* and *Vestispora*. Figure 12 presents a summary of the stratigraphic intervals delineated by these taxa.

It is hoped that continued work in Appalachian palyno-

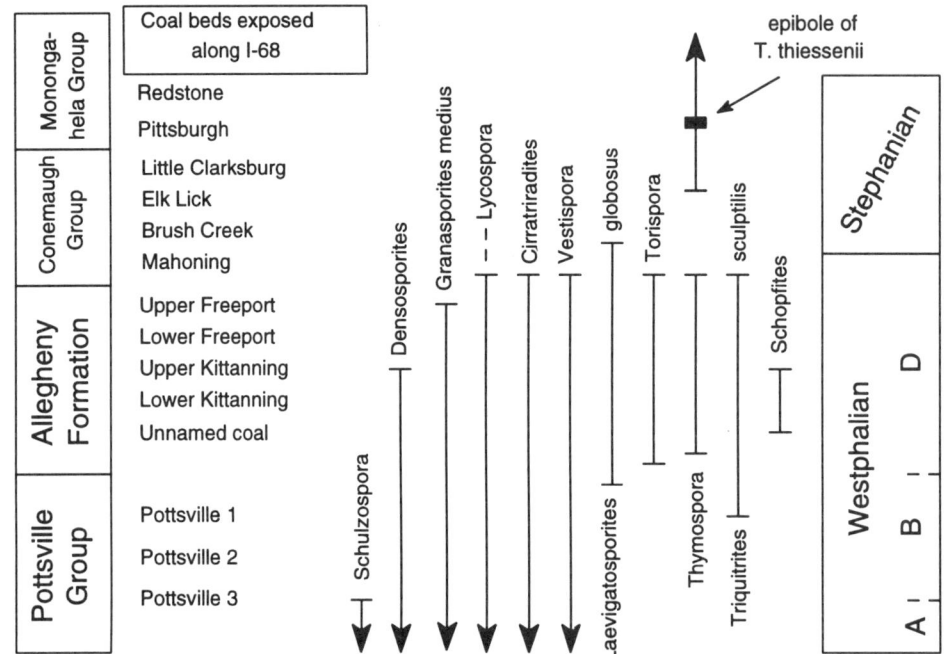

Figure 10. Extinction of miospores in the Interstate 68–Chestnut Ridge section in northern West Virginia. The Middle–Upper Pennsylvanian boundary approximately coincides with the Westphalian-Stephanian boundary, which occurs between the Mahoning and Brush Creek coal beds. Note that many spores end their ranges prior to this level.

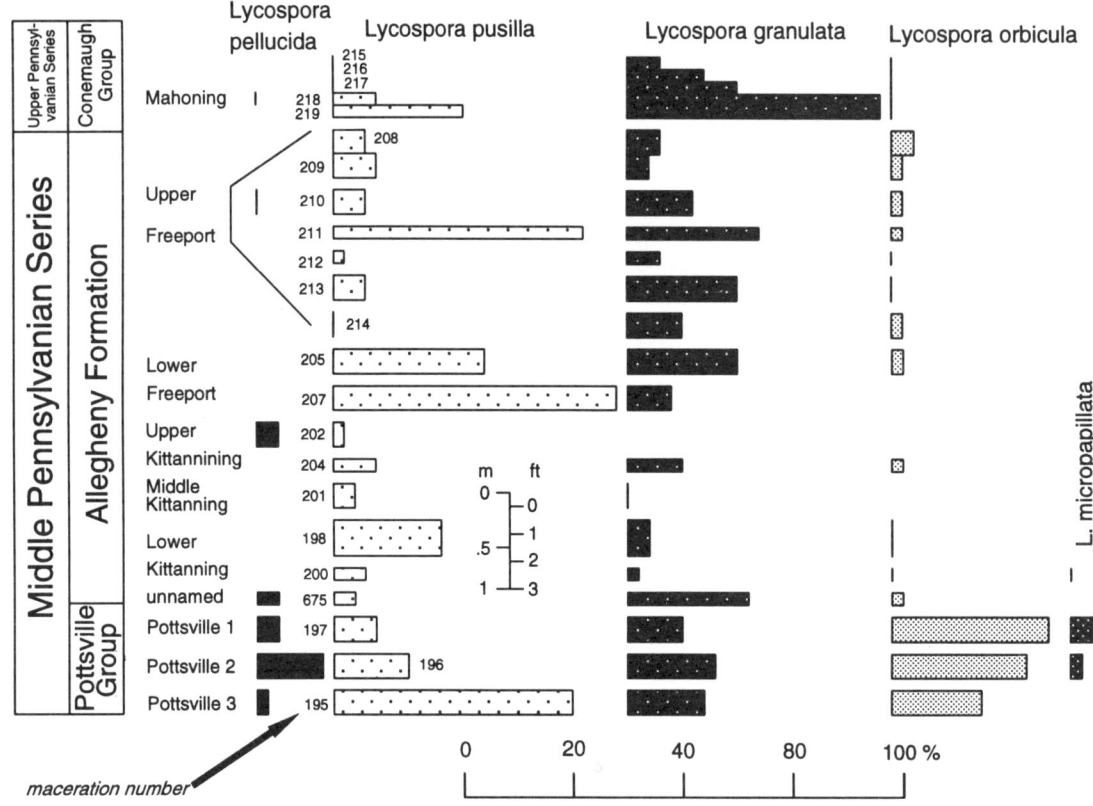

Figure 11. *Lycospora* species distribution in Middle and Upper Pennsylvanian coal beds exposed along Interstate 68 on the west flank of the Chestnut Ridge anticline in northern West Virginia.

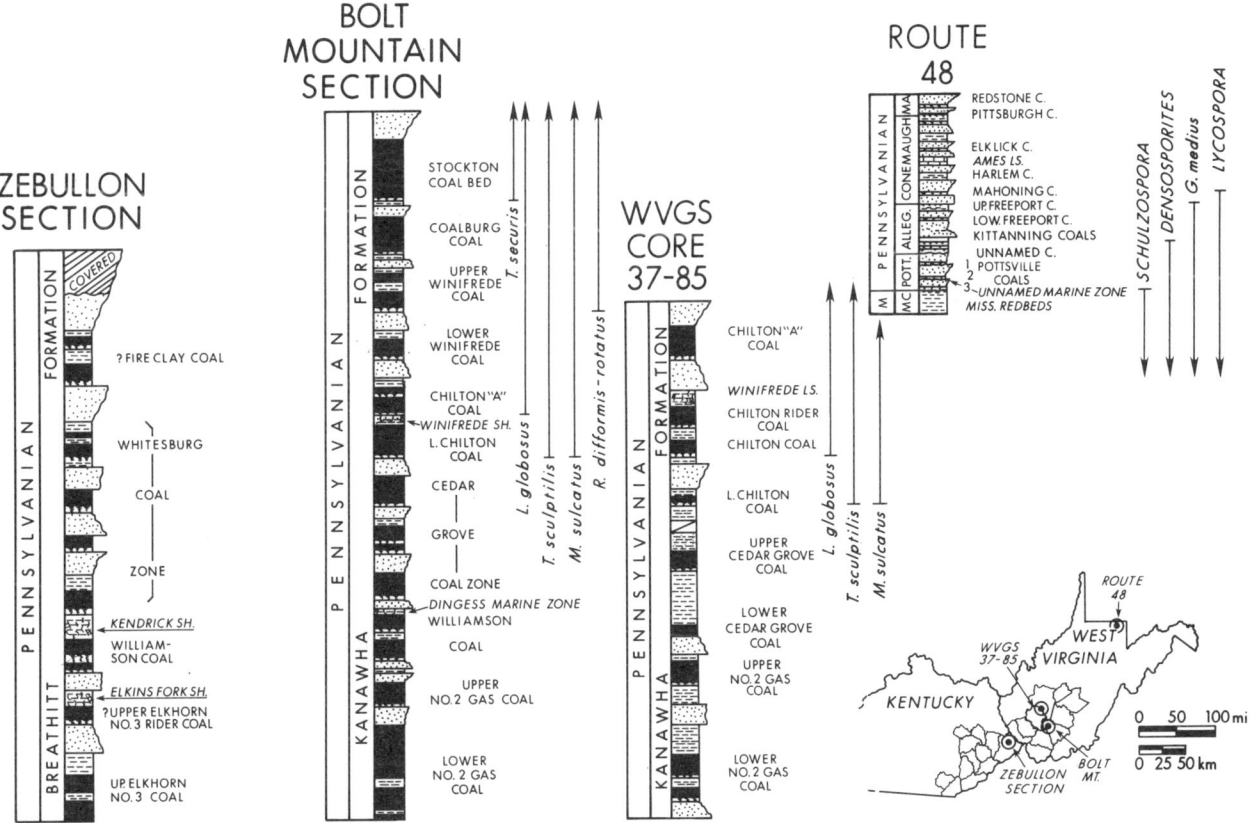

Figure 12. Summary of stratigraphically useful Middle and Upper Pennsylvanian miospore taxa (M = Mississippian, MC = Mauch Chunk Group, Pott. = Pottsville Group, Alleg. = Allegheny Formation, MA = Monongahela Group). Additional range zone termini for the Chestnut Ridge–Route 48 section are listed in Figure 11.

stratigraphy will result in not only range zone refinement of the forms listed above, but also recognition of other stratigraphically useful taxa. As such, this report, along with those of Kosanke (1973, 1984, 1988a–c), should be considered a foundation on which to build a more comprehensive palynostratigraphic data base for the Appalachian basin.

ACKNOWLEDGMENTS

I thank Alan Kaiser and Mitch Blake, West Virginia Geological and Economic Survey, Morgantown, for their assistance in the location and collection of samples, and Bob Kosanke, U.S. Geological Survey, and Russ Peppers, Illinois State Geological Survey, for critical reviews of the manuscript. This work represents a portion of my dissertation work at West Virginia University, Morgantown.

REFERENCES

ASTM, 1989, Preparing coal samples for analysis by reflected light, *in* Annual book of ASTM standards: Gaseous fuels—Coal and coke, v. 05.05, Philadelphia, American Society for Testing and Materials, p. 284–287.

Barss, M. S. and Williams, G. L., 1973, Palynology and nannofossil processing techniques: Canadian Geological Survey Paper 73–26, 25 p.

Butterworth, M. A., 1969, Microfloras of the Upper Carboniferous: Congrès International de Stratigraphie et de Géologie du Carbonifère, 6th, Sheffield, 1967: Compte Rendu, v. 1, p. 59.

Butterworth, M. A., and Smith, A.H.V., 1976, The age of the British upper coal measures with reference to their miospore content: Review of Paleobotany and Palynology, v. 22, p. 281–306.

Clayton, G., and 6 others, 1977, Carboniferous miospores of western Europe—Illustration and zonation: Mededelingen Rijks Geologische Dienst, v. 29, 71 p.

Clendening, J. A., 1974, Palynological evidence for a Pennsylvanian age assignment of the Dunkard Group in the Appalachian basin: West Virginia Geological and Economic Survey Coal Geology Bulletin no. 3, 107 p.

Cropp, F. W., 1960, Pennsylvanian spore floras from the Warrior basin, Mississippi and Alabama: Journal of Paleontology, v. 34, no. 2, p. 359–367.

Cropp, F. W., 1963, Pennsylvanian spore succession in Tennessee: Journal of Paleontology, v. 37, no. 4, p. 900–916.

Cross, A. T., 1947, Spore floras of the Pennsylvanian of West Virginia and Kentucky: Journal of Geology, v. 55, no. 3, p. 285–308.

Denton, G. H., 1957, Correlation of lower Allegheny coal beds of Columbiana County, Ohio and adjacent areas [M.S. thesis]: Morgantown, West Virginia University, 124 p.

Doher, I. L., 1980, Palynomorph preparation procedures currently used in the paleontology and stratigraphy laboratories, United States Geological Survey: U.S. Geological Survey Circular 830, 29 p.

Donaldson, A. C. and Eble, C. F., 1989, Morgantown area stops, *in* Cecil, C. B. and Eble, C. F., eds., Carboniferous geology of the eastern United

States, St. Louis, Missouri, to Washington, D.C.., June 28–July 8, 1989, Field trip guidebook T143 for the 28th International Geological Congress: Washington, D.C., American Geophysical Union, p. 133–142.

Eble, C. F., 1986, Miospore assemblages of selected coal beds in northern West Virginia, central Appalachian basin, and their stratigraphic implications: Geological Society of America Abstracts with Program, v. 18, no. 6, p. 592.

Eble, C. F., 1988, Palynology and paleoecology of a Middle Pennsylvanian coal bed from the central Appalachian basin [Ph.D. thesis]: Morgantown, West Virginia University, 495 p.

Eble, C. F. and Gillespie, W. H., 1984, Characteristic small spores of the Coalburg coal (Upper Pottsville, Pennsylvanian) in West Virginia: Proceedings of the West Virginia Academy of Science, v. 56, p. 104–123.

Eble, C. F., and Gillespie, W. H., 1986, Palynological studies of the upper Kanawha Formation (Pottsville, Pennsylvanian) in West Virginia: Compass, v. 63, no.2, p. 58–65.

Eble, C. F, and Gillespie, W. H., 1989, Palynology of selected Pennsylvanian coal beds from the central and southern Appalachian basin: Correlation and stratigraphic implications, in Englund, K. J., ed., Characteristics of the mid-Carboniferous boundary and associated coal-bearing rocks in the central and southern Appalachian basin: Washington, D.C., American Geophysical Union, Field trip guidebook T352B, 28th International Geological Congress, p. 61–66.

Englund K. J., Arndt, H. H. and Henry, T. W., 1979, Proposed Pennsylvanian System stratotype, Virginia and West Virginia: American Geological Institute Selected Guidebook Series no. 1, 136 p.

Fonner, R. F., Reynolds, J. H., Cole, G. A., Jake, T. R., Fedorko, N. and Connell, D. A., 1981, Geology along the West Virginia portion of U.S. Route 48: West Virginia Geological and Economic Survey Publication WV-14, 65 p.

Frederiksen, N. O., 1961, Sporomorphae of the Brookville Seam near Brookville, Pennsylvania [M.S. thesis]: State College, Pennsylvania State University, 273 p.

Kosanke, R. M., 1950, Pennsylvanian spores of Illinois and their use in correlation: Illinois State Geological Survey Bulletin 74, 128 p.

Kosanke, R. M., 1973, Palynological studies of the coals of the Princess District in northeastern Kentucky: U.S. Geological Survey Professional Paper 839, 24 p.

Kosanke, R. M., 1982, Mississippian-Pennsylvanian boundary, in the United States based on palynomorphs, in Ramsbottom, W.H.C., Sanders, W. B., and Owens, B., Biostratigraphic data for a mid-Carboniferous boundary: Leeds, England, Subcommission on Carboniferous Stratigraphy, p. 27–35.

Kosanke, R. M., 1984, Palynology of selected coal beds in the proposed Pennsylvanian stratotype of West Virginia: U.S. Geological Survey Professional Paper 1318, 44 p.

Kosanke, R. M., 1988a, Palynological studies of Middle Pennsylvanian coal beds of the proposed Pennsylvanian System stratotype in West Virginia: U.S. Geological Survey Professional Paper 1455, 73 p.

Kosanke, R. M., 1988b, Palynological studies of Lower Pennsylvanian coal beds and adjacent strata of the proposed Pennsylvanian System stratotype in Virginia and West Virginia: U.S. Geological Survey Professional Paper 1479, 17 p.

Kosanke, R. M., 1988c, Palynological analyses of Upper Pennsylvanian coal beds and adjacent strata from the proposed Pennsylvanian System stratotype in West Virginia: U.S. Geological Survey Professional Paper 1486, 24 p.

Owens, B., 1984, Miospore zonation of the Carboniferous: Congrès International de Stratigraphie et de Géologie du Carbonifère, 9th, Washington, D.C., and Champaign-Urbana, Illinois, 1979: Compte Rendu, v. 2, p. 90–102.

Owens, B., Lobozialk, S., and Teteriuk, V. K., 1978, Palynological subdivision of the Dinantian to Westphalian deposits of northwest Europe and the Donetz basin of the USSR: Palynology, v. 2, p. 69–91.

Patterson, T. R., and Fishbein, E., 1989, Re-examination of the statistical methods used to determine the number of point counts needed for micropaleontological quantitative research: Journal of Paleontology, v. 63, no. 2, p. 245–248.

Peppers, R. A., 1964, Spores in strata of Late Pennsylvanian cyclothems in the Illinois basin: Illinois State Geological Survey Bulletin 90, 89 p.

Peppers, R. A., 1970, Correlation and palynology of coals in the Carbondale and Spoon Formations (Pennsylvanian) of the northeastern part of the Illinois basin: Illinois State Geological Survey Bulletin 93, 173 p.

Peppers, R. A., 1979, Development of coal-forming floras during the early part of the Pennsylvanian in the Illinois basin, in Palmer, J. E., and Dutcher, R. R., eds., Depositional and structural history of the Illinois basin, pt. 2—Invited papers: Illinois State Geological Survey Guidebook Series 15a, p. 8–15.

Peppers, R. A., 1984, Comparison of miospore assemblages in the Pennsylvanian System of the Illinois basin with those in the Upper Carboniferous of western Europe: Congrès International de Stratigraphie et de Géologie du Carbonifère, 9th, Washington, D.C., and Champaign-Urbana, Illinois, 1979: Compte Rendu, v. 2, p. 483–502.

Phillips, T. L., and Peppers, R. A., 1984, Changing patterns of Pennsylvanian coal-swamp vegetation and implications of climate control on occurrence: International Journal of Coal Geology, v. 3, p. 205–255.

Phillips, T. L., Peppers, R. A., Avcin, M. J., and Laughnan, P. F., 1974, Fossil plants and coal: patterns of change in Pennsylvanian swamps of the Illinois basin: Science, v. 184, p. 1367–1369.

Phillips, T. L., Peppers, R. A., and DiMichele, W. A., 1985, Stratigraphic and interregional changes in Pennsylvanian coal-swamp vegetation and implications of climate control on occurrence: International Journal of Coal Geology, v. 5, nos. 1-2, p. 43–109.

Ravn, R. L., 1979, An introduction to the stratigraphic palynology of the Cherokee Group (Pennsylvanian) coals of Iowa, Iowa Geological Survey Technical Paper no. 6, 117 p.

Ravn, R. L., 1986, Palynostratigraphy of the Lower and Middle Pennsylvanian coals of Iowa: Iowa Geological Survey Technical Paper no. 7, 245 p.

Ravn, R. L., and Fitzgerald, D. L., 1982, A Morrowan (Upper Carboniferous) miospore flora from eastern Iowa, U.S.A.: Paleontographica, v. 183B, p. 108–172.

Ravn, R. L., Butterworth, M. A., Peppers, R. A., and Phillips, T. L., 1986, Proposed synonymy of *Granasporites* Alpern 1959 emend. and *Cappasporites* Urban emend. Chadwick 1983, miospore genera from the Upper Carboniferous of Europe and North America: Pollen et Spores, v. 28, nos. 3-4, p. 421–434.

Schemel, M. P., 1957, Small spore assemblages of mid-Pennsylvanian coals of West Virginia and adjacent areas [Ph.D. thesis]: Morgantown, West Virginia University, 222 p.

Smith, A.H.V., and Butterworth, M. A., 1967, Miospores in the coal seams of the Carboniferous of Great Britain: Paleontological Society Special Papers in Paleontology, London, no. 1, 324 p.

Thiessen, R., and Staud, J. N., 1923, Correlation of coal beds in the Monongahela Formation of Ohio, Pennsylvania and West Virginia: Coal Mining Investigations, Carnegie Institute Technical Bulletin 9, 64 p.

Thiessen, R., and Wilson, F. E., 1924, Correlation of coal beds of the Allegheny Formation of western Pennsylvania and West Virginia: Coal Mining Investigations, Carnegie Institute Technical Bulletin 10, 61 p.

Upshaw, C. F., 1967, Pennsylvanian palynology and age relationships, in Ferm, J. C., et al., eds., A field guide to Carboniferous detrital rocks in northern Alabama: Tuscaloosa, Alabama Geological Society, p. 16–20.

MANUSCRIPT ACCEPTED BY THE SOCIETY FEBRUARY 1, 1994

Facies analysis of Middle Pennsylvanian marine units, southern West Virginia

Ronald L. Martino
Department of Geology, Marshall University, Huntington, West Virginia 25755

ABSTRACT

At least 10 marine-influenced stratigraphic intervals have been distinguished in the Kanawha Formation. Marine units range from 3 to 34 m in thickness and have been divided into component sedimentary facies based on lithology, body and trace fossils, sedimentary structures, paleocurrents, and geometry. Offshore facies consist of dark gray laminated shales, whereas nearshore and littoral deposits are typified by interlaminated to thinly interbedded very fine sandstone, siltstone, and shale. Current ripple bedding (flaser, wavy, lenticular) and ripple cross-lamination are widely developed. Tidal influence was significant and is reflected by rhythmic textures and structures and bipolar paleocurrents.

Phosphatic brachiopods and burrowing bivalves predominate in nearshore and littoral facies, while calcareous brachiopods dominate offshore facies and are often accompanied by bivalves, bryozoans, echinoderms, cephalopods, and corals. Primary faunal distribution was likely to have been controlled mainly by substrate, salinity, and dissolved oxygen.

Two trace fossil assemblages are common. The *Olivellites* assemblage has a high diversity and is represented by both infaunal burrows and surface tracks and trails developed in the nearshore zone and low- to midtidal flats. The *Phycodes-Zoophycos* assemblage is a low-diversity assemblage of infaunal deposit feeders formed in upper tidal flats, restricted bays, and tidal creeks.

Lowstands of sea level caused incisement of fluvial channels, whereas rising sea level led to expansion of tidal plains and estuaries. Shoreface retreat produced ravinement surfaces and transgressive lags that now separate coastal plain and marginal marine facies from shallow subtidal facies. Coastal progradation began during highstands after estuarine sediment sinks had filled and may have intensified during falling sea level.

INTRODUCTION

Stratigraphic framework

The Kanawha and Allegheny Formations comprise the Middle Pennsylvanian Series in West Virginia (Fig. 1). The Kanawha Formation represents about 80% of this interval (Arndt, 1979). Marine units are restricted to the Kanawha Formation in the study area. The formation is 366 m thick in its type area and extends from the base of the Lower Douglass coal bed to the top of the Kanawha black flint (Arndt, 1979). Middle Pennsylvanian strata of southern West Virginia are characterized by coal-bearing sequences punctuated by numerous transgressive intervals. Equivalent stratigraphic intervals occur within the Breathitt Formation of eastern Kentucky and the Upper Pottsville Group of Ohio and Pennsylvania (Fig. 1) (Cardwell et al., 1968).

Geologic setting

The Middle Pennsylvanian paleogeography of central Appalachian basin has been described as one characterized by

---

Martino, R. L., 1994, Facies analysis of Middle Pennsylvanian marine units, southern West Virginia, *in* Rice, C. L., ed., Elements of Pennsylvanian Stratigraphy, Central Appalachian Basin: Boulder, Colorado, Geological Society of America Special Paper 294.

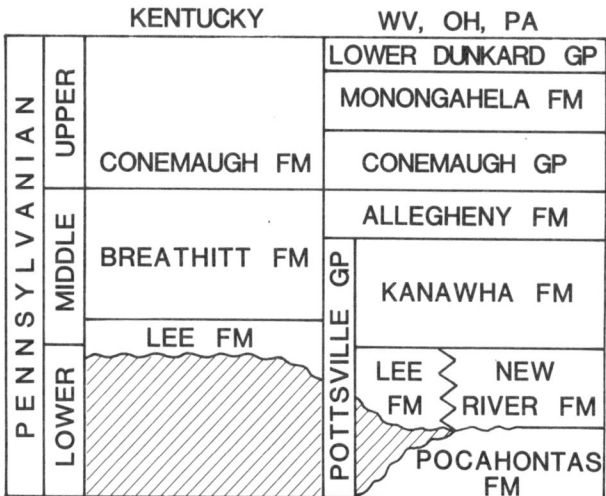

Figure 1. Stratigraphic framework for study area and vicinity (modified from Cardwell et al., 1968, and Englund and Randall, 1981).

fluvially dominated deltas that prograded northwestward from the rising Appalachian Mountains across West Virginia and eastern Kentucky into an interior sea (Fig. 2) (Horne and Ferm, 1978; Flores and Arndt, 1979; Donaldson and Schumaker, 1979; Donaldson et al., 1985). Superimposed on this broad regressive pattern were many pulses of marine inundation. Early workers reported as many as 12 marine units in the Kanawha Formation (Fayette County, West Virginia: Hennen and Teets, 1919), whereas Henry and Gordon (1979) have confirmed seven zones in Fayette County in the proposed Pennsylvanian System stratotype. The proposed stratotype has fewer marine units than are found in some other Pennsylvanian sections in the basin.

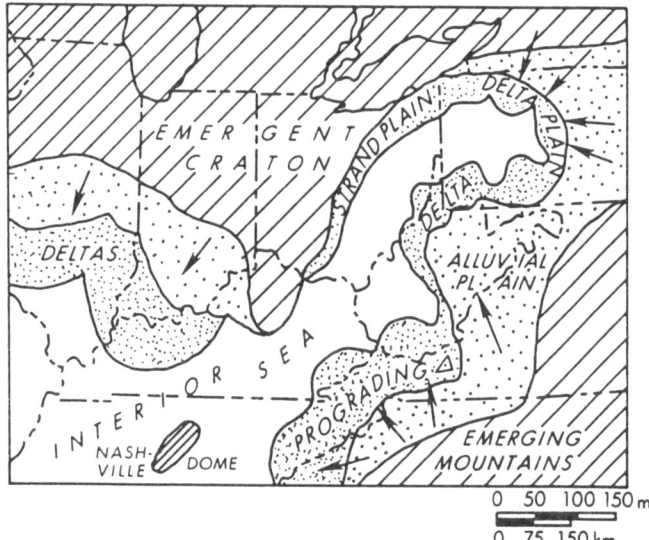

Figure 2. Middle Pennsylvanian regional paleogeography (after Donaldson and Eble, 1989; modified from Donaldson and Schumaker, 1979).

Delta switching and tectonically induced subsidence of the foreland basin are two suggested causes for the many transgressions (Horne and Ferm, 1978; Donaldson and Schumaker, 1979; Tankard, 1986; Klein and Willard, 1989). Others have attributed the transgressions largely to eustatic sea-level fluctuations (Busch and Rollins, 1984; Chesnut and Cobb, 1989). Donaldson (1991) has suggested that most of the Kanawha marine zones were part of minor allocycles that were probably glacioeustatic in origin, although autocycles may have locally occurred within allocycles.

Depositional models for the Kanawha and related Pennsylvanian coal-bearing strata have undergone considerable evolution over the years. The early work of Udden (1912) was reorganized and reinterpreted in the cyclothem model (Weller, 1930; Wanless and Weller, 1932). The cyclothem concept and the related stratigraphic sequence were reevaluated in the Allegheny model of Ferm and Williams (1965; modified by Ferm, 1975, and Ferm and Weisenfluh, 1989). The Allegheny model has dominated discussion for the last two decades. Whereas the cyclothem model tended to relate vertical facies sequences to depositional environments in Pennsylvanian coal-bearing strata of the Illinois basin, the Allegheny model tried to illustrate lateral relationships of facies in a regional context. The latter model emphasized the similarity of facies relationships with a deposystem comparable to that of the modern Mississippi delta.

Both the cyclothem and the Allegheny depositional models were developed for areas on the relatively stable part of the North American craton in western Pennsylvania, Ohio, and Illinois. The applicability of the Allegheny model to Pennsylvanian strata in the Pocahontas basin in southeastern West Virginia and southwestern Virginia was initially taken for granted by many workers (Donaldson, 1974; Horne and Ferm, 1978; Arndt, 1979). Rapid and differential subsidence of the Pocahontas foreland basin, however, may have caused drastic changes in the arrangement of depositional environments and facies from that predicted in the Allegheny model (Ferm and Weisenfluh, 1989).

According to the Allegheny model, marine units of the Kanawha and Breathitt Formations have been interpreted as lower delta plain bay fills (Horne and Ferm, 1978; Horne et al., 1978; Arndt, 1979; Flores and Arndt, 1979). These were thought to intergrade seaward (northwest) with orthoquartzitic barrier deposits and landward (southeast) with upper delta plain lithic sandstones and related facies. The Kanawha/Breathitt bay fills are described as coarsening-upward sequences of shale and siltstone 15 to 55 m thick that occur between coals (Baganz et al., 1975; Horne et al., 1978). In the Allegheny model, bay filling was accomplished mainly by direct fluvial processes (crevassing and mouth bar deposition) in a manner comparable to the modern river-dominated Mississippi delta. Lithic sandstones were described as fluvial or deltaic, whereas quartz-rich sandstones were interpreted to have been originally lithic sands that were subsequently modified by removal of labile constituents during reworking by waves and tides. In the model, tidal influence was restricted to barrier and back barrier environments, and tides were considered to have had little effect on deposition in the coastal zone of the lower delta plain (Horne et al., 1978). However, strata previously interpreted in this model as lower delta plain interdistributary bay fills are now viewed as prodelta and delta front facies (Ferm and Weisenfluh, 1989) that might be better described as "sea fills" (Chesnut, 1989).

Donaldson's (1979) generalized model for Pennsylvanian coal-bearing strata in West Virginia, Ohio, and Pennsylvania differs from the Allegheny model in two ways. First, widespread barrier island and back barrier facies are absent, and local barrier bars occasionally developed only as the result of delta switching. Second, and more important, Donaldson's model includes the presence of a tidal plain in areas of delta lobe abandonment and along interdeltaic portions of the coast. Although not specifically designed to explain the Kanawha Formation of southern West Virginia, this model more closely resembles the depositional setting interpreted here.

Cecil et al. (1985) have suggested that strata including those of the Kanawha Formation accumulated along a tropical, everwet, coastal lowland with domed peat swamps. They speculated that the Kanawha coastal zone may have been dominated by estuarine conditions having a meso- to macrotidal range.

Approach

This study is based on stratigraphic sections measured at 46 localities in Kanawha and Boone Counties, West Virginia (Fig. 3). Each section was divided into sedimentary facies distinguished by lithology, sedimentary structures, body fossils, trace fossils, paleocurrents, and facies geometry. Ten marine-influenced stratigraphic intervals have been identified in the Kanawha Formation in northwestern Boone County, West Virginia (Fig. 4). Marine units are analyzed using an integrated approach that combines sedimentology, paleocurrents, body fossils, and trace fossils that are known to be associated with marine-influenced Pennsylvanian strata (Miller, 1984; Miller and Jackson, 1984; Miller and Knox, 1985).

Among the earliest applications of trace fossils to the interpretation of coal-bearing strata was the work of Adolph Seilacher (1964) in paralic Carboniferous strata of Germany. Seilacher was able to use trace fossils and body fossils and their associated lithofacies to identify freshwater, brackish water, and marine deposits.

Trace fossil assemblages are now known to be temporally and geographically recurrent throughout Phanerozoic strata. They develop wherever the necessary ecologic conditions are repeated (Frey and Pemberton, 1984). Controlling factors include salinity, substrate, hydraulic energy, turbidity, rate of deposition, dissolved oxygen, food supply, capabilities of the tracemaker, and preservation potential (Frey and Pemberton, 1984; Frey et al., 1990). Additional controlling factors, which are apparent in modern environments but difficult to evaluate from the rock record, include competition and predation (Mill-

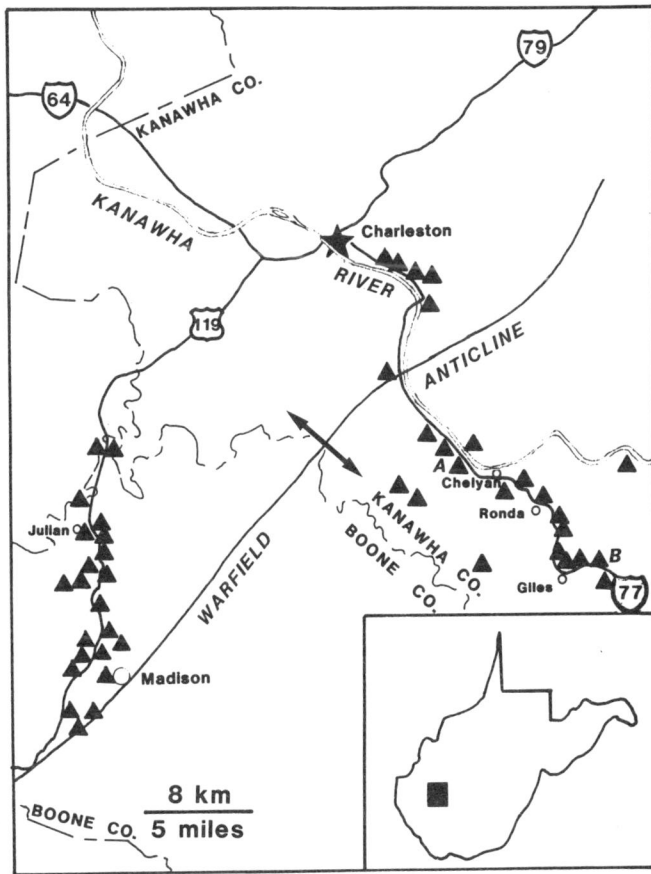

Figure 3. Inset map showing study area in southern West Virginia, and expanded view of portions of Kanawha and Boone Counties with 46 outcrop localities. Most outcrops are roadcuts along U.S. Route 119 and Interstate Highway I-77 (West Virginia Turnpike). A and B indicate the endpoints of the stratigraphic cross-section along I-77 in Figure 7.

er and Byers, 1989). Recent examples of the use of trace fossils to help recognize Pennsylvanian marine zones include Miller and Knox (Tennessee, 1985), Eagar et al. (Great Britain, 1985), Martino (West Virginia, 1988, 1989, 1991), and Maples and Suttner (Colorado, 1990).

The objectives of this chapter are to describe and evaluate the sedimentologic and paleontologic character of those facies that comprise marine-influenced stratigraphic intervals in the Kanawha Formation, and to demonstrate a greater influence of tidal processes and environments along the Kanawha coastal zone than has previously been realized.

PALEOENVIRONMENTS AND ASSOCIATED SEDIMENTARY FACIES

Depositional framework

Marine-influenced strata of the Kanawha Formation are interpreted within the context of the idealized depositional framework in Figures 5 and 6. The transgressive or regressive

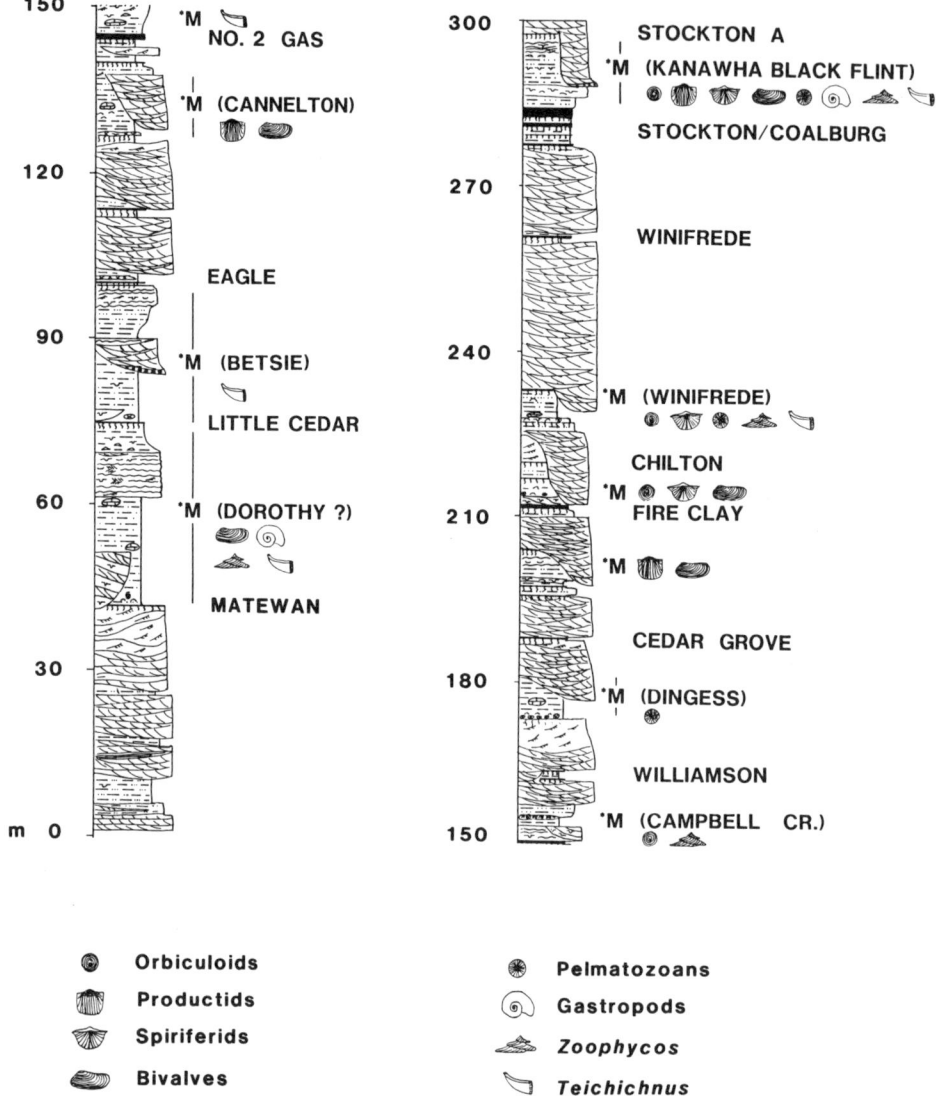

Figure 4. Composite section through the Kanawha Formation based on exposures along U.S. Route 119 near Madison, West Virginia. Ten marine-influenced stratigraphic intervals (*M) have been identified based on body and trace fossils, sedimentary structures, and paleocurrents. The names of major coal beds are indicated; named marine zones are in parentheses. The relations of stratigraphic terms used in this study with those of the Boone County [West Virginia] Report (Krebs and Teets, 1915) are shown in Table 1.

nature and the morphology of the shoreline were determined by the interplay between eustasy, basin subsidence, and sediment supply.

Lowstands of sea level caused incisement of fluvial drainage lines and soil development in interfluves as the result of sediment bypassing. Subsequent rising sea level led to aggradation of alluvial deposystems, valley filling, and the development of estuarine deposystems and associated tidal flats. These coastal facies were truncated by ravinement surfaces as the shoreface retreated. During the transgressive maximum, sediment supply to the shelf was temporarily reduced, and condensed sections developed. Much of the nearshore zone was muddy (similar to the Late Devonian Catskill shoreline of Walker and Harms, 1975). During highstands, estuaries became filled and coastal progradation subsequently occurred. Where point sources of sediment supply to the coast were significant, prograding deltaic systems developed. Falling sea level may have intensified regressive events, ultimately causing incision of alluvial channels and development of soil.

Wave energy along the coast was likely to have been low due to the paleolatitudinal position of the central Appalachian basin, which was approximately 3° to 5° south of the equator (Scotese, 1986). This position lies in intertropical convergence zone where warm moist air rises and winds tend to be weak

TABLE 1. CORRELATION OF SELECTED MEMBERS AND BEDS OF THE KANAWHA FORMATION*	
Boone County Report	This Chapter
Homewood sandstone	(Lower Allegheny Formation)
Black flint horizon	Kanawha black flint
Coalburg coal	Winifrede coal
Buffalo Creek coal	Chilton A Coal?
Winifrede coal	Fire Clay coal
Hernshaw coal	Cedar Grove coal
Dingess limestone	Dingess shale
Williamson coal	Williamson coal
Seth limestone	Campbell Creek limestone
Cedar Grove, Alma coals	No. 2 Gas coal
No. 2 Gas coal	Eagle coal
	Betsie shale
Eagle shale and limestone	Dorothy shale

*As used in the Boone County [West Virginia] Report (Krebs and Teets, 1915) in the Scott and northern Washington districts, and equivalent terminology used in this chapter

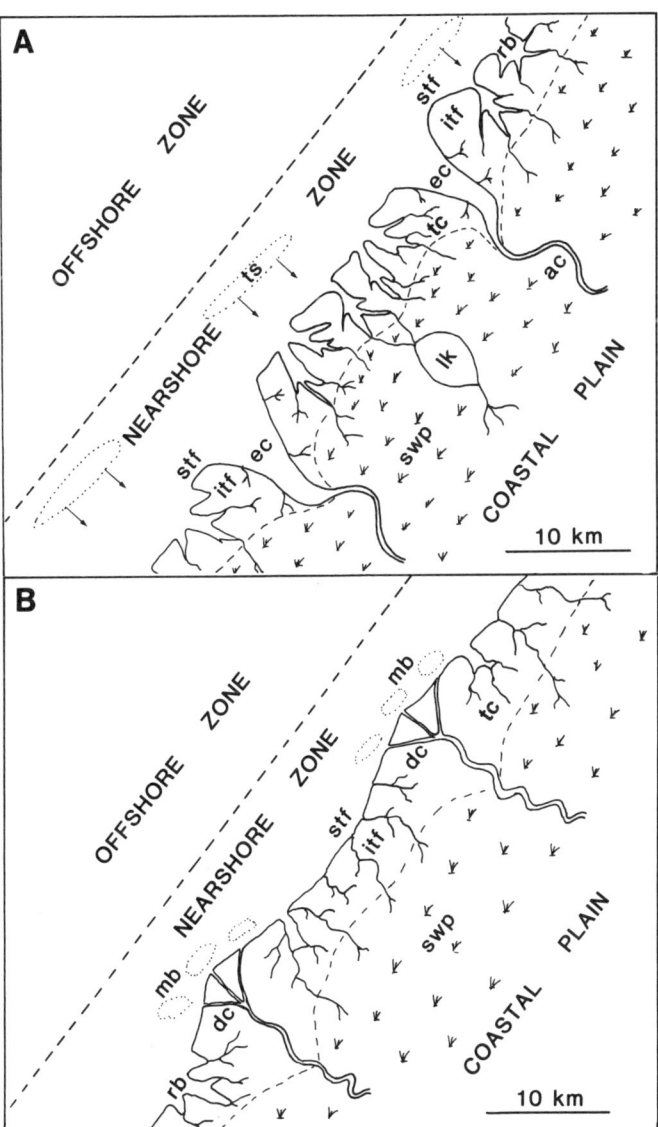

Figure 5. Proposed depositional framework for Kanawha Formation (modified from Donaldson, 1979, and Tankard, 1986). A, Transgressive coastal zone having estuarine channels (ec), tidal creeks (tc), intertidal flats (itf), subtidal flats (stf), restricted bays (rb), transgressive sands reworked during shoreface retreat (ts), alluvial channels (ac), coastal swamps (swp), and lakes (lk). B, Regressive coastal zone formed during highstands after estuaries have filled. Environments include deltaic distributary channels (dc), mouth bars (mb), and other environments as in A.

and doldrums occur. Wind-driven waves would have been further limited by the small fetch across the seaway (Fig. 2).

Opinion is divided as to the potential range of tides in epeiric seas. Shaw (1964) and Mazzullo and Friedman (1975) maintained that most epeiric seas were tideless, whereas more recent workers (Klein, 1977a; Klein and Ryer, 1978; Slingerland, 1986) have contended that many were tide-dominated. Mathematical modeling by Slingerland (1986) for the Late Devonian Catskill Sea indicated that shallow epicontinental seas (maximum depth, 45 m or less) are likely to be microtidal, but that intermediate depths (150 m maximum) can cause low macrotidal conditions. Low macrotidal conditions were predicted for the Catskill shelf in the mid-Atlantic states (Slingerland, 1986).

Shelf width and shoreline geometry influence the tidal range along coastal embayments (Hayes, 1975). Tidal ranges in the central portion of an embayment are likely to be greatest due to two factors: the funneling effect of the embayment on the tidal wave, and the frictional effects on the tidal wave exerted by the wider shelf in the central part of the embayment, which induces shoaling and augmentation. The influence of shoreline geometry on tidal range is exemplified by the German North Sea Bight, where microtidal ranges prevail along the outer edges of the embayment and increase to macrotidal at its center (Hayes, 1975).

Nonmarine facies

Coal beds, seat rock, and associated carbonaceous shale largely represent freshwater swamp accumulations and lake deposits, respectively. It is possible that some types of Pennsylvanian vegetation (for example, *Cordaites* and some lycopsids) may have been tolerant of brackish water conditions and, therefore, may have been similar to mangrove vegetation in modern tropical swamps (Gillespie et al., 1978; Gastaldo, 1986). Many of the Kanawha coals have been interpreted as the product of domed peats similar to those forming today in coastal Malaysia (Cecil et al., 1985).

Alluvial channels are represented by trough cross-stratified, channelform sandstones that are typically 7 to 10 m thick (single story). This facies is exemplified by channel sandstones above the No. 2 Gas coal and below the Cedar Grove

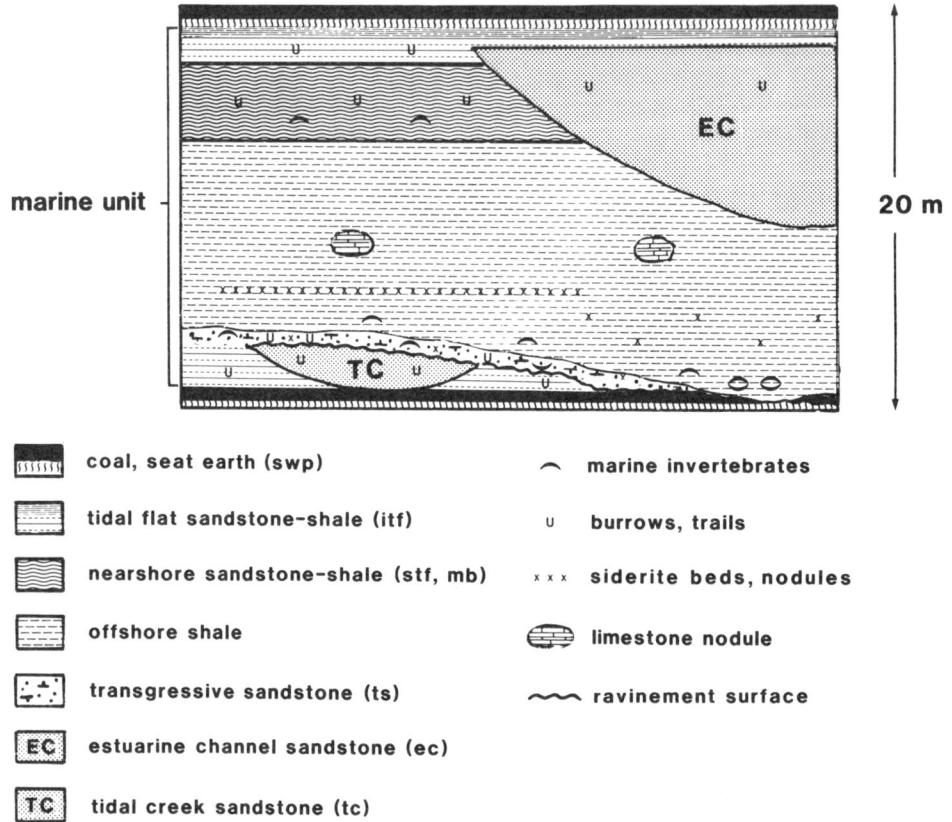

Figure 6. Idealized Kanawha marine unit with typical arrangement of component facies. Transgressive facies sequences 1, 2, and 3 are discussed in text under section on Facies architecture.

coal along Interstate Highway I-77 in Kanawha County, West Virginia (Fig. 7). Multistory channel sandstones as much as 35 m thick are locally present. Good examples of the latter sandstones are below the Coalburg coal along I-77 east of Giles, West Virginia, and along U.S. Route 119 at Julian, West Virginia (Figs. 3, 4). They are similar to alluvial sandstones described by Horne et al. (1978) and commonly occur with sufficient erosional relief to partially or totally cut out underlying marine units. Further discussion of facies is limited to those that preserve evidence of marine influence.

Estuarine channels

Estuaries have been defined in a variety of ways by different workers (Fairbridge, 1980; Davis, 1983; Dalrymple et al., 1992), but within the context of this study, the term is constrained to the marine-influenced lower reaches of a river or a deltaic distributary. Estuarine deposystems are divisible into (1) the lower (outer) estuary, which is marine-dominated and freely connected to the sea; (2) the middle estuary, which is defined as a zone of pronounced mixing of marine and freshwater; and (3) the upper (inner) estuary, which is dominated by riverine processes and characterized by fresh water although affected by daily tides (Fairbridge, 1980; Dalrymple et al., 1992). Two types of estuaries that are likely to have developed during Kanawha deposition include coastal plain estuaries (low relief, funnel shaped, flaring into open sea) and deltaic estuaries (e.g., ephemeral distributaries).

Estuarine channel fills of the Kanawha Formation are similar in size and gross appearance to the channel fills of fluvially dominated deltaic distributaries. They are typically 7- to 10-m-thick, single story channel fills and consist of very fine to fine-grained, micaceous lithic sandstone. Compound cross-stratification is commonly present. Features that help distinguish tidal or estuarine channels from the fluvial counterparts include rhythmic alternation of textures and structures, reactivation surfaces within cross-bed sets, bioturbation, and bipolar paleocurrents (Klein, 1977b; Frey and Howard, 1986). Good examples occur between the Gilbert and Little Cedar coals near Madison, West Virginia, and between the Winifrede shale and Winifrede coal southeast of Charleston, West Virginia (Figs. 3, 4, 8, 9C). Similar estuarine channel fills have been described from coeval strata of the Breathitt Formation in eastern Kentucky (Greb and Chesnut, 1992; Martino and Sanderson, 1993).

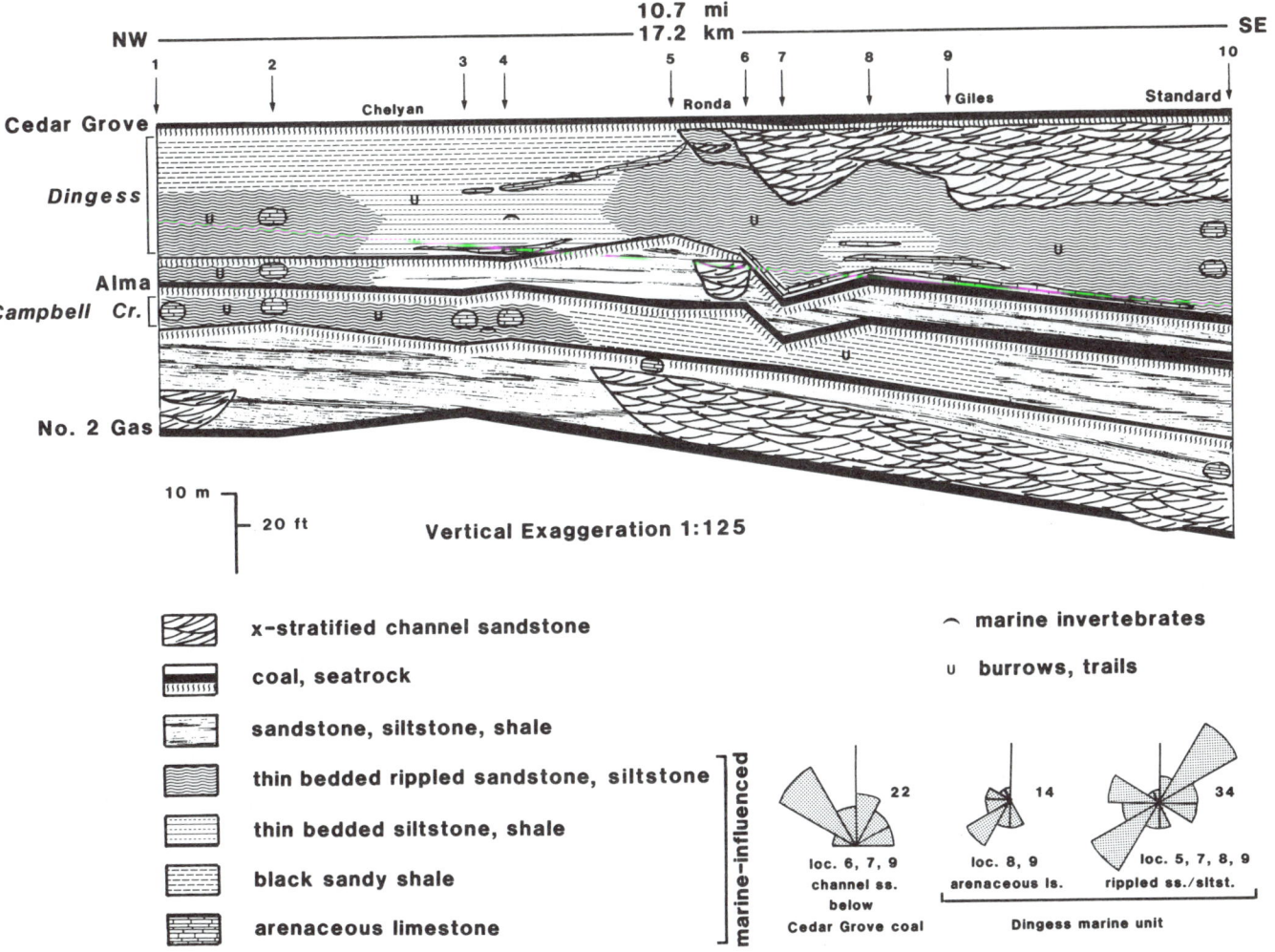

Figure 7. Stratigraphic cross section along Interstate Highway I-77 (A-B in Fig. 3) between the No. 2 Gas and Cedar Grove coals showing component facies of the Campbell Creek and Dingess marine units (indicated in italics and brackets at left margin). Outcrop locations are shown by arrows at top. The northeast-southwest orientation is parallel to the depositional dip. The Campbell Creek marine interval grades from lower tidal flat/nearshore rippled sands at Chelyan, West Virginia, southeastward into upper tidal flat black sandy shales at Ronda, West Virginia, and then into nonmarine overbank deposits at locality 10. The Dingess marine unit is laterally persistent and includes an extensive ravinement surface and transgressive lag above the Williamson coal zone. Paleocurrents based on cross-laminations indicated an average north-northwest flow direction in the channel sandstone below the Cedar Grove coal, and bipolar northeast-southwest flow in facies of the Dingess marine unit. The No. 2 Gas coal splits into two benches in the Chelyan area. The upper bench was locally mined as the Peerless coal (Krebs and Teets, 1914). It is this upper bench that is labeled as the No. 2 Gas.

Tidal creeks

Tidal creek deposits are usually channelform bodies 1 to 3 m thick and about 10 to 50 m wide. They are characteristically filled with very fine grained, bioturbated, sideritic sandstone and commonly overlie coal beds. Low-angle (less than 10°) lateral accretion surfaces are generally distinguishable and resemble those produced by tidal creek point bars. Ripple bedding (mostly flaser and wavy) is generally preserved where not destroyed by burrowing. Trace fossils typically found in marine-influenced facies are common (e.g., *Zoophycos* and *Helminthopsis*). The channel fills are overlain by thin-bedded sideritic siltstone and shale and are laterally equivalent to ripple-bedded sandstone and siltstone or dark gray to black shale. Good examples of this facies crop out on I-77 at Standard, West Virginia, between the No. 2 Gas and Powellton coals, and on U.S. Route 119 near Madison, West Virginia, above the Little Cedar coal and at the base of the Campbell Creek marine zone (Figs. 3, 4).

In some cases, the channels are filled with dark gray

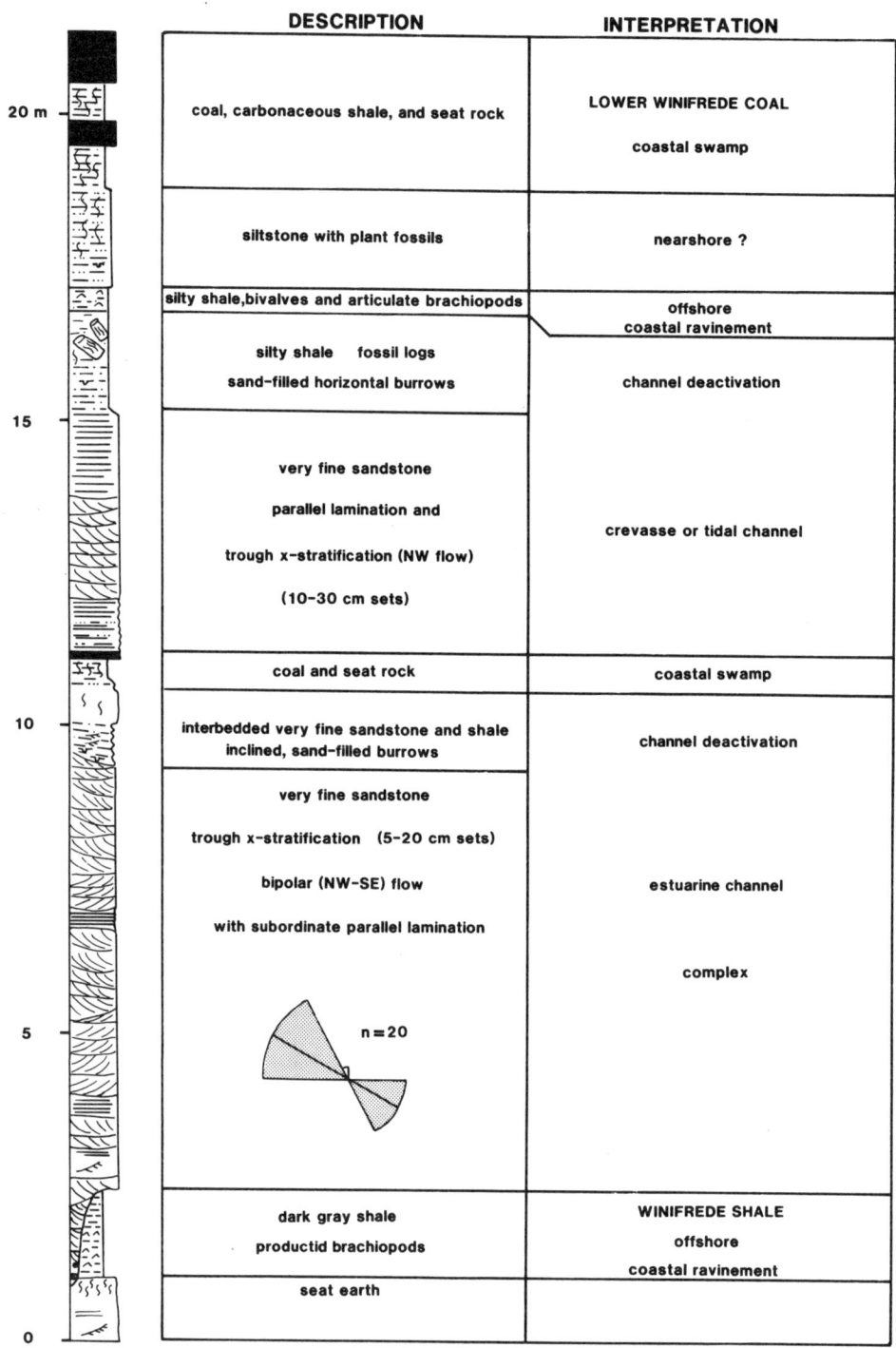

Figure 8. Stratigraphic section of outcrop located about 6.5 km southeast of Charleston, West Virginia, at the mouth of Campbell Creek. Channel sandstones at this locality show evidence of tidal influence in the form of bipolar paleocurrents, rhythmic sandstone/siltstone laminations, and burrowing, which illustrate a close association with facies containing marine invertebrates.

Figure 9. A, Outcrop section at Ronda, West Virginia, along Interstate Highway I-77. Dark mudstone at base represents the upper tidal flat facies of Campbell Creek marine unit. The overlying Williamson coal zone is split by overbank deposits from a fluviodeltaic distributary about 100 m east of this location. Thin-bedded sandstone and siltstone above the Williamson coal represents lower tidal flat and nearshore sandstones of the Dingess marine unit. This marine zone is truncated by a complex of estuarine and fluviodeltaic channels that is in turn capped by the Cedar Grove coal about 30 m above the base of the section (top of photo). Stakes at base are about 1 m high. B, Limestone nodule enclosed in thin-bedded, burrowed sandstone and siltstone showing compaction. Outcrop is from lower tidal flat/nearshore facies of Campbell Creek marine unit located along I-77 about 0.8 km east of Chelyan, West Virginia. Jacobs staff is 1.5 m. C, Wavy and flaser current ripple bedding and cross-lamination from estuarine channel facies at top of Dorothy marine unit. Exposure is located (along W. Va. Route 17) just north of Madison (see Fig. 3). Cross-lamination indicates bipolar flow (northwest-southeast). D, Marine zone above the Fire Clay coal in U.S. Route 119 roadcut at Julian, West Virginia. Scale is 30 cm. Light gray, burrowed, sparsely fossiliferous siltstone above coal is interpreted as tidal creek facies. It is truncated locally by a thin ravinement lag (sideritic burrowed sandstone), which in turn is overlain by fossiliferous, dark gray shale with siderite nodules. Fossils include orbiculoids, productids, spiriferids, and bivalves. E and F, Arenaceous, fossiliferous limestones associated with ravinement at base of Dingess shale. Outcrop along I-77 south of Cabin Creek, West Virginia. Fossils include abundant pelmatozoan plates along with spiriferid and productid brachiopods. Cross-stratification is common in this interval with a dominantly southwestward transport direction.

sandy shale or mudstone, or with thin-bedded sandy siltstone with large-scale lateral accretion surfaces 1.5 to 1.8 m in height; burrowing is sparse and includes *Planolites* and *Conostichus*. Body fossils including orbiculoid and spiriferid brachiopods are sometimes present.

Although tidal creeks and crevasse channels in the lower delta plain may form an intergradational continuum, the tide-dominated character of the channels is suggested by bioturbation, rhythmic internal structures and textures, marine trace and body fossils, and the apparent absence of genetically related distributary channels at the same stratigraphic level. The channels are similar to those that develop during the initial stages of transgression of the lower delta plain of the Mississippi Delta (Donaldson, 1979), although they should be expected in other transgressive coastal settings.

Tidal flats

Much of what previously has been considered bay fill splays and river-dominated mouth bars in the Kanawha Formation (Arndt, 1979; Flores and Arndt, 1979) may be more accurately interpreted as tidal sand and mud flats associated with shoals within the embayment or along its margin. Sand flats of the low to middle intertidal zone and shallow subtidal zone (nearshore zone) are represented by thin-bedded, rippled, very fine to fine-grained sandstone and siltstone. Asymmetric ripples predominate within wavy, flaser, and lenticular ripple bedding. Ripple cross-lamination is also abundant and associated with common herringbone cosets. Interference ripple bedding is also locally developed. Siltstone partings contain abundant mica and fine plant detritus, while *Calamites* and other plant fossils are sometimes preserved in the thin sandstones (Fig. 10A). Trace fossils are sparse to abundant and are preserved mainly along or between bedding planes. Ripple cross-lamination commonly indicates a bipolar paleocurrent pattern, as exemplified in the Dingess marine zone along I-77 between Rhonda and Giles, West Virginia (Fig. 7). Unimodal and polymodal patterns are also locally developed.

The heterolithic facies of the tidal flats commonly grades laterally and vertically into dark gray shale (upper tidal flat) that, in turn, is overlain by seat earth and coal. These facies transitions are demonstrated from northwest to southeast along I-77 in the Campbell Creek marine zone (Fig. 7). The dark shale is thin bedded to thinly laminated. Body and trace fossils are generally absent; exceptions occur where discontinuous, thin, horizontal sand laminations and lenticular ripples are present. Burrows in the sandy shale include *Planolites, Zoophycos,* and *Teichichnus* (Fig. 10).

The range of ripple bedding types reflects the alternation of weak, generally multidirectional traction currents and quiescence in an environment where both sand and mud were available. The trace fossils associated with this facies are characteristic of shallow marine and coastal environments (Chamberlain, 1978; Miller and Knox, 1985). Trace fossils include those formed by deposit-feeding worms (*Asterosoma, Teichichnus, Helminthopsis, Zoophycos, Rosselia,* and *Planolites*), creeping and grazing snails (*Scolicia, Curvolithus, Aulichnites,* and transversely ridged surface trails), and crawling and burrowing arthropods (*Olivellites?, Ancorichnus, Tasmanadia,* and *Petalichnus*). These trace fossils, in addition to the abundance of plant detritus, the vertical and lateral facies relations, and the multidirectional paleocurrents, indicate a depositional setting where sedimentation was influenced by coastal marine currents.

Plausible environmental interpretations for this facies include tidal flats as well as distributary mouth bars. The mouth bar interpretation, so widely invoked by previous workers, while consistent with many of the internal characteristics of the heterolithic facies, does not always provide a fully satisfactory explanation of the sedimentary features because of vertical and lateral facies relations. Coarsening-upward grain size trends are commonly not found as would be expected in distributary mouth bar facies (Horne et al., 1978). The mouth bar environment is usually too unstable to allow the development and preservation of abundant trace fossils and plant litter. In addition, there is commonly an apparent absence of genetically related distributary channels.

The dark gray shale facies that generally overlies the heterolithic facies and directly underlies coals is interpreted to have been deposited in upper tidal mud flats that intervened between more seaward sand flats (or sand flats along the more axial portions of the estuary) and coastal swamps that bordered the intertidal zone (Figs. 5, 6). The predominance of clay-sized material and thin, parallel laminations indicate that this facies accumulated under low-energy conditions because the fine-grained sediments were deposited mainly from suspension. The environment appears to have lacked indigenous macrorganisms, which suggests that stressful ecologic conditions prevailed. These conditions may have resulted from frequent subaerial exposure, high turbidity, or fluctuating salinity. The discontinuous and continuous, paper-thin sandstone laminations that are locally present closely resemble structures described as "pin-stripe" tidal bedding; this structure commonly develops in midtidal flats (Klein, 1977b).

Rare trace fossils (*Zoophycos, Teichichnus, Phycodes,* and *Planolites*) are preserved within sandier portions of the facies and were produced by infaunal deposit feeders (Chamberlain, 1971; Hantzschel, 1975). The restriction of trace fossils to rare sandy intervals suggests that more hospitable conditions temporarily prevailed during or following episodes of slightly increased energy, which were perhaps associated with unusually high tides (spring or storm). Such tides could have resulted in more prolonged inundation of the upper tidal flat and improved salinity, at least temporarily.

Nearshore zone

The nearshore zone extends from low tide to storm wave base. It includes shallow subtidal areas of interdistributary bays and the delta front, as well as interdeltaic or nondeltaic

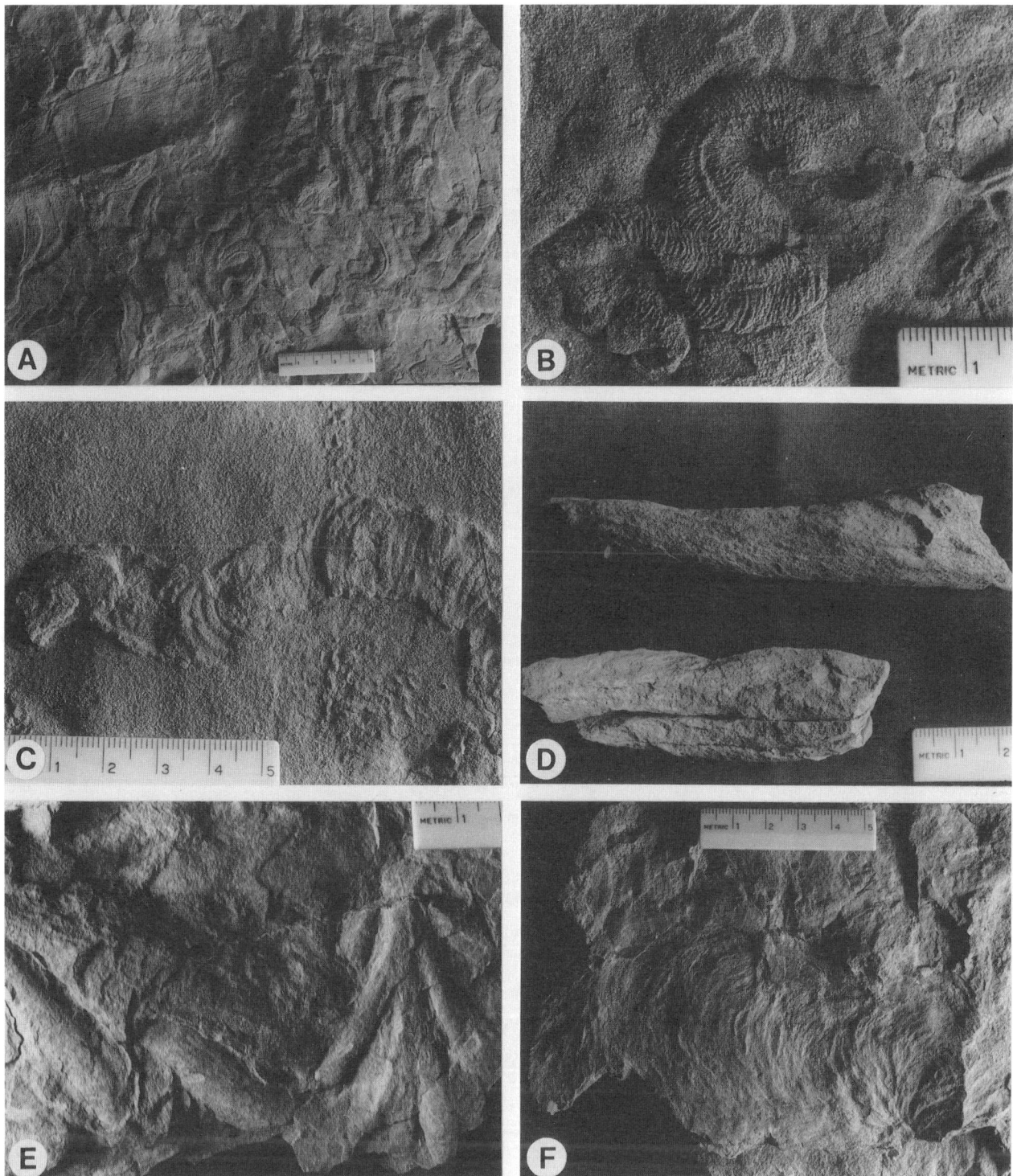

Figure 10. Trace fossils from the Dingess marine unit. All scales in centimeters. A, Top view of ripple bedded sandstone covered with trace fossil *Olivellites* from lower tidal flat/nearshore facies northeast of Chelyan, West Virginia. *Calamites* (upper left) and other plant fossils are common. B, *Olivellites* from same interval as in A. C, Surface trails preserved as casts from sole of rippled sandstone, located on Interstate Highway I-77 near Chelyan. Traces include transversely ridged surface trail ("snail trail") and arthropod trackway. D, *Teichichnus* trace fossil from ripple-bedded sandstone and siltstone facies at Giles, West Virginia, along I-77. E, *Phycodes* trace fossil from thin-bedded siltstone and shale facies about 1 km east of Chelyan. F, *Zoophycos* trace fossil from base of black sandy shale facies north of Ronda, West Virginia, on I-77.

nearshore (that is, shoreface and transition zone) deposits (Fig. 5). The Kanawha nearshore deposits consist mainly of very fine grained sandstone that is thinly interbedded with siltstone and shale. Body fossils are sparsely distributed and include *Lingula* and *Orbiculoidea*; these inarticulate brachiopods were tolerant of salinity fluctuations and pulses of rapid sediment influx (Gastaldo et al., 1989). Burrowing bivalves such as *Wilkingea* are also characteristic of nearshore facies.

A bed of very fine grained sandstone commonly overlies coal beds near the base of transgressive sequences. These sandstones contain abundant siderite nodules and are widespread but locally discontinuous. They are commonly richly fossiliferous and contain crinoid ossicles and columns, productid and spiriferid brachiopods, gastropods, and rugose corals (Fig. 9E, F). This facies is well-developed at the base of the Dingess marine unit along I-77 between Chelyan and Giles, West Virginia (Figs. 4, 7). In this area, the sandstone is locally cross-stratified, has as much as 1 m of relief, and partially truncates the underlying Williamson coal in some places. Elsewhere, thin bioturbated sandstones containing pebble-size siderite concretions intervene between the tops of channel sandstones or coals and overlying marine zones. The tendency for these thin, erosive based sideritic sandstones to occur above coals or distributary/estuarine channel fills and at the base of marine zones indicates that they are associated with ravinement.

The transgressive lag at the base of the Dingess marine unit is overlain by ripple-bedded sandstones deposited under the influence of bipolar, northwest-southeast flow and scour-fill sandy limestones deposited by southeastward currents (Fig. 7). These facies may represent nearshore deposits that were dominated by shore-parallel tidal flow. Similar tidal currents parallel the coast at the mouth of the Klang Delta (Coleman and Wright, 1975). A series of laterally coalescing channel fills at the top of the Dingess marine unit along I-77 is interpreted to represent tidal creek, estuarine, and fluviodeltaic channels that were telescoped as a result of north-northwest coastal progradation (Fig. 7).

Evidence of wave-generated currents in the nearshore deposits of the Kanawha Formation is exceedingly rare. Factors that may have limited wave energy along Kanawha coastal zone are discussed above in the section "Depositional framework. Along low-energy nearshore zones, bioturbation tends to destroy primary structures to within 2 m of low tide (Howard and Frey, 1981). Wave-formed structures may also have been inhibited by the prevalence of mud (rather than sand) in many nearshore areas.

Exceptional occurrences of wave-generated features occur in nearshore facies in the Kanawha black flint marine unit along U.S. Route 119 north of Julian (Figs. 3, 4). Hummocky cross-stratified, very fine grained sandstones are commonly capped by oscillation ripples. Small-scale (1- to 10-cm sets) trough cross-lamination is also common in this interval. Associated trace fossils include *Olivellites*, *Helminthopsis*, and *Neonereites*. This 1.1-m-thick interval is directly overlain by seat rock of the Stockton A coal bed (Fig. 4). Hummocky cross-stratification is generally attributed to the interplay between unidirectional and oscillatory currents generated by storms between fair weather and storm wave base (Swift et al., 1983). Tankard (1986) reported similar, thin (1-m-thick) sheet sandstones with hummocky cross-stratification associated with the Magoffin (=Winifrede) and Kendrick (=Dingess) marine units in eastern Kentucky. He interpreted them as caused by periodic reworking of bay margin sediments during storms.

Offshore zone

Offshore deposits are typically represented by dark gray to black shale. These shale units usually overlie a thin transgressive lag deposit and commonly contain micritic limestone nodules and thin beds and nodules of siderite (Fig. 6). The black shale units are as much as 22 m thick in outcrop (Dorothy shale along U.S. Route 119, 5 km southwest of Madison; Figs. 3, 4; Table 1).

Stenohaline fossil assemblages where present are typically localized near the base of these shale units, which suggests the assemblages were developed when sediment had become temporarily trapped in estuaries formed in response to rising sea level. A rise in sea level would have resulted in clear water conditions and low sedimentation rates in offshore areas. Offshore shales such as the Winifrede shale (=Magoffin of Kentucky) contain a diverse macrofauna consisting of calcareous brachiopods, cephalopods, bivalves, gastropods, and echinoderms (Dennis and Lawrence, 1979; Henry and Gordon, 1979; Martino and Adkins, 1986).

Although facies relations indicate an offshore setting for many of these transgressive shale units, the normal marine fauna that would be expected is not generally present. Factors that may have inhibited the development of offshore benthos include limited oxygen, high turbidity, and possibly abnormal salinity.

Oxygen-deficient bottom waters may result when these become isolated from well-oxygenated surface waters by strong density gradients associated with pycnoclines. Pycnoclines may result from thermal stratification or salinity stratification below brackish surface waters. Storms may temporarily destroy this stratification and result in short-lived increased oxygen levels in bottom waters (Wignall, 1991). Absolute water depths for black shale formation have been estimated as low as a few tens of meters. During the early stages of transgression in epicontinental seas, rapid expansion of oxygen-deficient "puddles" of deep water occurs in response to subsidence, sea-level rise, and reduced sediment supply. In initially shallow-water areas, these transgressive black shales rest on condensed, basal transgressive lags or unconformities (Wignall, 1991).

The paucity of macrofauna in the upper parts of offshore silty shales may be the result of increased turbidity and higher sedimentation rates that occurred during highstand progradation of the coastal zone.

Restricted marine or brackish water conditions in the seaway (Fig. 2) may have resulted from the northwestward progradation of the coast (Donaldson et al., 1985) toward a tectonic forebulge in eastern Kentucky (Tankard, 1986). Sea-lake conditions resulted from constriction of the seaway during the late Middle Pennsylvanian in Ohio and western Pennsylvania (Donaldson et al., 1985).

In addition to these factors that limit the initial distribution of body fossils, diagenetic factors may further reduce or eliminate macrofossils. Postmortem dissolution of calcareous skeletal material by acidic pore waters may be total (Oomkens, 1974; Goldring et al., 1978; Flessa and Brown, 1983).

Bioturbation in offshore shales is locally evident, but distinct trace fossils are generally absent. Only *Helminthopsis* and *Zoophycos* are rarely present. Anoxic conditions may have contributed to the development of faunally barren stratigraphic intervals, but poor preservation potential afforded by uniform lithology may also have been important.

DISCUSSION

Facies architecture

The Kanawha marine units of this study range from 3 to 34 m in thickness; truncation by overlying alluvial and fluviodeltaic channel sandstone substantially reduced the original thickness of some units. Commonly, the entire marine unit was locally removed. Examples of this removal occur in the Kanawha black flint and Winifrede marine units along I-77 near Charleston and near Julian West Virginia, on U.S. Route 119. The unnamed marine units above the Cedar Grove, and Fire Clay coals are also completely truncated near Julian on U.S. Route 119 (Figs. 3, 4; Table 1).

Three vertical facies sequences may occur during transgression that are illustrated in Figure 6. In the first facies sequence (left side, Fig. 6), the onset of transgression is evidenced by the development of tidal creek and tidal flat facies that succeed freshwater peats (coals). These intertidal deposits in turn are overlain by a thin transgressive sandstone containing abundant siderite nodules. Mud ripups, marine fossils, and trace fossils are also common. Transgressive sheet sandstones have sharp bases that are commonly erosional and have as much as 1 m of relief; these characteristics suggest that the sandstones are associated with ravinement during shoreface retreat. The transgressive sandstones are overlain by shale containing siderite concretions, limestone nodules, and sparsely distributed marine fossils that were deposited below wave base in the offshore zone. Examples of this facies sequence occur in the Campbell Creek marine unit and in the unnamed marine unit above the Fire Clay coal (Fig. 9D) along U.S. Route 119 near Julian.

A second facies sequence (Fig. 6, middle) occurs where the transgressive sandstone directly overlies a coal bed. The intertidal interval is absent. The second sequence is widely developed at the base of the Dingess marine unit between Chelyan and Giles (Figs. 7, 9).

A third transgressive facies sequence (Fig. 6, right side) occurs where tidal and nearshore transgressive facies are not preserved, and shale and limestone beds containing marine fossils directly overlie coal or carbonaceous shale roof rock. In some examples, an intervening surface of erosional truncation is evident above the coal bed that has locally downcut into the underlying seat rock. The base of the Winifrede shale at its type section illustrates this latter facies sequence.

The preservation of a relatively complete transgressive sequence including intertidal deposits between coals (freshwater) and subtidal facies (Fig. 6, left side) is favored by a high rate of sediment supply and rapid subsidence or relative sea-level rise (Fischer, 1961; Swift, 1968; Demarest and Kraft, 1987). Where relative sea level rises slowly, particularly along sediment-starved portions of the coast, the transgressive sequence will be more attenuated, and the transgressive sand will directly overlie the pretransgressive surface (Fig. 6, middle; Fischer, 1961; Swift, 1968; Elliott, 1986). Rapid relative sea-level rise, combined with a limited sediment supply, is envisioned as a mechanism for sequence 3 (Fig. 6, right side). These conditions would have been likely in southeastern West Virginia, along the basin axis during tectonically induced subsidence. Limited sediment supply, clear water, normal marine salinity, and adequate oxygen would have promoted the development of highly fossiliferous shale and limestone beds.

The regressive portion of the marine units usually includes sequences that coarsen upward from offshore shale and mudstone to nearshore bioturbated sandstone/siltstone or thinly interbedded sandstone, siltstone, and shale (Fig. 6). Channel fills are common near the tops of marine units and are related to distributary or estuarine channels and tidal creeks; these latter deposits are distinguished by scale, paleocurrents, and internal textures, structures, and trace fossils.

The facies architecture of marine units portrayed in Figure 6 uses coal beds as bounding terrestrial facies that enclose the marine-influenced strata deposited during a transgressive-regressive cycle. Where estuarine deposits occur as part of valley-fill deposits, they are likely to overlie fluvial channel facies and to be truncated by ravinement surfaces (Dalrymple et al., 1992).

Faunal associations

Onshore-offshore trends in benthic faunal assemblages have been identified in several studies of Pennsylvanian marine units (Pennsylvania: Williams, 1960; Ferm and Williams, 1965; Kentucky: Dennis and Lawrence, 1979; Illinois: Baird et al., 1985; Alabama: Gibson and Gastaldo, 1987; Gastaldo et al., 1989; Bennington, 1991). In general, nearshore sandstone and siltstone are characterized by the lingulid-molluscan association (Bretsky, 1969). The assemblage is dominated by abundant, infaunal sediment-feeding nuculid bivalves and suspension-feeding lingulid brachiopods, along with epifaunal grazing bellerophontid snails. Subordinate nearshore elements in the Pennsylvanian include pectinoid and myalinid bivalves and spiriferid brachiopods (Bretsky, 1969). In marine zones of the Allegheny Formation in Pennsylvania, phosphatic brachiopods (*Lingula* and *Orbiculoides*) and pectinids are associated with nearshore or restricted marine sideritic shales (Williams, 1960; Ferm and Williams, 1965).

Offshore Pennsylvanian assemblages in muddy substrates are generally brachiopod-dominated and characterized by productids and chonetids in association with athyroid brachiopods, crinoids, bryozoans, and bivalves (Bretsky, 1969). In marine zones of the Allegheny Formation of Pennsylvania, offshore facies are distinguished by calcareous brachiopods (*Marginifera, Chonetes, Mesolobus, Linoproductus, Composita*), cephalopods (*Pseudoorthoceras*), bivalves (*Allorisma, Myalina, Nuculana,* and corals (*Lophophyllum*) (Williams, 1960; Ferm and Williams, 1965).

Articulate brachiopods, gastropods, pelmatozoans, and protobranch bivalves dominate offshore, open marine zones in the Lower Pennsylvanian of Alabama (Gastaldo et al., 1989) and Illinois (Baird et al., 1985), whereas bivalves and inarticulate brachiopods dominate nearshore and marginal marine facies. In addition, abundant shrimp, medusae, and polychaetes are preserved in the nearshore facies of the Mazonian Delta complex of Illinois (Baird et al., 1985).

Although the bathymetric zones in the preceding section are presumed to correspond to environments that trend from nearshore areas of silt and sand, rapid deposition and erosion, and variable salinity to offshore muds, slower sedimentation, less turbid water, and more stable salinity, exceptions undoubtedly occur. Substrate is a critical factor in determining the distribution of benthic communities. In protected nearshore areas of slow mud deposition and narrow fluctuation in physical and chemical conditions, offshore benthic fauna may dominate the assemblage (Bretsky, 1969). On the other hand, where transgressing waters are less than normal marine, a distinct seaward shift of macrofauna may occur, and shallow-water species are displaced to deeper habitats (Remane and Scheiper, 1971; Ekdale et al., 1984).

Anomalies may also occur in the distribution of trace fossils as a result of nonuniform onshore-offshore grain-size trends. Ichnotaxa typically developed in offshore shelf facies may also be found in muddy nearshore or lagoonal/estuarine settings, whereas ichnotaxa typical of nearshore and littoral zones have been found in offshore sandy shelf and slope settings (Frey and Pemberton, 1984). Despite these exceptions, the body and trace fossil associations are useful paleoenvironmental tools, particularly where used as part of an integrated sedimentologic-paleontologic analysis.

The general trend toward increasing diversity from nearshore to offshore facies is interrupted at the transgressive maximum in some marine units of the Kanawha and Breathitt Formations. In the Winifrede marine zone (and its equivalent, Magoffin in Kentucky), the transgressive maximum is represented by a carbonaceous shale containing either a restricted molluscan fauna or no indigenous benthos (Bennington, 1991). Such faunal associations with transgressive maximum have been reported in Pennsylvanian cyclothems elsewhere in North America (Boardman et al., 1984). Small phosphatic nodules are sometimes present in the thin black shale intervals of the Magoffin/Winifrede marine zone that represent transgressive maximums. Dysaerobic or anaerobic conditions are likely to have prevailed during these depositional intervals.

Trace fossil assemblages

Trace fossils are controlled by the morphology and behavior of the tracemaker(s). Thus, their environmental significance depends on the range of the tracemakers and their response to ecologic conditions. As a result of these environmental constraints, assemblages of trace fossils tend to occur in specific depositional settings. An advantage of using trace fossil rather than body fossil assemblages in determining depositional environments is that trace fossils are normally in situ, and therefore not subject to transport.

Within the Kanawha Formation, two distinct intergradational trace fossil assemblages were distinguished in the Campbell Creek and Dingess marine units (see Figs. 10, 11; Martino, 1989); these assemblages recur in most other marine-influenced stratigraphic intervals. The first assemblage, the *Olivellites* assemblage, is a high abundance/high-diversity association of traces made by gastropods, arthropods, and worms (Martino, 1989). Endogenic feeding burrows and subordinate surface trails are represented. The assemblage is developed in thinly interbedded sandstone, siltstone, and shale that are interpreted as facies of nearshore (shoreface, transition zone, and delta front) and low- to midtidal flat environments.

The second assemblage, the *Phycodes-Zoophycos* assemblage, is associated with a lower energy setting as indicated by associated physical structures and textures (Fig. 11). The assemblage is characteristic of dark gray sandy shale and mudstone. It has a lower diversity and lower abundance than the *Olivellites* assemblage. The assemblage consists almost exclusively of endogenic deposit-feeding burrows. The low diversity is suggestive of environmental extremes such as low oxygen, fluctuating salinity, or subaerial exposure. Quiet, restricted coastal embayments and upper tidal flats are the probable paleoenvironments depending on facies relations. Although four ichnogenera are characteristic of the *Phycodes-Zoophycos* assemblage in the Campbell Creek marine unit (Martino, 1989), *Teichichnus* and *Planolites* dominate the assemblage where it occurs in other Kanawha marine units.

Tidal creek channel fills typically have a variation of the *Phycodes-Zoophycos* assemblage that also contains *Skolithos*-like vertical tubes. The assemblage developed in a stressful setting having fluctuating energy and salinity. Mud-loving benthos that produce feeding structures (*Zoophycos*, for example) are found side by side with sand-loving organisms that produce dwellings (exemplified by *Skolithos*; Ekdale et al., 1984). Bjerstedt (1987) described a similar association of dwellings (*Skolithos* and *Cylindrichnus*) and deposit feeding traces (*Teichichnus*) from estuarine point bar facies of the Devonian-Mississippian Price Formation in eastern West Virginia.

Highly burrowed tidal channel sandstone and siltstone as much as 10 m thick have been reported from the Lower Pennsylvanian Fentress Formation of Tennessee (Miller and Knox, 1985). Common trace fossils included *Zoophycos, Rosselia, Conostichus,* and *Planolites*. A subtidal lagoonal setting was suggested for this facies (Miller and Knox, 1985).

These two recurrent trace fossil assemblages represent the result of ongoing analysis; additional ichnogenera are likely to be found and new assemblages may be recognized.

Regional stratigraphy of marine zones

Pennsylvanian transgressions including those that produced the Magoffin and older marine units entered eastern Kentucky and southern West Virginia from the west, southwest, and south along the axis of the subsiding Appalachian basin; later Pennsylvanian transgressions entered the basin from the west and north (Rice et al., 1979; Chesnut, 1991). Marine units in the Breathitt Formation below the Magoffin are more numerous, of greater thickness and lateral continuity, are more fossiliferous, and have greater faunal diversity in eastern and southeastern Kentucky (Rice et al., 1979).

Middle Pennsylvanian marine units vary in their extent from basin-wide marker beds such as the Dingess and Winifrede units to more local ones (Blake et al., this volume). The limited areal extent of a marine unit may actually reflect the local nature of a transgression (for example, as that which results from delta lobe abandonment). Alternatively, dissection of a widespread but thin marine unit by coastal channel systems during regression may lead to preservation of several discontinuous marine units. The distinction between these conditions of formation is important because it affects the way in which the history of the basin is interpreted. At least some

OLIVELLITES ASSEMBLAGE

Ichnogenus	Occurrence	Tracemaker / Ethology	Facies	Paleoenvironment
Olivellites	A	arthropods	interbedded sandstone & siltstone	lower-mid tidal flat / nearshore zone
Tasmanadia	R	crawling & burrowing		
Petalichnus	R			
Teichichnus	C	worms deposit-feeding		
Planolites	C			
Asterosoma	R			
Ancorichnus	R			
Helminthopsis	R			
Rosselia	R			
Aulichnites	C	snails creeping & grazing	ripple bedding	
Scolicia	R			
Subphyllochordata	R			
transversely ridged surface trails	R			

PHYCODES - ZOOPHYCOS ASSEMBLAGE

Ichnogenus	Occurrence	Tracemaker / Ethology	Facies	Paleoenvironment
Phycodes	R-C	worms deposit-feeding	dark gray sandy shale & mudstone	upper tidal flat & restricted bay
Zoophycos				
Teichichnus				
Planolites				
Conostichus	R	sea anemone resting / dwelling		

Figure 11. Trace fossil assemblages from marine units of the Kanawha Formation. Tracemakers, ethology, facies associations and interpreted paleoenvironments are indicated for each assemblage.

of the local marine units may be extended throughout the region by using an integrated approach utilizing trace fossils and sedimentary structures as well as body fossils.

SUMMARY AND CONCLUSIONS

Marine-influenced strata of the Kanawha Formation are represented by a mosaic of facies reflecting deposition in offshore and nearshore zones of a shallow sea, as well as in tidal flats, tidal creeks, and estuarine channels. It is likely that the character of the coastal zone varied: low stands of sea level caused incisement of fluvial channels; rising sea level drowned most of the coastal zone and led to the expansion of tidal plains and estuaries. The influence of tides is reflected by rhythmic alternation of textures and structures through a range of high- and low-energy depositional settings and is demonstrated in many places by bipolar paleocurrent data.

As the coast retreated, shoreface erosion typically produced ravinement surfaces that were covered by transgressive lags, which now separate coastal and marginal marine facies from shallow subtidal facies. Coarsening-upward, regressive facies typically developed during highstands in the upper part of marine units after estuaries had filled with sediment. Marine units are capped by fining-upward sequences formed in estuarine channels and prograding tidal flats.

The distribution of body and trace fossils shows facies zonation. Substrate (type/stability), turbidity, dissolved oxygen level, salinity (level/stability), and food supply are likely to have been important factors that controlled the environmental range of organisms.

Calcareous benthos were discouraged in nearshore and intertidal areas characterized by silts and sands, rapid deposition and erosion, and variable salinity; phosphatic brachiopods, however, are locally preserved. In these environments, soft-bodied organisms (arthropods, medusae, annelids) and gastropods left behind the diverse *Olivellites* trace fossil assemblage. Mud-dominated areas in restricted embayments or upper tidal flats, which were associated with low oxygen, diluted salinity, and possibly exposure, are characterized by the low-diversity *Phycodes-Zoophycos* assemblage.

In mud-dominated offshore areas where deposition was slower and waters were clear and of more stable salinity, calcareous brachiopods dominated along with subordinate bivalves, bryozoans, echinoderms, cephalopods, and corals. In some instances, limited oxygen and possible salinity reduction of the seaway due to constriction may have been causes of stress that limited the development of offshore benthic communities. Although bioturbation is locally evident in offshore facies, distinct trace fossils are rare, probably due to limited oxygen and to the uniform lithology and corresponding low preservation potential.

The relative importance of autocyclic and allocyclic influences on Pennsylvanian depositional patterns the central Appalachian basin cannot be fully understood unless the total number and extent of marine units is clarified. Although this study represents progress toward such an objective, further analysis of equivalent strata elsewhere in the basin will be necessary before these influences can be properly evaluated.

ACKNOWLEDGMENTS

Various phases of this study were supported by summer research grants from the Marshall University Graduate School, the West Virginia Geological and Economic Survey, and the Petroleum Research Fund administered by the American Chemical Society. Discussions with B. M. Blake and A. Keiser of the Coal Resources Division, West Virginia Geologic and Economic Survey, were especially helpful in regional stratigraphic correlation of coals beds and marine zones. Data provided by B. Bennington, Virginia Polytechnic Institute, and several discussions in the field helped to clarify the relationships between faunal assemblages and water depth in the Winifrede and Magoffin marine zones.

REFERENCES CITED

Arndt, H. H., 1979, Middle Pennsylvanian Series in the proposed Pennsylvanian System stratotype, in Englund, K. J., Arndt, H. H., and Henry, T. W., eds., Proposed Pennsylvanian System stratotype, Virginia and West Virginia: American Geological Institute Selected Guidebook Series 1, p. 73–80.

Baganz, B. P., Horne, J. C., and Ferm, J. C., 1975, Carboniferous and Recent Mississippi lower delta plains—A comparison: Gulf Coast Association of Geological Societies, v. 37, p. 556–591.

Baird, G. C., Shabica, C. W., Anderson, J. L., and Richardson, E. S., Jr., 1985, Biota of a Pennsylvanian muddy coast—Habitats within the Mazonian Delta complex, northeast Illinois: Journal of Paleontology, v. 59, no. 2, p. 253–281.

Bennington, J. B., 1991, Paleoenvironmental analysis of the Magoffin marine zone transgression, Middle Pennsylvanian Breathitt Formation, eastern Kentucky: Geological Society of America Abstracts with Programs, v. 23, no. 1, p. 8.

Bjerstedt, T. W., 1987, Trace fossils indicating estuarine deposystems for the Devonian-Mississippian Cloyd Conglomerate Member, Price Formation, central Appalachians: Palaois, v. 2, p. 339–349.

Boardman, D. R., II, Mapes, R. H., Yancey, T. E., and Malinky, J. M., 1984, A new model for depth-related allogenic community succession within North American Pennsylvanian cyclothems and implications on the black shale problem, in Hyne, N.J., ed., Limestones of the Mid-continent: Tulsa Geological Society Special Publication 2, p. 141–182.

Bretsky, P. W., 1969, Evolution of Paleozoic benthonic marine invertebrate communities: Paleogeography, Paleoclimatology, Paleoecology, v. 6, p. 45–59.

Busch, R. M., and Rollins, H. B., 1984, Correlation of Carboniferous strata using a hierarchy of transgressive-regressive units: Geology, v. 12, p. 471–474.

Cardwell, D. H., Erwin, R. B., and Woodward, H. P., 1968, Geologic map of West Virginia: West Virginia Geological and Economic Survey, scale 1:250,000.

Cecil, C. B., Stanton, R. W., Neuzil, S. G., Dulong, F. T., Ruppert, L. F., and Pierce, B. F., 1985, Paleoclimate controls on Late Paleozoic sedimentation and peat formation in the central Appalachian Basin, USA: International Journal of Coal Geology, v. 5, p. 195–230.

Chamberlain, C. K., 1971, Morphology and ethology of trace fossils from the

Ouachita Mountains, southeastern Oklahoma: Journal of Paleontology, v. 45, p. 212–246.

Chamberlain, C. K., 1978, A guidebook to the trace fossils and paleoecology of the Ouachita Mountains: Tulsa, Oklahoma, Society of Economic Paleontologists and Mineralogists, 68 p.

Chesnut, D. R., 1989, Pennsylvanian rocks of the eastern Kentucky coal field, in Cecil, C. B., and Eble, C., eds., Carboniferous geology of the eastern United States: St. Louis, Missouri, to Washington, D.C., June 28–July 8, 1989, Field trip guidebook T143 for the 28th International Geologic Congress: Washington, D.C., American Geophysical Union, p. 57–60.

Chesnut, D. R., 1991, Marine transgressions in the central Appalachian Basin during the Pennsylvanian Period: Geological Society of America Abstracts with Programs, v. 23, p. 16.

Chesnut, D. R., and Cobb, J. C., 1989, Cycles in the Pennsylvanian rocks of the central Appalachian Basin: Geological Society of America Abstracts with Programs, v. 21., no. 6, p. A52.

Coleman, J. M., and Wright, L. D., 1975, Modern river deltas— Variability of process and sand bodies, in Broussard, M. L., ed., Deltas—Models for exploration: Houston, Texas, Houston Geological Society, p. 99–150.

Dalrymple, R. W., Zaitlin, B. A., and Boyd, R., 1992, Estuarine facies models—Conceptual basis and stratigraphic implications: Journal of Sedimentary Petrology, v. 62, no. 6, p. 1130–1146.

Davis, R. A., Jr., 1983, Depositional systems: Englewood Cliffs, New Jersey, Prentice-Hall, 669 p.

Demarest, J. M., and Kraft, J. C., 1987, Stratigraphic record of Quaternary sea levels—Implications for more ancient strata, in Nummedal, D., Pilkey, O. H., and Howard, J. D., eds., Sea level fluctuations and coastal evolution: Society of Economic Paleontologists and Mineralogists Special Publication 41, p. 223–239.

Dennis, A. M., and Lawrence, D. R., 1979, Macrofauna and fossil preservation in the Magoffin marine zone, Pennsylvanian Breathitt Formation of eastern Kentucky: Southeastern Geology, v. 20, no. 3, p. 181–192.

Donaldson, A. C., 1974, Pennsylvanian sedimentation of central Appalachians, in Briggs, G., ed., Carboniferous of the southeastern United States: Geological Society of America Special Paper 148, p. 47–78.

Donaldson, A. C., 1979, Origin of coal seam discontinuities, in Donaldson, A. C., Presley, W. W., and Renton, J. J., eds., Carboniferous coal short course guidebook: West Virginia Geological and Economic Survey Bulletin B-37-1, p. 102–132.

Donaldson, A. C., 1991, Causes of Pennsylvanian cyclicity in the Appalachian basin: Geological Society of American Abstracts with Programs, v. 23, p. 22–23.

Donaldson, A.C., and Eble, C. F., 1989, Morgantown area stops, in Cecil, C. B., and Eble, C. F., eds., Carboniferous geology of the eastern United States: 28th International Geological Congress Field Trip Guidebook T 143, p. 104–111.

Donaldson, A. C., and Schumaker, R. C., 1979, Late Paleozoic molasse of the central Appalachians, in Donaldson, A. C., Presley, M. W., and Renton, J. J., eds., Carboniferous coal guidebook: West Virginia Geological and Economic Survey Bulletin B-37-3, p. 1–42.

Donaldson, A. C., Renton, J. J., and Presley, M. W., 1985, Pennsylvanian deposystems and paleoclimates of the Appalachians: International Journal of Coal Geology, v. 5, p. 167–193.

Eagar, R.M.C., Baines, J. G., Collinson, J. D., Hardy, P. G., Okolo, S. A., and Pollard, J. E., 1985, Trace fossils and their occurrence in Silesian (Mid-Carboniferous) deltaic sediments of the central Pennine Basin, in Curran, H. A., ed., Biogenic structures—Their use in interpreting depositional environments: Society of Economic Paleontologists and Mineralogists Special Publication 35, p. 99–149.

Ekdale, A. A., Bromley, R. G., and Pemberton, S. G., 1984, Ichnology—Trace fossils in sedimentology and stratigraphy: Society of Economic Paleontologists and Mineralogists Short Course 15, 317 p.

Elliott, T., 1986, Siliciclastic shorelines, in Reading, H. G., ed., Sedimentary environments and facies: Oxford, Blackwell Scientific Publications, p. 155-188.

Englund, K. J., and Randall, A. H., III, 1981, Stratigraphy of the Upper Mississippian and Lower Pennsylvanian Series in the eastern Appalachians, in Roberts, T. G., ed., Stratigraphy, sedimentology: Geological Society of America Cincinnati Field Trip Guidebooks, v. 1, p. 154–158.

Fairbridge, R. W., 1980, The estuary—Its definition and geodynamic cycle, in Olausson, E., and Cato, I., eds., Chemistry and biochemistry of estuaries: New York, Wiley, p. 1–35.

Ferm, J. C., 1975, Pennsylvanian cyclothems in the Appalachian Plateau—A retrospective view, in McKee, E. D., et al., eds., Paleotectonic investigation of the Pennsylvanian System in the United States, Part II: U.S. Geological Survey Professional Paper 853, p. 57–64.

Ferm, J. C., and Weisenfluh, G. A., 1989, Evolution of some depositional models in Late Carboniferous rocks of the Appalachian coal fields: International Journal of Coal Geology, v. 12, p. 259–292.

Ferm, J. C., and Williams, E. G., 1965, Characteristics of a Carboniferous marine invasion in western Pennsylvania: Journal of Sedimentary Petrology, v. 35, p. 319–330.

Fischer, A. G., 1961, Stratigraphic record of transgressing seas in the light of sedimentation on the Atlantic coast of New Jersey: American Association of Petroleum Geologists Bulletin, v. 45, p. 1656–1666.

Flessa, K. W., and Brown, T. J., 1983, Selective solution of macroinvertebrate calcareous hard parts—A laboratory study: Lethaia, v. 16, p. 193–205.

Flores, R. M., and Arndt, H. H., 1979, Depositional environments of the Middle Pennsylvanian Series in proposed Pennsylvanian stratotype, in Englund, K. J., Arndt, H. H., and Henry, T. W., eds., Proposed Pennsylvanian System stratotype, Virginia and West Virginia: American Geological Institute Selected Guidebook Series 1, p. 115–122.

Frey, R. W., and Howard, J. D., 1986, Mesotidal estuarine sequences—A perspective from the Georgia Bight: Journal of Sedimentary Petrology, v. 56, p. 911–924.

Frey, R. W. and Pemberton, S. G., 1984, Trace fossil facies models, in Walker, R. G., ed., Facies models: Geological Association of Canada, Reprint Series 1, p. 189–207.

Frey, R. W., Pemberton, S. G., and Saunders, T.D.A., 1990, Ichnofacies and bathymetry—A passive relationship: Journal of Paleontology, v. 64, no. 1, p. 155–158.

Gastaldo, R. A., 1986, Standing lycopod forests in northern Alabama: Paleoclimatology, Paleogeography, Paleoecology, v. 53, p. 191–212.

Gastaldo, R. A., Gibson, M. A., and Gray, T. D., 1989, An Appalachian-sourced deltaic sequence, northern Alabama, USA—Biofacies-lithofacies relationships and interpreted community patterns: International Journal of Coal Geology, v. 12, p. 225–257.

Gibson, M. A., and Gastaldo, R. A., 1987, Invertebrate paleoecology of the Upper Cliff coal interval (Pennsylvanian), Plateau Coal Field, northern Alabama: Journal of Paleontology, v. 61, p. 439–450.

Gillespie, W. H., Clendening, J. A., and Pfefferkorn, H. W., 1978, Plant fossils of West Virginia: West Virginia Geological and Economic Survey Educational Series ED-3A, 172 p.

Goldring, R., Bosence, D.W.J., and Blake, T., 1978, Estuarine sedimentation in the Eocene of southern England: Sedimentology, v. 25, p. 861–876.

Greb, S. R., and Chesnut, D. R., Jr., 1992, Transgressive channel filling in the Breathitt Formation (Upper Carboniferous), eastern Kentucky coal field, USA: Sedimentary Geology, v. 75, p. 209–221.

Hantzschel, W., 1975, Trace fossils and Problematica (second edition), in Teichert, C., ed., Treatise on invertebrate paleontology, Part W, Miscellanea, Supplement 1: Boulder, Colorado, Geological Society of America (and University of Kansas Press), 269 p.

Hayes, M. O., 1975, Morphology of sand accumulation in estuaries: An introduction to the symposium, in Cronin, L. E., ed., Estuarine research: New York Academic Press, v. II, p. 3–21.

Hennen, R. V., and Teets, D. D., 1919, Fayette County: West Virginia Geological and Economic Survey [County Report], 1002 p.

Henry, T. W., and Gordon, M., Jr., 1979, Late Devonian through early Permian(?) invertebrate faunas in proposed Pennsylvanian System stratotype area, in Englund, K. J., Arndt, H. H., and Henry, T. W. eds., Proposed Pennsylvanian System stratotype, Virginia and West Virginia: American Geological Institute Selected Guidebook Series 1, p. 97–104.

Horne, J. C., and Ferm, J. C., 1978, Carboniferous depositional environments—Eastern Kentucky and southern West Virginia: Columbia, University of South Carolina, Guidebook, 151 p.

Horne, J. C., Ferm, J. C., Caruccio, F. T., and Baganz, B. P., 1978, Depositional models in coal exploration and mine planning in the Appalachian region: American Association of Petroleum Geologists Bulletin, v. 62, p. 2379–2411.

Howard, J. D., and Frey, R. W., 1981, Depositional facies of high energy beach-to-offshore sequence—Comparison with low-energy sequence: American Association of Petroleum Geologists Bulletin, v. 65, p. 807–830.

Klein, G. V., 1977a, Tidal circulation model for deposition of clastic sediment in epeiric and mioclinal shelf seas: Sedimentary Geology v. 18, p. 1–12.

Klein, G. D., 1977b, Clastic tidal facies: Champaign, Illinois, Continuing Education Publication Co., 147 p.

Klein, G. D., and Ryer, P. A., 1978, Tidal circulation patterns in Precambrian, Paleozoic, and Cretaceous epeiric and mioclinal shelf seas: Geological Society of America Bulletin, v. 89, p. 1050–1058.

Klein, G. D., and Willard, D. A., 1989, Origin of the Pennsylvanian coal-bearing cyclothems of North America: Geology, v. 17, p. 152–155.

Krebs, C. E., and Teets, D. D., Jr., 1914, Kanawha County: West Virginia Geological Survey [County Report], 679 p.

Krebs, C. E., and Teets, D. D., 1915, Boone County: West Virginia Geological Survey [County Report], 648 p.

Maples, C. G., and Suttner, L. J., 1990, Trace fossils and marine-nonmarine cyclicity in the Fountain Formation (Pennsylvanian: Morrowan-Atokan) near Manitou Springs, Colorado: Journal of Paleontology, v. 64, p. 859–880.

Martino, R. L., 1988, The Campbells Creek marine zone—Its extent, component facies, and relation to coals of the Kanawha Formation in southern Kanawha County, West Virginia [abs.]: American Association of Petroleum Geologists Bulletin, v. 72, p. 967.

Martino, R. L., 1989, Trace fossils from marginal marine facies of the Kanawha Formation (Middle Pennsylvanian), West Virginia: Journal of Paleontology, v. 63, no. 4, p. 389–403.

Martino, R. L., 1991, Facies analysis of Middle Pennsylvanian marine units of the Kanawha Formation, southern West Virginia: Geological Society of America Abstracts with Programs, v. 23, p. 99.

Martino, R. L., and Adkins, K., 1986, Fauna of the Winifrede limestone (Middle Pennsylvanian), upper Kanawha Formation Boone County, West Virginia: Proceedings of the West Virginia Academy of Science Abstracts, v. 58, no. 1, p. 26.

Martino, R. L., and Sanderson, D. D., 1993, Fourier and autocorrelation analysis of estuarine tidal rhythmites, lower Breathitt Formation (Pennsylvanian), eastern Kentucky, USA: Journal of Sedimentary Petrology, v. 63, no. 1, p. 105–119.

Mazzullo, S. J., and Friedman, G. M., 1975, Conceptual model of tidally influenced deposition on margins of epeiric seas—Lower Ordovician (Canadian) of eastern New York and southwestern Vermont: American Association of Petroleum Geologists Bulletin, v. 59, p. 2123–2141.

Miller, M. F., 1984, Distribution of biogenic structures in Paleozoic nonmarine and marine margin sequences—An actualistic model: Journal of Paleontology, v. 58, p. 550–570.

Miller, M. F., and Byers, C. W., 1989, Effective use of biogenic structures [abs.]: 28th International Geological Congress Abstracts, v. 2, p. 438.

Miller, M. F., and Jackson, S. R., 1984, Biogenic structures as salinity indicators—Lower Pennsylvanian coal-bearing sequences, northern Cumberland Plateau, Tennessee: Geological Society of America, Abstracts with Programs, v. 16, no. 2, p. 180.

Miller, M. F., and Knox, L. W., 1985, Biogenic structures and depositional environments of a Lower Pennsylvanian coal-bearing sequence, northern Cumberland Plateau, Tennessee, USA, in Curran, H. A., ed., Biogenic structures, Their use in interpreting depositional environments: Society of Economic Paleontologists and Mineralogists Special Publication 35, p. 67–97.

Oomkens, E., 1974, Lithofacies relations in the late Quaternary Niger Delta complex: Sedimentology, v. 21, p. 195–222.

Remane, A., and Scheiper, C., 1971, Biology of brackish water: New York, Wiley, 372 p.

Rice, C. L., Sable, E. G., Dever, G. R., Jr., and Kehn, T. M., 1979, The Mississippian and Pennsylvanian (Carboniferous) Systems in the United States—Kentucky: U.S. Geological Survey Professional Paper 1110-F, 32 p.

Scotese, C. R., 1986, Atlas of Paleozoic space maps—Paleoceanographic mapping project: Austin, University of Texas Institute for Geophysics Technical Report No. 66, p. 1–23.

Seilacher, A., 1964, Biogenic sedimentary structures, in Imbrie, J., and Newell, N. D., eds., Approaches to paleoecology: New York, Wiley, p. 296–312.

Shaw, D. P., 1964, Time in stratigraphy: New York, McGraw-Hill, 365 p.

Slingerland, R., 1986, Numerical computation of co-oscillating paleotides in the Catskill epeiric sea of eastern North America: Sedimentology, v. 33, p. 487–497.

Swift, D.J.P., 1968, Coastal erosion and transgressive stratigraphy: Journal of Geology, v. 76, p. 444–456.

Swift, D.J.P., Figueiredo, A.G., Jr., Freedland, G.L., and Oertel, G.F., 1983, Hummocky cross-stratification and megaripples—a geologic double standard?: Journal of Sedimentary Petrology, v. 53, p. 1295–1317.

Tankard, A.J., 1986, Depositional response to foreland deformation in the Carboniferous of eastern Kentucky: American Association of Petroleum Geologists Bulletin, v. 70, p. 853–868.

Udden, J. A., 1912, Geology and mineral resources of the Peoria quadrangle, Illinois: U.S. Geological Survey Bulletin 506, 103 p.

Walker, R. G., and Harms, J. C., 1975, Shorelines of weak tidal activity—Upper Devonian Catskill Formation, central Pennsylvania, in Ginsburg, R. N., ed., Tidal deposits: New York, Springer-Verlag, p. 103–108.

Wanless, H. R., and Weller, J. M., 1932, Correlation and extent of Pennsylvanian cyclothems: Geological Society of America Bulletin, v. 43, p. 1003–1016.

Weller, M. J., 1930, Cyclic sedimentation of the Pennsylvanian Period and its significance: Journal of Geology, v. 38, p. 97–135.

Wignall, P. B., 1991, Model for transgressive black shales?: Geology, v. 19, p. 167–170.

Williams, E.G ., 1960, Marine and fresh water fossiliferous beds in the Pottsville and Allegheny Groups of western Pennsylvanian: Journal of Paleontology, v. 34, no. 5, p. 908–922.

Manuscript Accepted by the Society February 1, 1994

The Pennsylvanian Fire Clay tonstein of the Appalachian basin— Its distribution, biostratigraphy, and mineralogy

Charles L. Rice and Harvey E. Belkin
U.S. Geological Survey, National Center, MS 926, Reston, Virginia 22092
Thomas W. Henry and Robert E. Zartman
U.S. Geological Survey, DFC, Box 25046, Denver, Colorado 80225
Michael J. Kunk
U.S. Geological Survey, National Center, MS 981, Reston, Virginia 22092

ABSTRACT

The Middle Pennsylvanian Fire Clay tonstein, mostly kaolinite and minor accessory minerals, is an altered and lithified volcanic ash preserved as a thin, isochronous layer associated with the Fire Clay coal bed. Seven samples of the tonstein, taken along a 300-km traverse of the central Appalachian basin, contain cogenetic phenocrysts and trapped silicate-melt inclusions of a rhyolitic magma. The phenocrysts include beta-form quartz, apatite, zircon, sanidine, pyroxene, amphibole, monazite, garnet, biotite, and various sulfides. An inherited component of the zircons (determined from U-Pb isotope analyses) provides evidence that the source of the Fire Clay ash was Middle Proterozoic (Grenvillian) continental crust inboard of the active North American margin. $^{40}Ar/^{39}Ar$ plateau ages of seven sanidine samples from the tonstein have a mean age of 310.9 ± 0.8 Ma, which suggests that it is the product of a single, large-volume, high-silica, rhyolitic eruption possibly associated with one of the Hercynian granitic plutons in the Piedmont. Biostratigraphic analyses correlate the Fire Clay coal bed with a position just below the top of the Trace Creek Member of the Atoka Formation in the North American Midcontinent and near the Westphalian B-C boundary in western Europe.

INTRODUCTION

The flint clay parting or tonstein of the Fire Clay coal bed has been considered historically the best single lithostratigraphic marker in the Middle Pennsylvanian of the central Appalachian basin (Wanless, 1946). Although the tonstein is commonly only a few tenths of a meter thick, the "jackrock," as it was generally known to miners and coal geologists, made mining of the enclosing coal bed difficult or impossible in the early part of the 20th century. The cooperative geologic mapping program between the U.S. Geological Survey and the Kentucky Geological Survey, carried out between 1960 and 1978, demonstrated the continuity of the Fire Clay coal bed (and its equivalents) and its tonstein across most of the eastern Kentucky coal field on quadrangle maps at a scale of 1:24,000. The tonstein lies lithostratigraphically about midway between the bases of two widely recognized marine units, the Kendrick Shale Member and the Magoffin Member of the Breathitt Formation (Rice, 1986). Because these three units have been identified in many other areas of the central Appalachian basin, they have become the main elements of the Middle Pennsylvanian lithostratigraphic framework for much of that area (Fig. 1).

Field studies have identified three other coal beds having similar flint clay partings (tonsteins) in the middle part of Breathitt Formation in Kentucky (Rice, 1986), and earlier re-

---

Rice, C. L., Belkin, H. E., Henry, T. W., Zartman, R. E., and Kunk, M. J., 1994, The Pennsylvanian Fire Clay tonstein of the Appalachian basin—Its distribution, biostratigraphy, and mineralogy, *in* Rice, C. L., ed., Elements of Pennsylvanian Stratigraphy, Central Appalachian Basin: Boulder, Colorado, Geological Society of America Special Paper 294.

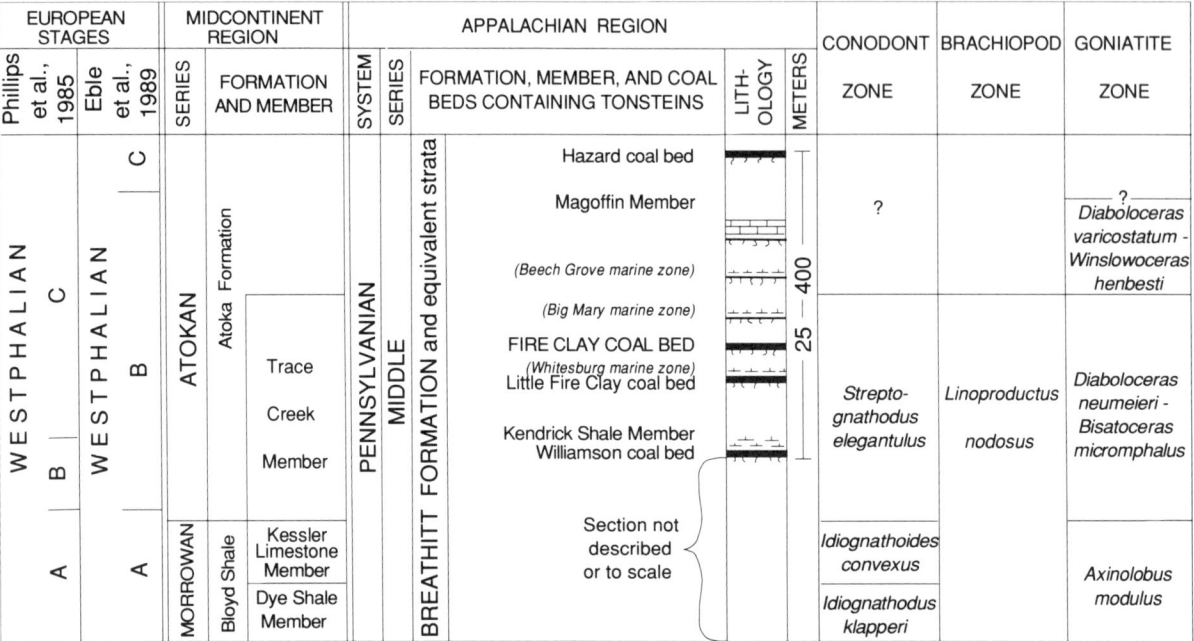

Figure 1. Middle Pennsylvanian stratigraphic section showing the correlation of the four coal beds that contain a tonstein parting in the central Appalachian basin to the Midcontinent stratigraphic section, the faunal zones of North America, and to the European stages as interpreted in terms of floral evidence. The Morrowan-Atokan boundary is shown as defined by Sutherland and Grayson (1978). Conodont zones are from Manger and Sutherland (1985).

ports show that these other tonsteins and enclosing coal beds were at times misidentified as the Fire Clay (Harris, 1963). Figure 1 shows the stratigraphic relations between the Fire Clay coal bed and other coal beds containing tonsteins with which it has been confused: from oldest to youngest, they are the Williamson, Little Fire Clay, and Hazard coal beds and their equivalents. These tonsteins and several others in the Pennsylvanian of the Appalachian basin have been catalogued and classified from earlier reports by Burger and Damberger (1985). Because of the wide extent and uniformity of the tonsteins, they have been considered "time lines," or isochrons, representing altered deposits of volcanic ash falls (Chesnut, 1985). To date, only the Fire Clay tonstein has proven to be distinctive and continuous enough to be useful for regional stratigraphic studies.

The Fire Clay tonstein is composed mostly of well-crystallized kaolinite, which contains from 3 to 5% accessory minerals. The previous identification of high-temperature sanidine (Seiders, 1965) and beta-form quartz (Robl and Bland, 1977) among the accessory minerals by microscopic examination of thin sections strongly suggests a volcanic origin for the tonstein. Those studies, however, did not fully resolve the problems of tonstein genesis or accessory mineral origin; some mineral grains were thought to be either detrital or, like the kaolinite, authigenic (Huddle and England, 1966).

Recent studies by Hess and Lippolt (1986) have demonstrated that sanidine from volcanic tonsteins in western Europe can be concentrated and dated by means of $^{40}Ar/^{39}Ar$ age spectra. Hess et al. (1988), applying the same method to two samples of a Middle Pennsylvanian tonstein from Kentucky, reported ages of 311.8 ± 1.6 Ma and 310.9 ± 8.3 Ma for sanidine and 310.6 ± 1.0 Ma for plagioclase(?). Although the specific locations of those samples from Kentucky were not identified, they would appear to represent "grab" samples of the Fire Clay tonstein.

This study of the Fire Clay tonstein provides a general definition of its distribution, its biostratigraphic position in the Middle Pennsylvanian, and its physical characteristics and mineralogy. Seven samples of the tonstein were collected across a large part of the basin to determine the concordance of its isotopic ages, and analyses of its mineralogy. The isotopic dating by the $^{40}Ar/^{39}Ar$ method of sanidine separated from the Fire Clay tonstein is presented in an accompanying paper (Kunk and Rice, this volume). The details of the mineralogy and petrogenesis are in preparation by Belkin and Rice and will only be summarized briefly herein. Preliminary results from U-Pb series radiometric dating also constrain the petrogenesis of the parent magma. Because this is the only Carboniferous horizon dated isotopically in North America, the present study includes detailed collection and collation of

DISTRIBUTION

Nomenclature

The Fire Clay coal bed is known by more than a dozen names in the Appalachian coal field. In addition to those listed in Table 1, it is commonly called Windrock, Wallins Creek, or, incorrectly, "Chilton" (see Blake et al., this volume). The tonstein has not been identified and is probably missing north of the Warfield anticline in West Virginia and Kentucky (see Fig. 2). It also appears to be missing in the northeasternmost part of Kentucky and locally along the northwestern part of its outcrop belt. In addition, the tonstein is generally absent in several linear areas, 15- to 20-km-wide west- and northwest-oriented, that cross the coal field at intervals of 50 to 80 km. In those areas, geologic reports (see, for example, Froelich and Stone, 1973; Brown, 1977) note the absence of the tonstein, and, locally, the Fire Clay coal bed. Those areas may represent well-developed meander belts of consequent, northwest-flowing, river systems that drained the Appalachian highlands and crossed the peat swamps at the time of the ash fall (see Fig. 14). Of particular interest to stratigraphers is the area just west of Rocky Face fault on the Cumberland overthrust sheet (the north-south tear fault about 15 km southeast of sample RK, Fig. 2), where the absence of the Fire Clay tonstein has long been noted (Wanless, 1946; Newell, 1975; Rice and Maughan, 1978). That area (Fig. 2) is offset from the autochthon by movement of the Pine Mountain thrust fault (Newell, 1975; Rice and Newell, 1975). In the absence of the tonstein, four separate systems of lithostratigraphic nomenclature have developed for the coal beds (two on the overthrust sheet and two on the autochthon). Similarly, the lack of the Fire Clay tonstein in the type area of the Kanawha Formation in West Virginia may account for many of the miscorrelations of coal beds and marine units from that area to other areas of the State (Blake et al., this volume).

Thickness

A search of geologic reports, combined with field reconnaissance, indicates that the thickness of the Fire Clay tonstein is commonly exaggerated, perhaps by 30 to as much as 80%. Commonly, other clay and "bone coal" partings have been included with the tonstein measurements in coal bed descriptions for economic reasons. The thickest occurrence of the Fire Clay parting found in our reconnaissance is about 31 cm. At that locality (on the Kentucky-Virginia state line about 10 km north of Pennington Gap, Virginia; see Fig. 2), only the upper and lower parts of the parting are flint clay; the intervening material is a plastic clay.

In Tennessee, the Fire Clay tonstein, which ranges from 0 to 20 cm thick, commonly occurs at the base of the Windrock coal according to Glenn (1925). Glenn noted that the floor of the coal contains "gently undulating sags" and "narrow ridges" that apparently affected the thickness of the tonstein as well as the overlying coal. Deposition of volcanic ash on an uneven surface or on a surface controlled by water currents within a shallow lake or lagoon (as suggested by Glenn for the structures noted above) could result in the reworking of the ash and account for the reported local thickness variations. Similarly, variations of topography within the peat mires such as domes, hammocks, and open-water channels and basins result in complicated flow patterns of surface water during periods of intense precipitation (Ingram, 1983). Thus, surface wash of unconsolidated volcanic ash probably accounts for much of the variations in thickness of the Fire Clay tonstein.

Maps of the distribution of the Fire Clay tonstein could be extremely valuable to resolve questions of provenance and to establish minimum estimates of erupted volume. Our sampling (Table 1) and extensive field work has convinced us that the

TABLE 1. TONSTEIN SAMPLE DATA*

Sample	7.5-Minute Quadrangle	Local Coal Bed Name	Latitude (N)	Longitude (W)	Elevation (m)	(ft)	Thickness (cm)	(in.)	$^{40}Ar/^{39}Ar$ Plateau Age (Ma)
RI-89	Ivydell, TN	Walnut Mountain	36°24'18"	84°10'48"	817	2,680	19	7.5	310.87 ± 0.55
RK-89	Kayjay, KY	Fire Clay (Dean)	36°43'19"	83°49'22"	501	1,645	16	6.5	310.71 ± 0.55
RH-89	Hoskinston, KY	Fire Clay (Hazard #4)	37°00'03"	83°23'53"	341	1,120	14	5.5	311.20 ± 0.55
RB-89	Blackey, KY	Fire Clay (Jack Rock)	37°13'06"	82°58'44"	369	1,210	11	4.5	310.25 ± 0.55
RFG-89	Flat Gap, VA	Phillips (No. 7)	37°01'23"	82°41'11"	847	2,780	13	5	311.38 ± 0.55
RV-89	Varney, KY	Fire Clay	37°37'45"	82°23'07"	315	1,035	10	4	310.27 ± 0.54
RS-89	Sylvester, WV	"Hernshaw"	38°03'03"	81°33'10"	287	940	13	5	311.30 ± 0.55

*Isotope age dates from Kunk and Rice (this volume), who state that the mean of the measured plateau ages, combined with the uncertainty constant "J," is 310.9 ± 0.8 Ma.

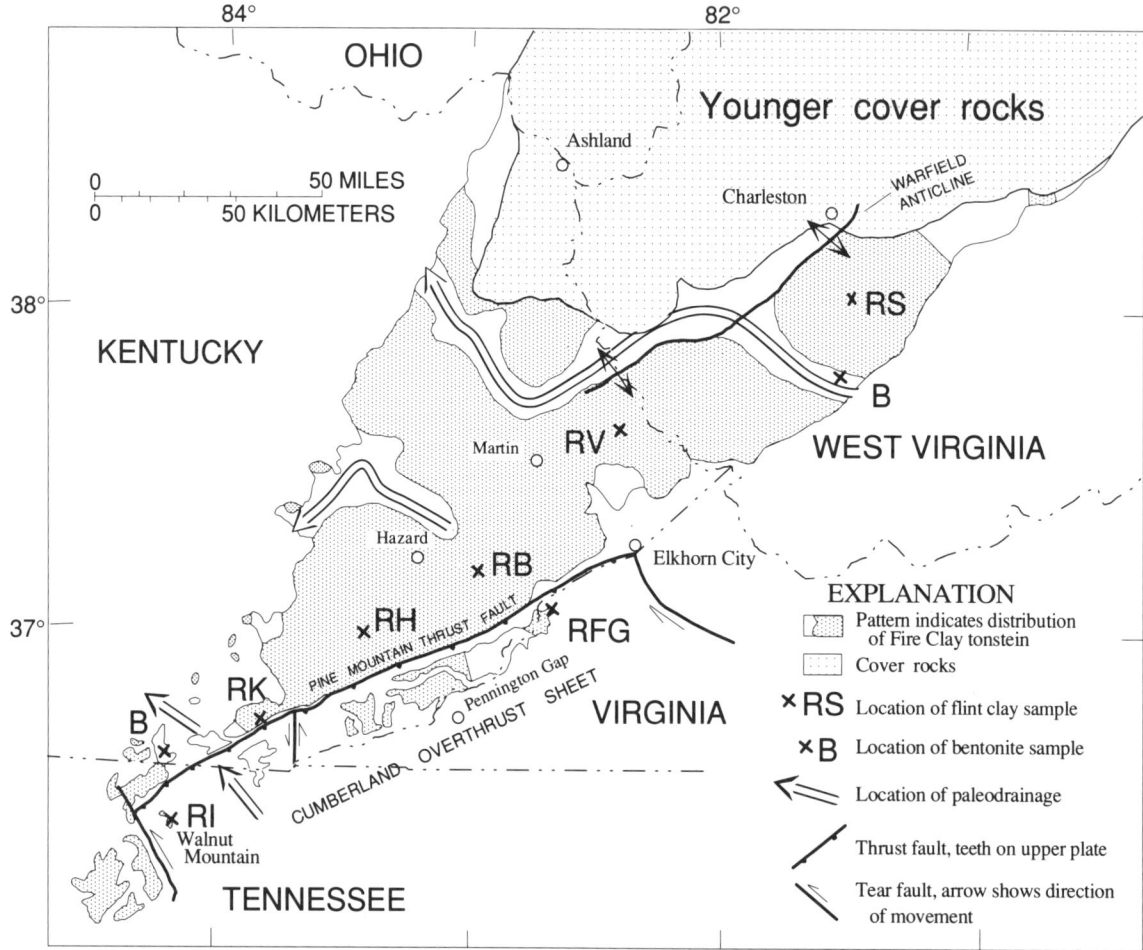

Figure 2. Sample locations and occurrence of the tonstein parting within the outcrop area of the Fire Clay coal bed. West Virginia data in part from B. M. Blake, Jr., and A. F. Keiser (oral communication, 1993). Samples of the Fire Clay ash fall represented by swelling clays are here termed "bentonites."

thickness of the Fire Clay tonstein is really nonsystematic both on a scale of outcrop and quadrangle. Such variability would seem to preclude meaningful isopach analysis. However, isopach maps of the thickness of the tonstein have been published by Lyons et al. (1992) and Chesnut (1985).

The Fire Clay tonstein isopach map published by Lyons et al. for the central Appalachian basin, indicates the tonstein is 0 to more than 10 in. (25 cm) thick. Lyons et al. gave no source of data for construction of their map, which disagrees with many tonstein thicknesses reported in U.S. Geological Survey geologic quadrangle (GQ) maps in Kentucky. For example, three areas of maximum thickness that, on their map, extend along a southwest-northeast line northwest of Pine Mountain in Kentucky, could not be confirmed by geologic reports or reconnaissance. One of those thickness maxima is shown to cross an area where mapping by Rice and Newell (1975) specifically demonstrated that the tonstein is absent. Additionally, tonstein samples RK and RH (Fig. 2; Table 1) were collected from two areas of the reported maximum thickness trend, yet the tonstein is half as thick as the distribution map of Lyons et al. would indicate. We believe that the great variation of thickness within small areas and the general lack of thickness data must compromise the accuracy of a detailed isopach map of the Fire Clay tonstein such as presented by Lyons et al. (1992).

Eaton (1964) suggested that as few as just 10 thickness measurements of an ash deposit are necessary to determine the direction of the wind currents that carried the ash. Where the topography is irregular and the ash deposit is reworked by slope wash, he suggested a method of moving averages in which data points are averaged for small areas of a sampling grid. In the absence of those kinds of point data, Chesnut (1985) used a method similar to that suggested by Eaton in which he showed the distribution of maximum thicknesses based on 7.5-min quadrangle maps in Kentucky and Tennessee. Those data were derived from GQ map reports and other published reports, as well as from field observations collected during resource studies conducted by the Kentucky

Geological Survey. Our experience suggests that all those reports are subject to thickness exaggeration.

Nevertheless, Chesnut's map, with which we generally agree, shows three belts of tonstein maximum thickness that decrease northward from southeastern Kentucky and Tennessee to southern Ohio and central West Virginia, where the tonstein is absent. Chesnut projected an elongate ash-fall pattern to account for the apparent thickness distribution of the Fire Clay tonstein; the vector of that wind pattern (reproduced in Fig. 13) is at about 135 degrees (almost opposite) to the wind vector suggested by Lyons et al. (1992). A strict application of Eaton's (1964) method of moving averages for construction of thickness isopachs would probably smooth out or even eliminate many of the thickness details upon which both published isopach maps are based.

Few significant physical or mineralogic differences were found among the seven samples of tonstein other than thickness. Sample RB contained few sanidine grains larger than 120-mesh screen (0.125-mm openings) and, overall, contained probably less than one-quarter of the concentration of sanidine found in the other samples. Sample RB may duplicate one of the tonstein samples analyzed from eastern Kentucky by Hess et al. (1988), the plagioclase(?) sample, which we speculate was a mixture of quartz and sanidine. The small differences between samples suggest the possibility that some mineral fractionation occurred locally in the ash clouds before their deposition. Varying convection currents might also contribute to the large reported local differences in tonstein thickness as it was deposited in the main part of the peat mire. Most mineral grains, with the exception of ilmenite and biotite, appear to be relatively unaltered by diagenesis (Fig. 6). In general, the southernmost and thickest tonstein samples contained more and larger volcanic phenocrysts.

The preservation of the Fire Clay tonstein is related to the distribution of Fire Clay peat mire. Both the coal bed and its tonstein are thin and discontinuous along the northwestern margin of the coal field, where subsidence of the foreland basin was not sufficient to allow for the development of thick peats or to ensure their preservation. The swamp seems to have occupied parts of an alluvial plain flanking a major fluvial system. That system may have extended down the regional paleoslope toward the present southwest along and east of the Cincinnati arch in Middle Pennsylvanian time (see Fig. 14). Neither the tonstein nor any correlative units in this part of the Middle Pennsylvanian section have been identified west of the Cincinnati arch in the Eastern Interior basin. If the ash fall extended into that area, conditions there were not favorable for its preservation.

BIOSTRATIGRAPHIC RELATIONS AND CORRELATIONS

The interval including the Williamson and Hazard coal beds, illustrated in Figure 1, contains as many as 15 coal beds. Correlation of either the long-ranging flora or palynomorphs of these coals to areas beyond the Appalachian basin in North America is difficult because strata of the same general age in the Eastern Interior basin contain few coal beds, and in the Western Interior basin, they contain none (Phillips et al., 1985). Correlation of this interval with the European coal-bearing section is based on the occurrence of extensive and overlapping palynomorph assemblages from coal samples that may represent site-specific environments. Thus, two recent analyses of the coal bed flora place the Westphalian B-C boundary either near the base of the Williamson to Hazard interval (Phillips et al., 1985) or near the top (Eble et al., 1989) (Fig. 1).

Most of the marine units of the Middle Pennsylvanian of the Appalachian basin are relatively thin and consist largely of shale and siltstone. The megafauna of these marine units is typically mollusk-dominated and long ranging. The distribution of species suggests that many parts of the marine embayments had restricted circulation or were brackish. Detailed mapping during the last two decades has allowed construction of a more complete biostratigraphic framework for this part of the Pennsylvanian than what had been possible in the past from the material collected from widely different parts of the Appalachian basin. It is now known that the strata of the interval between the Hazard and Williamson coal beds include five marine units, only one or two of which are generally well represented in any one section (Fig. 1). The position of the Fire Clay coal bed and its tonstein (as a recognizable marker bed about halfway between the bases of the Kendrick and Magoffin marine units) has been uniquely useful for the identification and separation of these marine units in thin and otherwise difficult-to-analyze stratigraphic sections. Extensive collections of invertebrates from this interval have been assembled within the last 10 yr. They provide a better understanding of the biostratigraphic relationships of the marine units in this part of the Breathitt Formation and make possible the correlation of these units with those of the Midcontinent, and, to a limited extent, with the European marine bands. The goniatites and brachiopods have been particularly useful in this endeavor (Table 2; Fig. 1).

The Kendrick Shale Member, which is best developed in Floyd, Pike, and Letcher Counties, Kentucky (Fig. 3), contains the youngest Pennsylvanian fauna typical of that found in the Trace Creek Member, which was reassigned from the Bloyd Shale to the Atoka Formation by Sutherland and Grayson (1978) (see Manger and Sutherland, 1985, for discussion). The goniatites from this member are especially biostratigraphically diagnostic and consist of species characteristic of the *Diaboloceras neumeieri–Bisatoceras micromphalus* Goniatite Assemblage Zone of Gordon (1970), which is restricted to the Trace Creek Member in the Ozarks (see Manger et al., 1992). In addition to one of the zonal name bearers (*D. neumeieri* Quinn & Carr), the Kendrick goniatites include *Dimorphoceratoides campbellae* Furnish & Knapp and *Gastrioceras occidentale* (Miller & Faber) (Furnish and Knapp, 1966). Those genera also permit correlation with the Westphalian B of western

TABLE 2. GONIATITE AND BRACHIOPOD FAUNAS IN PROXIMITY TO THE FIRE CLAY TONSTEIN*

Fauna	Marine Units†					Fauna	Marine Units†				
	K	W	BM	BG	M		K	W	BM	BG	M
Goniatites						*Spirifer goreii* Mather	X	–	–	–	–
Diaboloceras neumeieri Quinn and Carr	X	–	–	–	–	*Hustedia* cf. *H. miseri* Mather	X	–	–	–	–
Dimorphoceratoides campbellae Furnish and Knapp	X	–	–	–	–	*Cleiothyridina milleri* Sutherland and Harlow	X	–	–	–	–
Gastrioceras occidentale (Miller and Faber)	X	–	–	–	–	*Composita gibbosa* Mather (?)	X	–	–	–	–
Phaneroceras compressum (Hyatt)	X	X	–	–	X	*Composita ovata* Mather	X	–	–	–	–
Gastrioceras sp. A	–	–	–	–	X	*Desmoinesia* cf. *D. nambeensis* Sutherland and Harlow	X	X	X	–	–
Brachiopods						*Antiquatonia coloradoensis* (Girty)	X	–	–	–	X
Derbyia cf. *D. crassa* (Meek and Worthen)	X	–	–	–	–	*Composita subtilita* (Hall)	X	–	–	–	X
Derbyia cf. *D. bonita* Sutherland and Harlow	X	–	–	–	–	*Anthracospirifer chavezae* Sutherland and Harlow	–	–	–	–	X
Schizophoria altirostris (Mather)	X	–	–	–	–	*Derbyia bonita* Sutherland and Harlow	–	–	X	–	–
Schizophoria oklahomae Dunbar and Condra (?)	X	–	–	–	–	*Neochonetes? platynotus* (C. A. White)	–	–	X	–	–
Neochonetes? cf. *N.? platynotus* (C. A. White)	X	–	–	–	–	*Buxtonia* sp. (intermediate-sized form)	–	–	X	–	X
Linoproductus nodosus (Newberry)	X	–	–	–	–	*Linoproductus planiventralis* Hoare	–	–	–	X	X
Marginovatia pumila (Sutherland and Harlow)	X	–	–	–	–	*Derbyia crassa* (Meek and Hayden)	–	–	–	–	X
Sandia cf. *S. welleri* (Mather)	X	–	–	–	–	*Schizophoria* cf. *S. texana* Girty	–	–	–	–	X
Tesuquea formosa Sutherland and Harlow	X	–	–	–	–	*Sandia* n. sp. A (large form)	–	–	–	–	X
Buxtonia grandis Sutherland and Harlow	X	–	–	–	–	*Juresania* cf. *J. nebrascensis* (Dunbar and Condra)	–	–	–	–	X
Pulchratia? picuris Sutherland and Harlow	X	–	–	–	–	*Hustedia* cf. *H. rotunda* B. O. Lane	–	–	–	–	X
Pulchratia? pustulosa Sutherland and Harlow	X	–	–	–	–	*Cleiothyridina pecosi* (Marcou)	–	–	–	–	X
Anthracospirifer cf. *A. newberryi* Sutherland and Harlow	X	–	–	–	–						
Anthracospirifer tanoensis Sutherland and Harlow	X	–	–	–	–						

*Arranged by stratigraphic first occurrence.
†X = presence; K = Kendrick Shale Member (Breathitt Formation); W = Whitesburg marine zone; BM = Big Mary marine zone; BG = Beach Grove marine zone; M = Magoffin Member (Breathitt Formation) (see Fig. 1 for stratigraphic position).

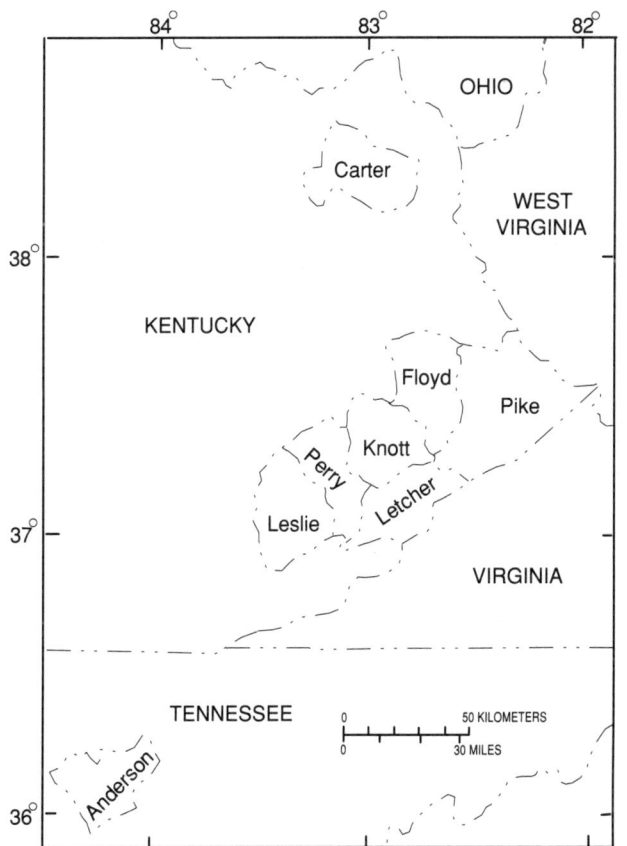

Figure 3. Counties in which marine fossil collections were made for this study.

Europe. Moreover, the species *D. neumeieri* (= *Rodiezmoeras bisati* Wagner-Gentis) is present in the Westphalian B of northern Spain.

A composite list of brachiopods from the Kendrick is shown in Table 2. Only a few of these species range into younger Breathitt strata. In general, these species are typical of the *Linoproductus nodosus* Brachiopod Assemblage Zone of Sutherland and Henry (1975, 1980) that includes the middle and upper parts of the Bloyd Shale of the type Morrowan Series, and the overlying Trace Creek Member, Atoka Formation, in the Ozarks (Fig. 1).

The cephalopods (goniatites) and brachiopods collected from the Whitesburg marine zone in Carter County in northeastern Kentucky (Fig. 3; see also Fig. 1) and the two brachiopod collections made from the roof shale of the Big Mary coal bed in Anderson County, Tennessee (Fig. 3), are related closely to faunas of the Kendrick Shale Member and are representative of the *Linoproductus nodosus* Zone as well.

The fauna of the Magoffin Member has a high diversity in the areas of Floyd, Knott, Perry, and Leslie Counties, Kentucky (Fig. 3), but unfortunately contains only a few nondiagnostic goniatites. The brachiopods from the Magoffin are more diagnostic and are markedly different at the species level from those of the Kendrick Shale Member. The brachiopod fauna also includes several species that are present in the faunas from the Atoka strata that overlie the Trace Creek Member in northwestern Arkansas. The Beech Grove marine zone, which is only about 10 m below the Magoffin Member (Fig. 1) in northeastern Tennessee, contains a few poorly preserved brachiopods, including *Linoproductus planiventralis*(?) Hoare, which is a common form in the Magoffin. Tentatively, these two marine units correlate with the post–Trace Creek sequence in the Atoka Formation of the Ozarks and to the upper part of the Westphalian B in Europe according to boundary definitions used by Eble et al. (1989) (Fig. 1).

MINERALOGY AND PHYSICAL CHARACTERISTICS

Seven samples of the Fire Clay tonstein were collected from well-identified localities across a 300-km traverse in the central part of the Appalachian basin (Table 1; Fig. 1). The samples were spaced to provide the widest basis for comparison of mineralogy and physical character and material for ^{40}Ar/^{39}Ar age dating (see Kunk and Rice, this volume). All tonstein samples were selected from partings within the coal bed in order to minimize contamination from waterborne clastic sediments from adjacent sedimentary deposits. We attempted to sample the thickest locally reported occurrences of the tonstein.

The Fire Clay tonstein is mostly cryptocrystalline, commonly aphanitic, and generally dark brownish gray. Kaolinite (see Fig. 4) is the main alteration product of the volcanic ejecta, where they fell into the Fire Clay peat mire. The nonclay portion of the tonstein is generally characterized by small (<1-mm diameter) water-clear crystals of quartz. As its common name "flint clay" implies, it is hard and has a conchoidal fracture and a waxy luster; thin edges are translucent. It weathers to blocky chips and develops a light blue patina much like that of chert. Flint clays do not slack or become plastic in water.

Two additional samples of the Fire Clay ash deposit were collected (West Virginia and Kentucky localities B; see Fig. 2), where the depositional environments were probably less acid and favored alteration of the volcanic glass to mixed-layer illite or bentonitic clay rather than kaolinite (Fig. 4). These clays are light gray and swell rapidly when wet. One sample, provided in 1992 by B. M. Blake, Jr., West Virginia Geological Survey, is from a coal company exploratory core drilled in southern West Virginia (Fig. 2). The other, also a core sample, was described as a bentonite by Bergeron et al. (1983) from the 95-m depth in core D-13, drilled in southern Kentucky (locality B, Fig. 2). Both samples contain volcanic phenocryst assemblages similar to those of the Fire Clay tonstein. Both samples appear to be enclosed by freshwater shale and siltstone, which are perhaps shallow lake deposits. Three X-ray diffractometer analyses of clay samples (Fig. 4) compare a typical kaolinitic Fire Clay tonstein (RH) with the compositions of the swelling clays, which are typically a mixture of illite and mixed-layer clays.

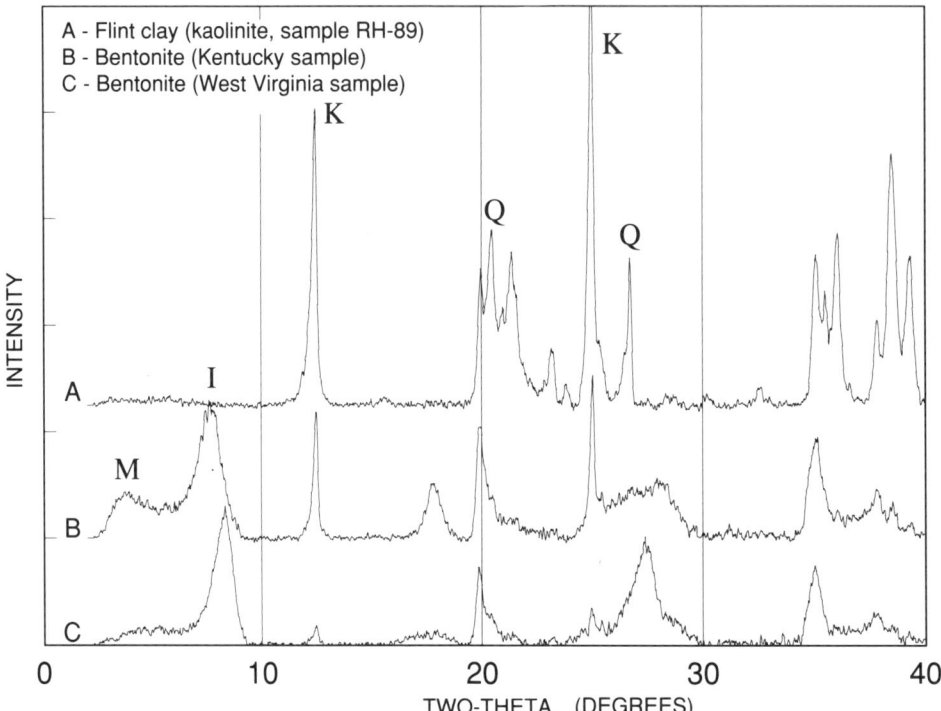

Figure 4. Smoothed X-ray diffractometer traces of oriented clay-size samples from the Fire Clay ash fall (see Fig. 2 for locations). The upper trace A is of well-crystallized kaolinite and some quartz; only the main peaks of kaolinite (K) and quartz (Q) are identified. Two main peaks of kaolinite can also be identified in traces B and C for swelling clay (bentonite), which are composed of different mixtures of illite (I) and mixed-layer (M) clays.

Of the Middle Pennsylvanian tonsteins of the Appalachian basin, the Fire Clay tonstein is the most areally continuous and the most consistent in thickness, commonly ranging from 5 to 20 cm. Other tonsteins, particularly those found in the middle and lower parts of the Middle Pennsylvanian, have far more limited areal extent and generally are very discontinuous, being only a few millimeters to rarely more than 5 cm thick. Where it is thickest, the Fire Clay tonstein locally is graded.

Most of the mineral grains of the Fire Clay tonstein are in the size range of fine to very fine sand (0.25 to 0.088 mm). The phenocrysts consist of quartz, sanidine, apatite, zircon, iron and titanium oxides, and minor to rare pyroxene, amphibole, monazite, garnet, biotite, and various sulfides (Figs. 5, 6). Several euhedral biotite plates were observed in silicate-melt inclusions hosted by beta-form quartz, but biotite is only very rarely found in the Fire Clay tonstein, probably being mostly destroyed or altered by humic acids present in the peat swamp. The corrosive nature of the humic acids is apparent also from an examination of the etched surfaces of most accessory crystals (e.g., garnet) (Fig. 6D, G) (H. E. Belkin and C. L. Rice, unpublished data, 1993). However, biotite is present in small but noticeable quantities in both swelling clay samples (B, Fig. 2), where the depositional environment was more favorable for its preservation.

Stevens (1976) estimated the amount of accessory minerals in the Fire Clay tonstein to be 6.5%, of which three-quarters was identified as quartz. The accessory minerals (caught on a 325-mesh screen) of the two swelling clay samples (B) each weighed slightly less than 3% of the total sample. Many of the beta-form quartz crystals are euhedral (about 10%), whereas others show rounding and embayment that resulted from partial resorption in the host magma (Fig. 5). Morphology and contained silicate-melt inclusions indicate that most quartz grains in the tonstein are phenocrysts, but 2 to 6% of the quartz grains appeared to be of nonigneous origin (Fig. 7) based on electron microprobe beam cathodoluminescence (Ruppert, 1987). These grains are not rounded and may be xenocrysts, associated with strata penetrated by the volcanic feeders. However, they could be detrital, because Pennsylvanian detrital grains of the region are commonly angular. Current laminae were observed in only one of the seven samples (RV).

Contrary to the suggestions of Huddle and Englund (1966) and Höehne (1957), none of the quartz grains appear to be authigenic (that is, have an alpha form). These authors interpreted border relations between quartz and kaolinite (viewed in thin sections) as evidence of repeated formation of kaolinite and solution and precipitation of quartz, in which one mineral replaced the other. Our data show that spherical or ellipsoidal

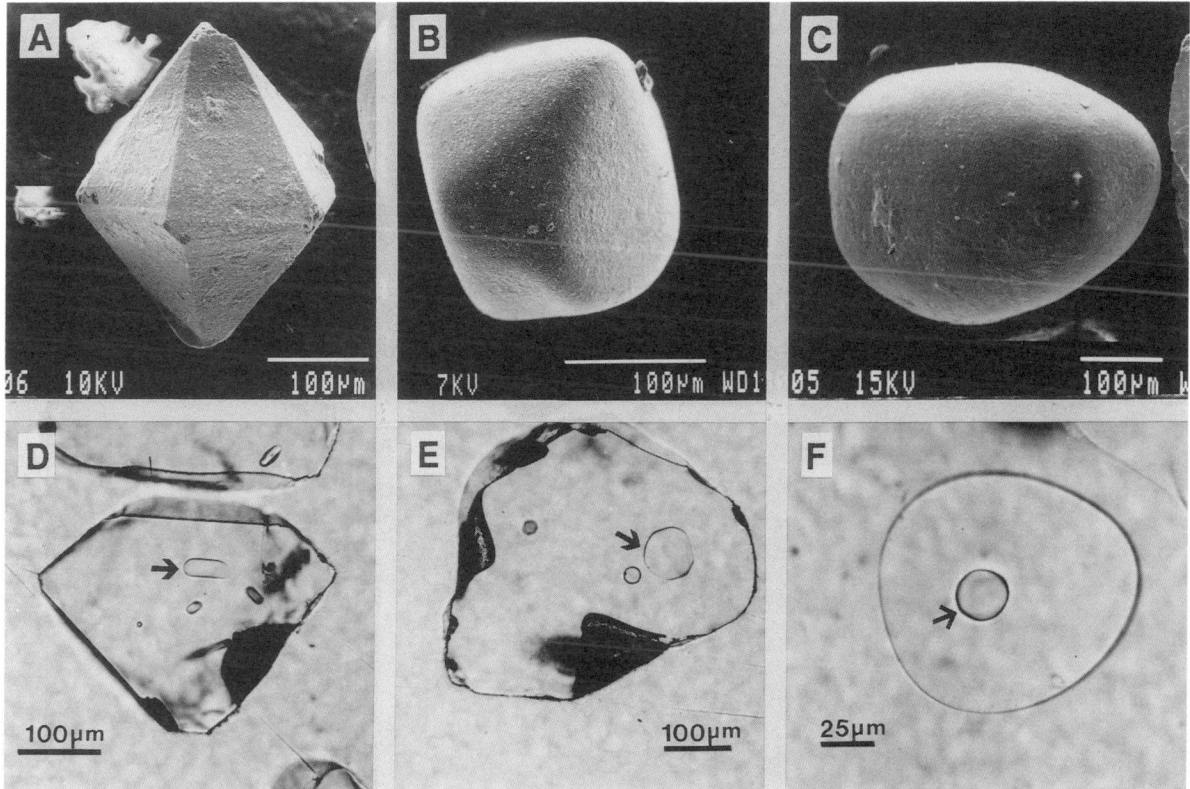

Figure 5. Beta-form quartz showing different degrees of magmatic resorption. A, B, and C, Scanning electron microscope (SEM) secondary electron images show increasing magmatic resorption from left to right. D, E, and F, Crystals photographed in transmitted light also show increasing magmatic resorption (from left to right) and contain typical silicate-melt inclusions (indicated by arrows).

quartz crystals (the glassy blebs and splinters of Huddle and Englund [1966]) were probably crystals in the process of being resorbed by the magma just before eruption. Most sanidine grains are clear, but some have tough coatings of clay that rarely penetrate into the crystals. In general, this clay appears to be an alteration of glass rather than authigenic alteration of the sanidine because the main portion of the crystals are clear.

Electron microprobe analyses of the beta-form quartz crystals from the samples show that they contain chemically indistinguishable silicate-melt inclusions (Fig. 5) within the limits of the technique (H. E. Belkin and C. L. Rice, unpublished data, 1993). These data suggest that the crystals are cogenetic phenocrysts of a rhyolitic magma (Belkin and Rice, 1989, 1990) (Fig. 8). A comparison of the composition of the silicate melts of the Fire Clay tonstein with those of a sample of the younger Hitchins tonstein—the upper part of Middle Pennsylvanian—(see Carlson, 1965) (Fig. 8) indicates that at least some of these tonsteins can be distinguished from one another by their magmatic chemistry, even though they are all of generally rhyolitic composition (Belkin and Rice, 1990). Microprobe analyses of phenocrysts show that the composition of sanidine and apatite grains is also the same for all samples of the Fire Clay tonstein. Garnets from the samples are essentially almandine, have a range of chemical composition (Fig. 9), and are typical of those described from acid volcanics (Gilbert and Rogers, 1989; see also Speer, 1981).

The accessory suite of minerals includes trace amounts of pyroxenes and amphiboles. On the basis of petrographic studies, none of these mineral grains appear to be detrital, although some could be xenocrysts. Some grains show moderate rounding that is probably due to partial resorption in the magma. The occurrence of different types of pyroxenes and amphiboles within single samples may represent the admixture of xenocrysts or may represent sampling of various pressure-temperature-compositional environments within a large preeruption magma chamber. The data for amphiboles from four tonstein samples (Fig. 10) indicate the difficulty of discriminating individual tonsteins on the basis of the chemistry of their trace minerals and suggest, at least for the Fire Clay tonstein, that processes of differentiation and mixing may have taken place before and during its eruption.

Zircon U-Pb systematics

A preliminary attempt to date the Fire Clay zircons by the U-Pb isotope method was unsuccessful, owing mainly to the presence of a significant component of older, inherited zircon

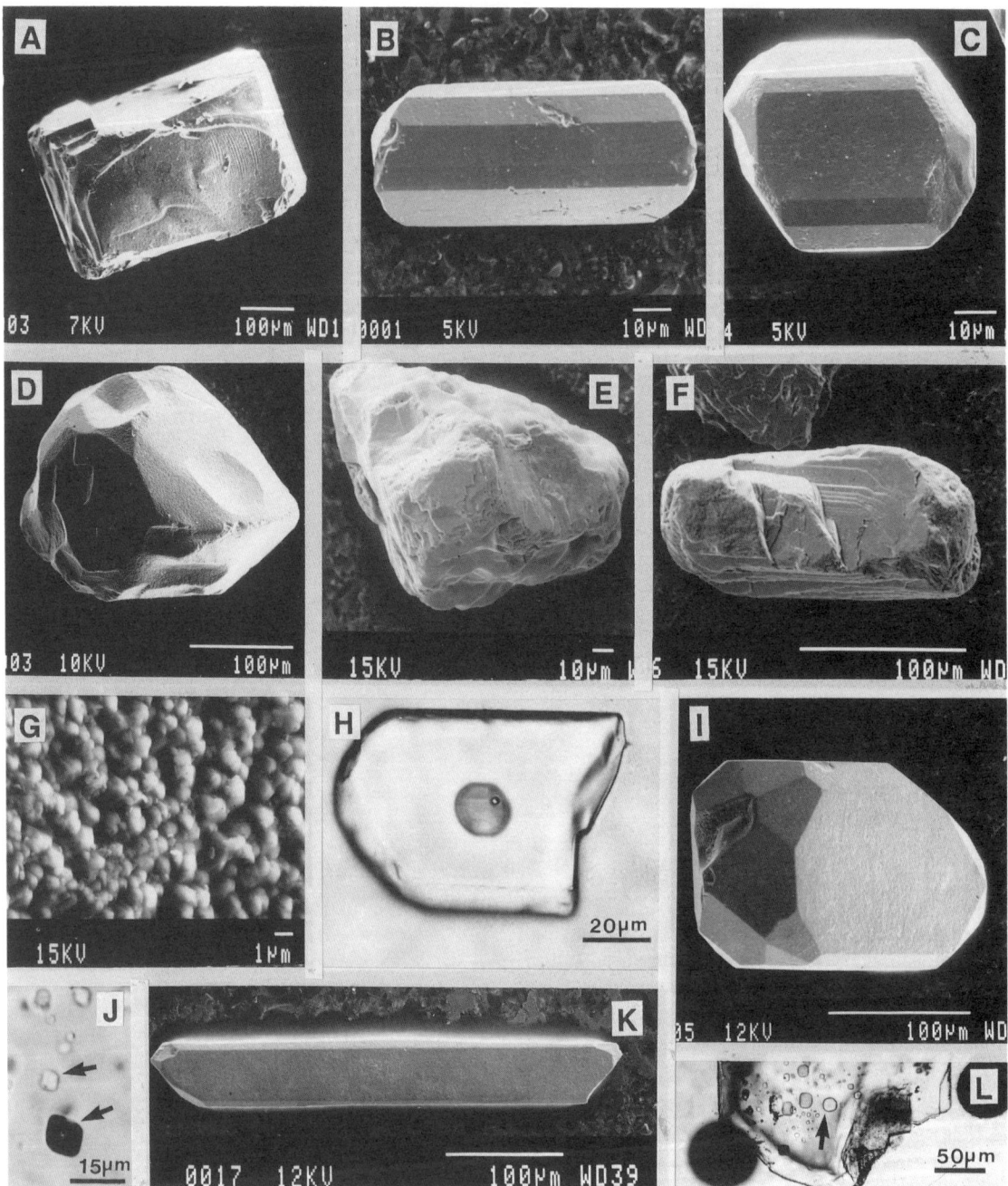

Figure 6. Accessory crystals from the Fire Clay tonstein, illustrating crystal shapes and fluid inclusions. Figure parts A through G, I, and K are SEM secondary electron images; figure parts H, J, and L are transmitted light photomicrographs. A, Sanidine crystal, sample RI. B, Apatite crystal, sample RI. C, Apatite crystal, sample RI. D, Garnet crystal, sample RFG. E, Monazite crystal, sample RK. F, Hornblende crystal, sample RK. G, Highly magnified surface of garnet (Fig. 6D) showing alteration texture. H, Broken apatite crystal containing a typical silicate-melt inclusion, sample RI. I, Zircon crystal, sample RK, J, Rare CO_2 fluid inclusion (lower arrow) in quartz containing silicate-melt inclusions (upper arrow), sample RB. K, Zircon crystal, sample RK. L, Sanidine from sample RI containing uncommon silicate-melt inclusions (arrow). See Table 1 and Figure 2 for sample locations.

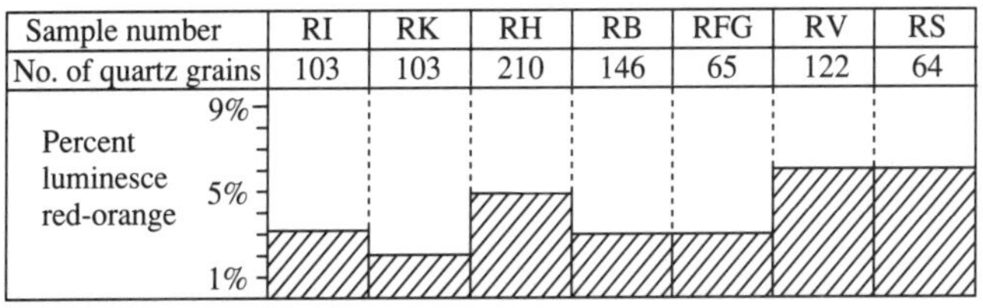

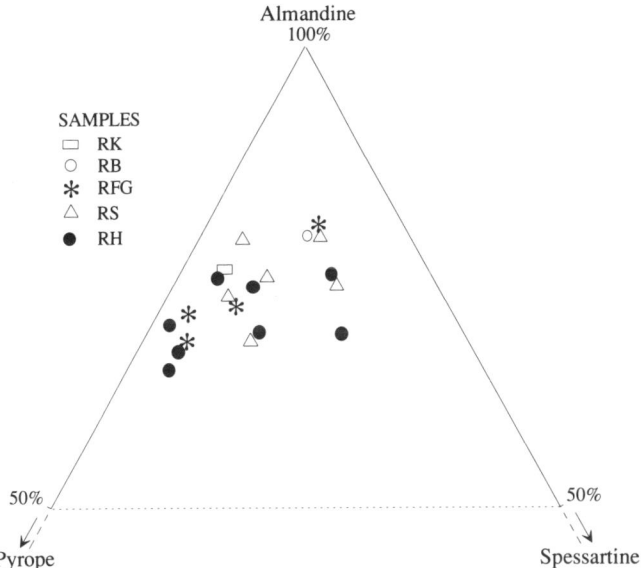

Figure 7. Percentage of accessory nonigneous quartz grains (shown by hachures) in tonstein samples, as indicated by red-orange cathodoluminescence in electron microprobe analyses (Ruppert, 1987). Samples arranged from southwest to northeast (left to right).

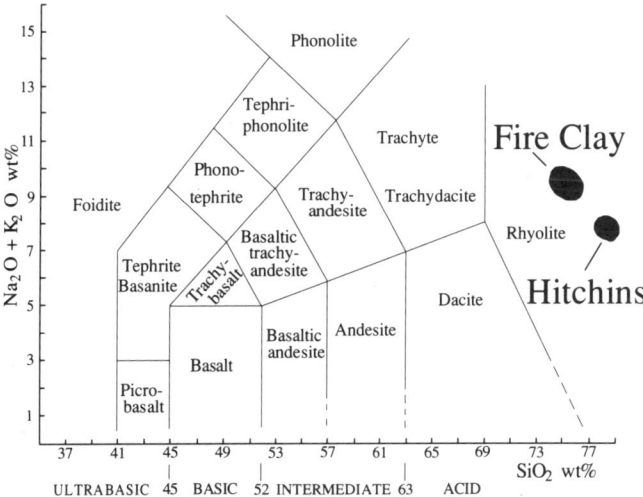

Figure 8. Comparison of silicate-melt composition of inclusions of the Pennsylvanian Fire Clay and Hitchins tonsteins by electron microprobe analysis using the total alkali-silica diagram of Le Bas et al. (1986).

Figure 9. Compositional range of garnets from the Fire Clay tonstein plotted as the molar percentage of the three main components. Grossular plus andradite are ≤5%; grossular ≤1%. See Table 1 and Figure 2 for sample locations.

cise volcanic (=chronostratigraphic) age to be extrapolated on a concordia diagram.

Despite an effort to choose only elongate, visually noncomposite grains—deemed most diagnostic of volcanogenic origin—the resultant U-Pb isotopic ages clearly demonstrate that the inherited core component was not eliminated from the Fire Clay zircons (Table 3). The four samples define an approximate linear array on a concordia diagram (Fig. 12) having lower and upper intercept ages of 344 ± 35 Ma and 1,224 ± 150 Ma, respectively (95% uncertainty confidence level, mean-square weighted deviation (MSWD) = 15). The scatter in the fit of a discordia line, apparent from the high MSWD, together with the lack of data control close to either intercept, accounts for the large uncertainties attached to both ages. Although the lower intercept age is ~10% older than that given by the $^{40}Ar/^{39}Ar$ method, the overlap of the uncertainties and the likely complexities in the U-Pb isotopic systematics of the zircon preclude a meaningful evaluation of this difference. Obviously, in this case, the U-Pb method cannot be used to place additional constraints on the precision of the $^{40}Ar/^{39}Ar$ plateau age spectra.

The upper intercept of the discordia line does appear to identify a late Middle Proterozoic age for the inherited component of the Fire Clay zircons (unless the upper intercept represents an average age of two or more different source terranes). The upper intercept age of 1,224 ± 150 Ma suggests that the magma giving rise to the Fire Clay tonstein was derived from, or involved with, rocks of Middle Proterozoic (Grenville) age. Interestingly, this tonstein is the only one from the Middle Pennsylvanian of the Appalachian basin investigated thus far

in the four samples investigated. Although insufficiently precise to confirm the $^{40}Ar/^{39}Ar$ plateau age spectra by an independent decay scheme, the U-Pb data nevertheless do provide useful insight into the age of the source rock for the magma from which the Fire Clay tonstein was derived. Zircons in the tonstein samples vary in size from 0.2 to <0.04 mm and in shape from stubby, nearly equidimensional to acicular grains having length-to-width ratios as great as 10:1. As Figure 11 shows, the Fire Clay zircons are not only highly zoned internally, but they may also reveal conspicuous core-overgrowth relationships. The composite nature of such grains suggests the incorporation of partially resorbed, preexisting zircon in the volcanic magma chamber—a phenomenon commonly producing hybrid ages from the mixture of inherited and newly crystallizing components. In such instances, selection of morphologically favorable zircon can sometimes significantly reduce the proportion of the inherited component in the sample to be analyzed (Zartman et al., 1994) and thereby allow a pre-

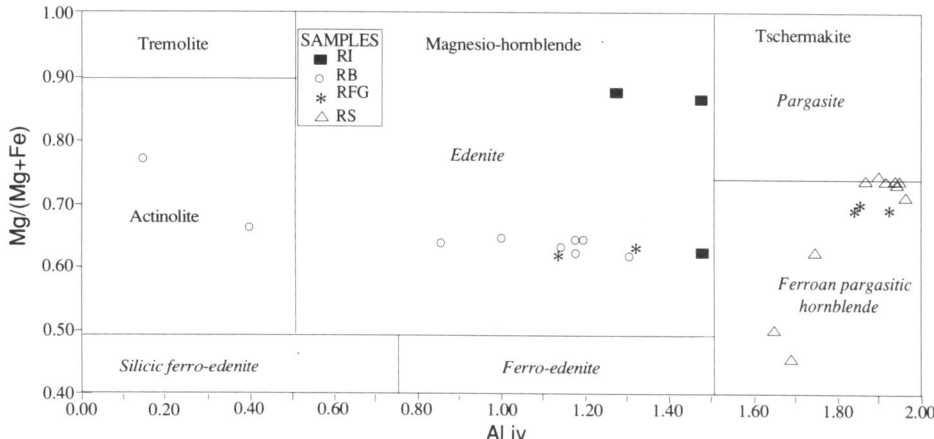

Figure 10. Range of some chemical elements for amphibole grains in tonstein samples RI, RB, RFG, and RS. See Table 1 and Figure 2 for sample locations.

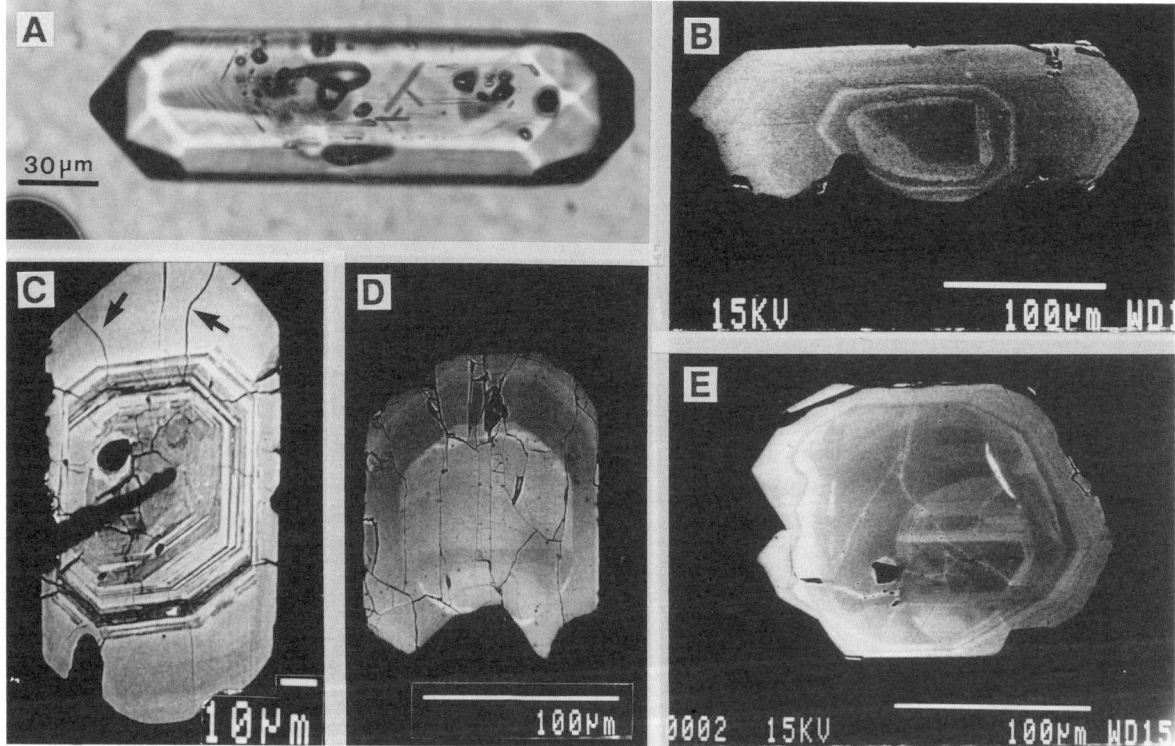

Figure 11. Zoned zircons from the Fire Clay tonstein showing conspicuous cores, silicate-melt and solid inclusions, and differential shrinkage cracks. Figure part A is transmitted light photomicrograph; figure parts B through E are SEM backscattered electron images. A, Zoned zircon having a core containing silicate-melt inclusions, apatite, and ilmenite solid inclusions, sample RB. B, Zircon showing simple zoning, sample RS. C, Zircon having a core sufficiently different in composition from subsequent growth to cause cracks (arrows) from differential shrinkage on cooling, sample RI. D, Zircon showing complex zoning (the cracks observed in this crystal were caused by the grinding process), sample RS. E, Zircon showing complex zoning, sample RS. See Table 1 and Figure 2 for sample locations.

TABLE 3. U-Th-Pb ISOTOPIC AGES OF ZIRCON FROM THE FIRE CLAY TONSTEIN*

Sample	Concentration† U	Th	Pb	Raw§ $^{206}Pb/^{204}Pb$	Isotopic Composition of Pb§ ^{204}Pb	^{206}Pb	^{207}Pb	^{208}Pb	Age $^{206}Pb/^{238}U$	$^{207}Pb/^{235}U$	$^{207}Pb/^{206}Pb$	$^{208}Pb/^{232}Th$
	ppm				atom %				m.y.**			
RB-89	197.4	82.8	18.46	2,912	0.0169	83.54	5.930	10.51	558 ± 5	624 ± 6	871 ± 6	494 ± 4
RS-89	219.7	94.5	18.21	774	0.0925	80.39	6.460	13.06	471 ± 5	518 ± 7	733 ± 7	413 ± 5
RI-89	215.4	n.d.	14.65	2,380	0.0221	82.39	5.155	12.48	405 ± 4	428 ± 4	555 ± 4	n.d.
RFG-89	186.9	87.4	14.63	567	0.1207	78.14	6.565	15.17	431 ± 5	469 ± 7	662 ± 9	399 ± 5

*n.d. = no data. Decay constants: ^{238}U = 1.55125 x 10^{-10} yr^{-1}; ^{235}U = 9.8485 x 10^{-10} yr^{-1}; ^{232}Th = 4.9375 x 10^{-11} yr^{-1}; $^{238}U/^{235}U$ = 137.88. Isotopic composition of common lead assumed to be ^{204}Pb:^{206}Pb:$^{207}Pb$$^{208}Pb$ = 1:18.24:15.61:38.08.
†Sample weights ranged from 103 to 145 mg.
§Corrected for fractionation and spike only.
**Corrected for fractionation, spike, and 20 pg laboratory Pb blank.

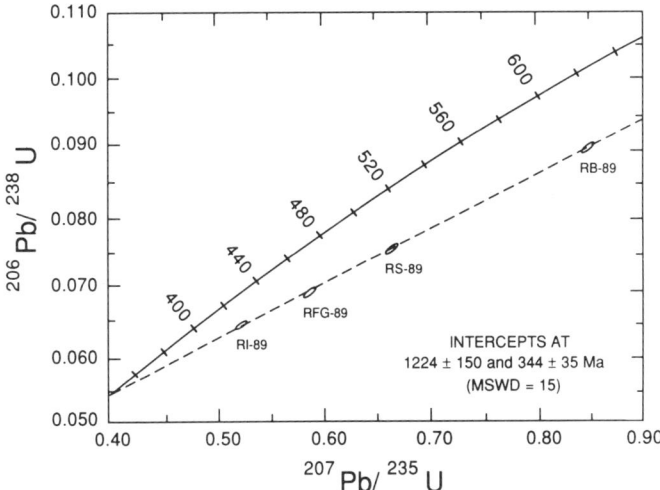

Figure 12. Concordia diagram for the Fire Clay tonstein.

that contains so large a fraction of inherited Precambrian zircon (R. E. Zartman, unpublished data, 1992). Together with chemical distinctions iterated below, the geochemical features of the zircon serve as potential discriminants of the various tonsteins that may aid in differentiating among lithostratigraphic units and in characterizing the source terranes on which the volcanos were sited.

The uranium concentrations of the zircons vary significantly between the several tonsteins that have been analyzed by the U-Pb method. Those of the Hitchins tonstein (upper Middle Pennsylvanian) have about two times (299 to 523 ppm for six samples) the uranium content of the Fire Clay tonstein (197 to 220 ppm for four samples), and those of the Upper Banner tonstein (lowermost Middle Pennsylvanian) have about three times (573 to 779 ppm for four samples) the content. Likewise, discernible differences in the U/Th ratio are noted among the Hitchins (1.29 to 1.56), Upper Banner (1.45 to 1.73), and Fire Clay (2.12 to 2.35) tonsteins as well.

Little Fire Clay tonstein

The stratigraphy of the Fire Clay tonstein is now so well known as a result of detailed geologic mapping that it is not likely to be confused with other tonsteins, except in areas where it is absent and the tonstein of the underlying Little Fire Clay (Upper Whitesburg) coal bed is well developed (for distribution of the Little Fire Clay tonstein, see Fig. 13). These conditions appear to occur only locally in areas northwest of Elkhorn City, Kentucky (Fig. 2), where the Little Fire Clay coal bed ranges from 6 to 15 m below the Fire Clay coal bed or its horizon.

In order to compare the mineralogy and chemistry of the two tonsteins, the Little Fire Clay tonstein was sampled in the roadcut on the western side Kentucky State Highway 80, about 100 m north of the crest of the hill about 1.5 km north of Martin, Kentucky (section illustrated by Cobb et al., 1981, Plate 16). At that locality, the Little Fire Clay coal bed (about 40 cm thick) underlies dark gray shale and siltstone about 6 m thick below a 1-m-thick Fire Clay coal bed that contains an almost black, coaly tonstein near its base. The Little Fire Clay tonstein, which directly underlies the coal bed, is dark brownish gray, about 8 cm thick, and, because it is intensively rooted, is friable and difficult to recognize; it grades down into a lighter colored plastic underclay. Most of the accessory grains of this sample are very small, mostly in the size range of very fine sand to coarse silt (0.074 to 0.044 mm). Accessory volcanic crystals consist of beta-form quartz, zircon, sanidine, apatite, amphibole, garnet, and a few euhedral crystals of tourmaline. In addition, about 30% of the quartz grains appear to be nonigneous, on the basis of their cathodoluminescence. The sample contained other apparent detrital grains, such as albite and white mica, and a large amount of pyrite. The composition of the silicate-melt inclusions of the beta-form quartz crystals of this one small sample of tonstein is not significantly different from that of the Fire Clay tonstein (H. E. Belkin and C. L. Rice, unpublished data, 1993).

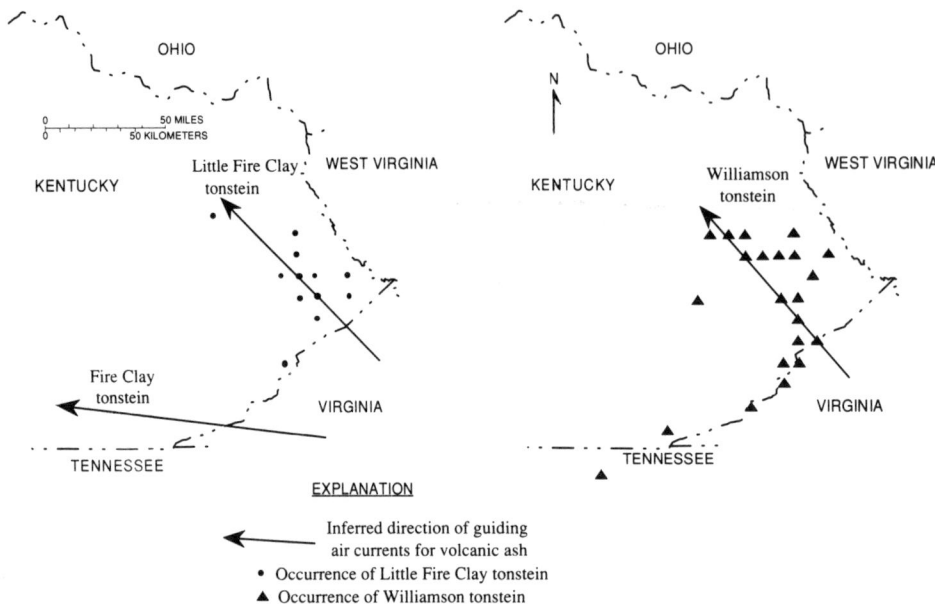

Figure 13. Inferred vectors of air currents controlling the distribution of some Middle Pennsylvanian tonsteins. Current vector for the Fire Clay tonstein is from Chesnut (1985). Also shown are the locations of reported occurrences of the Little Fire Clay tonstein (left) and the Williamson tonstein (right). Data points are centered on 7.5-min quadrangles.

DISCUSSION

Volcanic source

Two paleogeographic models have been proposed for the source area of the Fire Clay volcanic ash. In the first, Chesnut (1985) argued that the geographic position of the Fire Clay peat swamp in Middle Pennsylvanian time was about 7° south of the equator and rotated about 45° clockwise from its present position. He suggested that the ash plume was probably controlled by southeasterly winds that dominate the Intertropical Convergence Zone (Robinson, 1973) of that equatorial area. Chesnut also suggested that the volcanic source was east or southeast of eastern Kentucky, probably one of the Hercynian plutons of the magmatic arc identified by Sinha and Zietz (1982) in the Piedmont outcrop belt of southeastern North America (Fig. 14).

The second tectonic model is that of Lyons et al. (1992), who suggested that the Yucatan block of the Sabine plate, which collided with the southern margin of the North American plate during the late Paleozoic, was a possible source for the Fire Clay ash fall. On the basis of reports of late Paleozoic, Andean-type arc vulcanism and related andesitic tuffaceous deposits in the Yucatan block by Pindell and Dewey (1982), Lyons et al. (1992) postulated an associated back-arc epicontinental caldera system to explain the rhyolitic nature of the Fire Clay volcanic ash fall. Using the same paleogeographic reconstruction as Chesnut (1985), they argued that the Fire Clay tonstein thickens toward the Gulf of Mexico and the position of the Yucatan block during the Middle Pennsylvanian. They suggest that the volcanic ash would have been carried to the Appalachian foreland basin by westerly equatorial winds (counter to the prevailing easterly winds) based on continental reconstructions for the Middle Pennsylvanian.

Isotopic studies (such as those by Fullagar and Butler [1979] and Russell et al. [1981]) have identified a belt of Hercynian (330 to 260 Ma), granitoid plutons in the Piedmont of the central and southern Appalachians. Sinha and Zietz (1982) interpreted those plutons as parts of a continental magmatic volcanic arc (Fig. 14) that collided with the North American plate during the late Paleozoic. The belt of plutons apparently extends eastward beneath Coastal Plain deposits (see Horton et al., 1991). Our chemical data for silicate-melt inclusions of the Fire Clay tonstein generally fit well with those listed by Sinha et al. (1989, see Fig. 18) for the known granitoid plutons of that magmatic arc. But Sinha and Zietz (1982) showed that the SiO_2 and K_2O contents of the arc plutons tend to increase eastward and southward and thus suggest an association of the Fire Clay ash fall with the eastern part of the arc because of the high SiO_2 ($\geq 73\%$) and K_2O (>5%) contents of the silicate-melt inclusions.

As Figure 14 shows, the Fire Clay swamp appears to have occupied a position of between 3° and 8° south latitude during the Middle Pennsylvanian (see Kent and Keppie, 1988) and was probably in a zone of prevailing easterly winds that dominate the Intertropical Convergence Zone. Although both Eaton (1964) and Reiter (1967) showed that wind currents do vary from prevailing directions, according to seasons and altitude, counter winds are commonly weaker and therefore have a re-

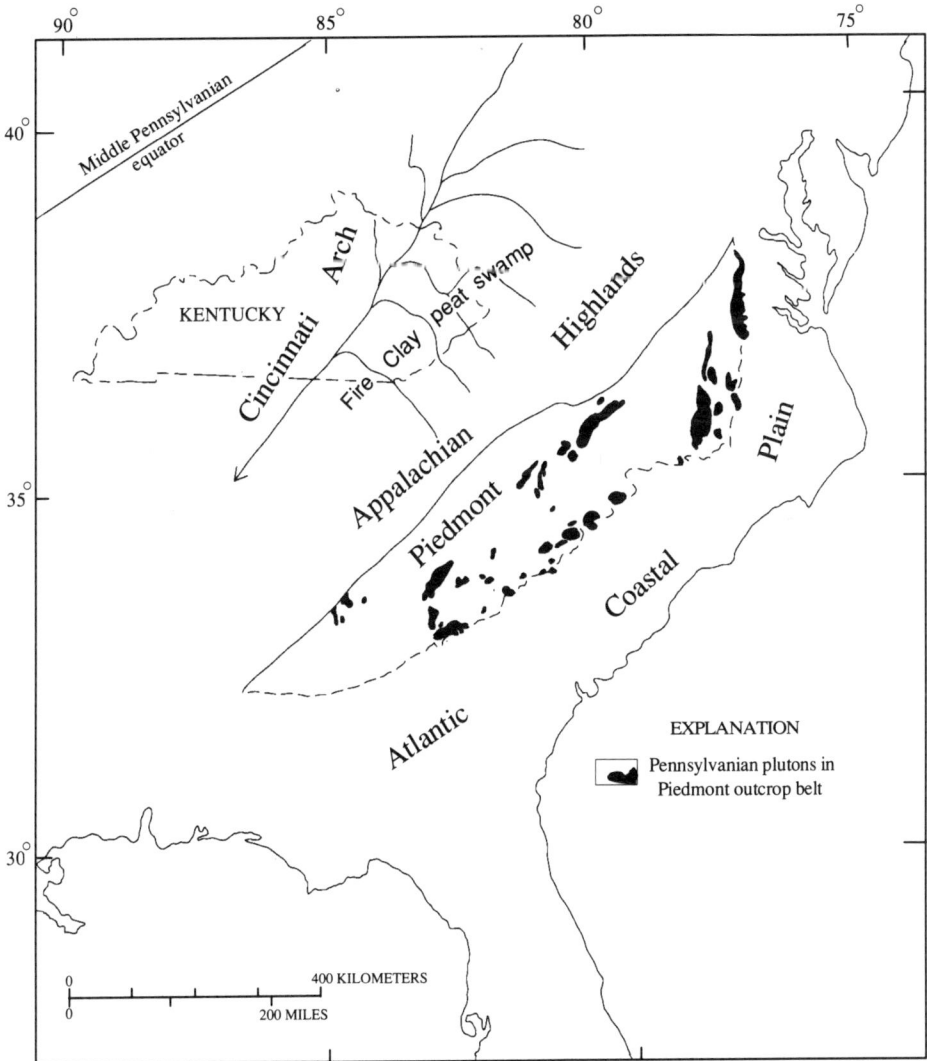

Figure 14. Map of the southeastern United States showing the present location of plutons of Pennsylvanian age in the Piedmont outcrop belt that may be the roots of source volcanoes for the Pennsylvanian ash falls and the Fire Clay drainage basin that probably lay between the Cincinnati arch and the Appalachian highlands at the time of eruption. Sources of Pennsylvanian plutonic data include Sinha et al. (1989) and Horton et al. (1991). Note the approximate position of the Middle Pennsylvanian equator (Kent and Keppie, 1988).

duced carrying capacity. The plutons of the Piedmont, which lay east and southeast of the Fire Clay swamps in the Middle Pennsylvanian, were generally in the prevailing zone of easterly winds; thus, we believe they were ready sources of volcanism for the Middle Pennsylvanian ash deposits in the central Appalachian basin.

An additional consideration should be made for Alleghanian deformation that is estimated to have produced crustal shortening across the Piedmont of at least 300 km (Rankin et al., 1989), which would increase the distance shown in Figure 14 between the volcanic sources and sites of deposition by perhaps as much as 75%. Yet, the latter consideration may not be an important factor with respect to the Fire Clay ash deposit. As previously noted, the zircons from the Fire Clay tonstein contain cores that were probably inherited from continental rocks of Middle Proterozoic age. Hence, the Fire Clay ash deposit is likely to represent emanations originating within the North American continent, where today, Middle Proterozoic (Grenvillian) rocks occur in the Blue Ridge (just west of the Piedmont, see Fig. 14) and as windows in the adjacent western Piedmont. The eastern margin of this Precambrian crust lies subparallel to, and extends some distance beneath, the Piedmont outcrop belt, as shown in Figure 14 (see Bartholomew, 1984, p. v–vii). If the volcanic source was situated somewhere along the then-active Pennsylvanian continental margin, it may have been as little as 400 km from the Fire Clay peat swamp.

Because the zircons of the Upper Banner and Hitchins tonsteins (as well as other sampled Pennsylvanian tonsteins) lack radiogenic inheritance (R. E. Zartman, unpublished data, 1993), they may derive from the magmatic arc of Sinha and Zietz (1982), which was farther east and outboard of the North American continent at that time. It is perhaps noteworthy that other physical and chemical properties of the Fire Clay tonstein mentioned in this chapter—such as wide areal distribution, relatively uniform thickness, silicate-melt composition, larger phenocryst grain size, low uranium concentration, and high U/Th—support a different provenance.

The known distribution of the other tonsteins of the Williamson to Hazard coal bed interval appears to extend from easternmost Kentucky toward the west-northwest and has not been reported northeast or southwest of that area. The distribution of the Williamson and Little Fire Clay tonsteins is shown in Figure 13 and appears to represent deposits narrowly controlled by high-altitude wind currents. The two most southwestern locations shown for the Williamson tonstein are questionable, isolated single occurrences. Data for the Hazard tonstein are not included in our maps because the tonstein has not been proven to be volcanic in origin; it does, however, seem to have a similar distribution as do the Williamson and Little Fire Clay tonsteins. The general pattern of distribution suggests that the Williamson and Little Fire Clay deposits derived also from Hercynian volcanism to the east and southeast.

The actual source of the Fire Clay ash fall may have been eroded or may not be presently exposed. Furthermore, since the volcanic magma chamber was probably stratified, no specific pluton can be identified as the Fire Clay source using our limited chemical data. However, additional analyses, such as Pb and Sr isotopes of the accessory minerals, may provide data to eventually "fingerprint" one of the plutons of the Piedmont as its source.

Tonsteins in Pennsylvanian stratigraphy

Because coal beds are discontinuous and tend to split, they generally cannot be traced regionally as single beds and are therefore commonly identified as members of coal zones. The wide distribution of the Fire Clay tonstein, however, records an instant of time in the development of the Fire Clay coal bed that can be traced across a large part of the central Appalachian basin. The distribution pattern indicated in Figure 2 probably cannot be duplicated by any other terrestrial deposit in the Pennsylvanian section (no other horizon can be confidently traced across such distances), but the pattern does demonstrate that single coal beds may be essentially isochronous (on the order of 10^4 to 10^5 yr) and have great regional continuity. Unfortunately, Figure 2 is a synthesis of incidental, incomplete, and mostly undocumented raw data found in general geologic reports that we have been able to check only by limited reconnaissance. The areas where the tonstein is indicated to be missing in Figure 2 correspond to areas of major nomenclatural changes from one part of the basin to another; this fact alone illustrates the utility and importance of the tonstein as a lithostratigraphic marker bed in areas where it is continuous.

The analyses of the phenocrysts and their melt inclusions from samples of all of the Pennsylvanian tonsteins across the Appalachian basin studied herein suggest that they can be used to help characterize and distinguish different tonsteins. Our investigations indicate that the Fire Clay tonstein is the only horizon in the Hazard to Williamson coal bed interval (Fig. 1) that contains a rich, relatively coarse mineral suite of volcanic phenocrysts throughout its area of occurrence. Reconnaissance related to this study suggests that the Little Fire Clay and Williamson tonsteins may locally contain datable pyroclastic minerals, such as sanidine, biotite, and zircon, that can be used to further extend and develop a reliable geochronologic time scale for the Carboniferous.

The four tonstein horizons shown in Figure 1 are easily distinguished one from another by physical stratigraphy; only the Little Fire Clay tonstein might be confused locally with the Fire Clay tonstein because of its proximity, but the former is best known from a relatively small area, where both generally occur together. The Westphalian B, which in eastern Kentucky represents a stratigraphic interval as thick as 800 m (based on boundaries correlated by Eble et al., 1989), is estimated by Hess and Lippolt (1986) to have a duration of only about 2 m.y. (see Fig. 1). If a constant rate of deposition of about 400 m/m.y. is assumed for the Hazard to Williamson interval, it is possible that radiometric ages for all the tonsteins (having the same accuracy as that of the Fire Clay tonstein, ±0.8 Ma) would overlap one another. However, the Fire Clay tonstein probably would still be distinguishable from the Hazard and Williamson tonsteins on the basis of size of their mineral grains, and on the small, but real, differences between their mineral populations and chemistries. For example, the isolated Walnut Mountain tonstein (sample RI, Fig. 2), has been miscorrelated previously with a tonstein in the Hazard coal zone (or its equivalents) about 80 m higher in the section (Rice, 1984). Physical and chemical characteristics of its phenocrysts (as well as its radiometric age) demonstrate that it is essentially identical to the other six Fire Clay tonstein samples, with which it is correlated.

SUMMARY

The Fire Clay tonstein is the first stratigraphic unit in the Carboniferous of North America to be precisely dated by radiometric means. Its stratigraphic position just below a major faunal break makes it particularly useful in the construction of a geochronologic time scale and for comparison of North American sections with those of other continents. A comparison of megafauna suggests that Atoka Formation strata younger than the Trace Creek Member are correlative with uppermost Westphalian B and lowermost Westphalian C, whereas Figure 1 shows that paleobotanists of this country have placed the

Westphalian B-C boundary at approximately equal distances either above or below the Fire Clay coal bed. Our mean $^{40}Ar/^{39}Ar$ plateau age of 310.9 ± 0.8 Ma (see Kunk and Rice, this volume) for the Fire Clay tonstein is consistent with the age of 311 Ma developed by Hess and Lippolt (1986) for the Westphalian B-C boundary based on $^{40}Ar/^{39}Ar$ ages of tonsteins in western Europe. Thus, the radiometric age data generally appear to support the invertebrate paleontologic data.

Much more work needs to be done with regard to the Fire Clay tonstein. It would be very useful to confirm the $^{40}Ar/^{39}Ar$ age results with U-Pb ages from the zircons. The distribution of grain sizes and minerals should be examined more carefully than this study was able to do. Perhaps isotopic studies of various mineral grains can help "fingerprint" possible volcanic sources in the Piedmont and thereby provide a better chronology for tectonic activity in that area. The wealth of material that has been found in the Fire Clay tonstein for mineral, geochemical, and isotopic studies should encourage a diligent search for other tonsteins in the Pennsylvanian section having similar potential for developing a better geochronologic scale and for expanding our understanding of this important tectonic period on the North American continental margin.

ACKNOWLEDGMENTS

We thank Victor M. Seiders, Robert P. Koeppen, Norman C. Hester, and Walter L. Manger for their careful review of the final versions of this paper and for providing many useful suggestions.

REFERENCES CITED

Bartholomew, M. J., ed., 1984, The Grenville event in the Appalachians and related topics: Geological Society of America Special Paper 194, 287 p.

Belkin, H. E., and Rice, C. L., 1989, A rhyolite ash origin for the Hazard No. 4 flint clay (Appalachian basin)—Evidence from silicate-melt inclusions: Geological Society of America Abstracts with Programs, v. 21, no. 6, p. A360.

Belkin, H. E., and Rice, C. L., 1990, The origin of some flint clays in the central Appalachian basin: Evidence from silicate-melt inclusions [abs.]: Biennial Pan-American Conference on Research On Fluid inclusions, 3d, Toronto, Ontario, 1990, Program and Abstracts, Pacrofi III, v. 3, p. 15.

Bergeron, R. R., Cobb, J. C., Slucher, E. R., and Smath, R. A., 1983, Geologic descriptions and coal analyses for 1982 coal drilling in the Daniel Boone National Forest, eastern Kentucky: Kentucky Geological Survey Information Circular 12, ser. 11, 243 p.

Brown, W. R., 1977, Geologic map of the Willard quadrangle, eastern Kentucky: U.S. Geological Survey Geologic Quadrangle Map GQ-1387, scale 1:24,000.

Burger, K., and Damberger, H. H., 1985, Tonsteins in the coalfields of Western Europe and North America: Congrès International de Stratigraphie et de Géologie du Carbonifère, 9th, Washington, D.C., and Champaign-Urbana, Illinois, 1979: Compte Rendu, v. 4, p. 433–448.

Carlson, J. E., 1965, Geology of the Rush quadrangle, Kentucky: U.S. Geological Survey Geologic Quadrangle Map GQ-408, scale 1:24,000.

Chesnut, D. R., Jr., 1985, Source of the volcanic ash deposit (flint clay) in the Fire Clay coal of the Appalachian basin: Congrès International de Stratigraphie et de Géologie du Carbonifère, 10th, Madrid, 1983: Compte Rendu, v. 1, p. 145–154.

Cobb, J. C., Chesnut, D. R., Jr., Hester, N. C., and Hower, J. C., 1981, Coal and coal-bearing rocks of eastern Kentucky (Geological Society of America field trip guidebook): Lexington, Kentucky Geological Survey, 169 p.

Eaton, G. P., 1964, Windblown volcanic ash—A possible index to polar wandering: Journal of Geology, v. 72, p. 1–35.

Eble, C. F., Grady, W. C., and Gillespie, W. H., 1989, Palynology, petrography and paleoecology of the Hernshaw–Fire Clay coal bed in the central Appalachian basin, in Cecil, C. B., and Eble, C. F., eds., Carboniferous geology of the eastern United States, St. Louis, Missouri, to Washington, D.C., June 28–July 8, 1989, Field trip guidebook T143 for the 28th International Geological Congress: Washington, D.C., American Geophysical Union, p. 133–142.

Froelich, A. J., and Stone, B. D., 1973, Geologic map of parts of the Benham and Appalachia quadrangles, Harlan and Letcher Counties, Kentucky: U.S. Geological Survey Geologic Quadrangle Map GQ-1059, scale 1:24,000.

Fullagar, P. D., and Butler, J. R., 1979, 325 to 265 m.y. old granitic plutons in the Piedmont of the southeastern Appalachians: American Journal of Science, v. 279, p. 161–185.

Furnish, W. M., and Knapp, W. D., 1966, Lower Pennsylvanian fauna from eastern Kentucky; Pt. 1, Ammonoids: Journal of Paleontology, v. 40, p. 296–308.

Gilbert, J. S., and Rogers, N. W., 1989, The significance of garnet in the Permo-Carboniferous volcanic rocks of the Pyrenees: Journal of the Geological Society, London, v. 146, p. 477–490.

Glenn, L. C., 1925, The northern Tennessee coal field: Tennessee Division of Geology Bulletin 33-B, 478 p.

Gordon, M., jr., 1970, Carboniferous ammonoid zones of the south-central and western United States: Congrès International de Stratigraphie et de Géologie du Carbonifère, 6th, Sheffield, 1967: Compte Rendu, v. 2, no. 6, p. 817–826.

Harris, L. D., 1963, Coal beds of the Big Sandy reserve district, in Huddle, J. W., Lyons, E. J., Smith, H. L., and Ferm, J. C., Coal reserves of eastern Kentucky: U.S. Geological Survey Bulletin 1120, p. 81–100.

Hess, J. C., and Lippolt, H. J., 1986, $^{40}Ar/^{39}Ar$ Ages of tonstein and tuff sanidines—New calibration points for the improvement of the Upper Carboniferous time scale: Chemical Geology (Isotope Geoscience Section), v. 59, p. 143–154.

Hess, J. C., Lippolt, H. J., and Burger, K., 1988, New time-scale calibration points in the Upper Carboniferous from Kentucky, Donetz Basin, Poland, and West Germany [abs.]: Besançon, France, 6th International Conference on Fission-Track Dating.

Höehne, K., 1957, Tonsteine in Kohlenflözen der Oststaaten von Nordamerika und Ostaustralien: Chemie Erde, v. 19, no. 2, p. 111–129.

Horton, J. W., Jr., Drake, A. A., Jr., Rankin, D. W., and Dallmeyer, R. D., 1991, Preliminary tectonostratigraphic terrane map of the central and southern Appalachians: U.S. Geological Survey Miscellaneous Investigations Series Map I-2163, scale 1:2,000,000.

Huddle, J. W., and Englund, K. J., 1966, Geology and coal reserves of the Kermit and Varney area, Kentucky: U.S. Geological Survey Professional Paper 507, 83 p., and map scale 1:24,000.

Ingram, H.A.P., 1983, Hydrology, in Gore, A.J.P., ed., Mires: Swamp, bog, fen and moor—Ecosystems of the world, Vol. 4A: New York, Elsevier, p. 67–158.

Kent, D. V., and Keppie, J. D., 1988, Silurian-Permian palaeocontinental reconstructions and circum-Atlantic tectonics, in Harris, A. L., and Fettes, D. J., eds., The Caledonian-Appalachian orogen: Geological Society of London Special Publications 38, p. 469–480.

Le Bas, M. J., Le Maitre, R. W., Streckeisen, A., and Zanettin, B., 1986, A chemical classification of volcanic rocks based on the total alkali-silica diagram: Journal of Petrology, v. 27, p. 745–750.

Lyons, P. C., and 7 others, 1992, An Appalachian isochron—A kaolinized

Carboniferous air-fall volcanic-ash deposit (tonstein): Geological Society of America Bulletin, v. 104, p. 1515–1527.

Manger, W. L., and Sutherland, P. K., 1985, Mid-Carboniferous biostratigraphic relations, southern midcontinent, North America: Congrès International de Stratigraphie et de Géologie du Carbonifère, 10th, Madrid, 1983: Compte Rendu, vol. 1, p. 383–389.

Manger, W. L., Miller, M. S., and Mapes, R. H., 1992, Age and correlation of the Gene Autry Shale, Ardmore basin, southern Oklahoma, in Sutherland, P. K., and Manger, W. L., eds., Recent advances in middle Carboniferous biostratigraphy—A symposium: Oklahoma Geological Survey Circular 94, p. 101–109.

Newell, W. L., 1975, Geologic map of the Frakes quadrangle and part of the Eagan quadrangle, southeastern Kentucky: U.S. Geological Survey Geologic Quadrangle Map GQ-1249, scale 1:24,000.

Phillips, T. L., Peppers, R. A., and DiMichele, W. A., 1985, Stratigraphic and interregional changes in Pennsylvanian coal-swamp vegetation—Environmental inferences: International Journal of Coal Geology, v. 5, p. 43–109.

Pindell, J., and Dewey, J. F., 1982, Permo-Triassic reconstruction of western Pangea and the evolution of the Gulf of Mexico/Caribbean region: Tectonics, v. 1, no. 2, p. 179–211.

Rankin, D. W., and 12 others, 1989, North American Continent–Ocean Transect Program Explanatory pamphlet for Transect E-4, central Kentucky to the Carolina trough: Geological Society of America Centennial/Ocean Transect 16, 41 p.

Reiter, E. R., 1967, Jet streams: New York, Doubleday, 189 p.

Rice, C. L., 1984, Stratigraphic framework and nomenclatural problems in the Pennsylvanian of the Cumberland overthrust sheet, Kentucky and Tennessee: Geological Society of America Bulletin, v. 95, p. 1475–1481.

Rice, C. L., 1986, Pennsylvanian System, in The geology of Kentucky—A text to accompany the geologic map of Kentucky: U.S. Geological Survey Professional Paper 1151-H, p. H31–H43.

Rice, C. L., and Maughan, E. K., 1978, Geologic map of the Kayjay and part of the Fork Ridge quadrangle, Bell and Knox Counties, Kentucky: U.S. Geological Survey Geologic Quadrangle Map GQ-1505, scale 1:24,000.

Rice, C. L., and Newell, W. L., 1975, Geologic map of the Saxton quadrangle and part of the Jellico East quadrangle, Whitley County, Kentucky: U.S. Geological Survey Geologic Quadrangle Map GQ-1264, scale 1:24,000.

Robinson, P. L., 1973, Paleoclimatology and continental drift, in Tarling, D. H., and Runcorn, S. K., Implications of Continental drift to the earth sciences, v. 1: University of Newcastle upon Tyne, England, NATO Advanced Study Institute, p. 449–476.

Robl, T. L., and Bland, A. E., 1977, The distribution of aluminum in shales associated with the major economic coal seams of eastern Kentucky, in Rose, T. G., ed., Proceedings, Kentucky Coal Refuse Disposal and Utilization Seminar, 3rd, May 11–12, 1977, Lexington: Lexington, University of Kentucky, p. 97–101.

Ruppert, L. F., 1987, Applications of cathodoluminescence of quartz and feldspar to sedimentary petrology: Scanning Microscopy, v. 1, no. 1, p. 63–72.

Russell, G. S., Russell, W. C., Speer, A. J., and Glover, L., 1981, Rb-Sr evidence of latest Precambrian to Cambrian and Alleghanian plutonism along the eastern margin of the sub-coastal plain Appalachians, North Carolina and Virginia: Geological Society of America Abstracts with Programs, v. 13, p. 543.

Seiders, V. M., 1965, Volcanic origin of flint clay in the Fire Clay coal bed, Breathitt Formation, eastern Kentucky, in Geological Survey research: U.S. Geological Survey Professional Paper 525-D, p. D52–D54.

Sinha, A. K., and Zietz, I., 1982, Geophysical and geochemical evidence for a Hercynian magmatic arc, Maryland to Georgia: Geology, v. 10, p. 593–596.

Sinha, A. K., Hund, E. A., and Hogan, J. P., 1989, Paleozoic accretionary history of the North American plate margin (central and southern Appalachians); Constraints from the age, origin, and distribution of granitic rocks, in Hillhouse, J. W., ed., Deep structure and past kinematics of accreted terranes: Virginia Polytechnic Institute and State University Geophysical Monograph 50, p. 219–238.

Speer, J. A., 1981, Petrology of cordierite- and almandine-bearing granitoid plutons of the southern Appalachian piedmont, U.S.A.: Canadian Mineralogist, v. 19, p. 35–46.

Stevens, S. S., 1976, Petrogenesis of a tonstein in the Appalachian basin [M.S. thesis]: Richmond, Eastern Kentucky University, 83 p.

Sutherland, P. K., and Grayson, R. C., Jr.,1978, Redefinition of the Morrow Series (Lower Pennsylvanian) in its type area in northwestern Arkansas: Geological Society of America Abstracts with Programs, v. 10, no. 7, p. 501.

Sutherland, P. K., and Henry, T. W., 1975, Brachiopod zonation of the Lower and Middle Pennsylvanian System in the western United States of America [abs.]: Congrès International de Stratigraphie et de Géologie Carbonifére, 8th, Moscow, 1975; p. 274–275.

Sutherland, P. K., and Henry, T. W., 1980, Brachiopod zonation of the Lower and Middle Pennsylvanian System in the central United States: Congrès International de Stratigraphie et de Géologie Carbonifére, 8th, Moscow, 1975: Compte Rendu, v. 6, p. 71–75, 2 figs.

Wanless, H. R., 1946, Pennsylvanian geology of a part of the southern Appalachian coal field: Geological Society of America Memoir 13, 162 p.

Zartman, R. E., Dyman, T. S., Tysdal, R. G., and Pearson, R. C., 1994, U-Pb ages of volcanogenic zircon from porcellanite beds in the Vaughn Member of the mid-Cretaceous Blackleaf Formation, southwestern Montana: U.S. Geological Survey Bulletin 2113-B.

MANUSCRIPT ACCEPTED BY THE SOCIETY FEBRUARY 1, 1994

High-precision $^{40}Ar/^{39}Ar$ age spectrum dating of sanidine from the Middle Pennsylvanian Fire Clay tonstein of the Appalachian basin

Michael J. Kunk
U.S. Geological Survey, National Center, MS 981, Reston, Virginia 22092
Charles L. Rice
U.S. Geological Survey, National Center, MS 926, Reston, Virginia 22092

ABSTRACT

$^{40}Ar/^{39}Ar$ plateau age spectra of seven sanidine samples from the Fire Clay tonstein (Middle Pennsylvanian), collected along a 300-km traverse in the Appalachian basin, range from 310.3 to 311.4 Ma. All plateau ages agree, within the limits of analytical precision, with their respective total gas ages. This agreement, together with the reproducibility between samples, suggests the analyzed samples did not contain any significant contaminant feldspar. The mean of these seven plateau ages, 310.9 ± 0.8 Ma, is interpreted to represent a precise numerical estimate of time of eruption and deposition of this tonstein and the coal bed in which it is found. The lack of any discernible difference between the age of two samples of the Fire Clay tonstein collected from east of the Pine Mountain thrust fault, along with the age of five samples from west of this fault, suggests that the Fire Clay tonstein has been reliably correlated with a tonstein on the Cumberland overthrust sheet. This correlation, together with the age data presented in this paper, indicates that the Pine Mountain thrust fault must be younger than the 310.9-Ma age obtained for the Fire Clay tonstein.

The Fire Clay tonstein is biostratigraphically correlated with the Trace Creek Shale Member of the Atoka Formation in the Midcontinent of North America and with a position near the Westphalian B-C boundary in Western Europe. Our age of 310.9 ± 0.8 Ma for the Westphalian B-C boundary represents a well-constrained point, useful for the numerical refinement of the geologic time scale.

INTRODUCTION

A "flint clay parting" in the Fire Clay coal bed has long been considered the best single stratigraphic marker in the Middle Pennsylvanian of the central Appalachian basin (Wanless, 1946) because of its broad regional extent and the relative ease with which it is identified in outcrop. This flint clay parting was first suggested to be of volcanic origin by Ashley (1928). Seiders (1965) and others (including Robl and Bland [1977] and Chesnut [1985]) confirmed this suggestion with studies that showed its phenocryst mineralogy to be consistent with that of a water-lain air-fall tuff that has been altered since deposition, mostly to kaolinite. For the purposes of this chapter, we refer to this flint clay parting as the Fire Clay tonstein; the term tonstein is used in the sense of Burger and Damberger (1985), whereas Fire Clay comes from the common name of the coal bed in which the tonstein is found. Although the term tonstein does not imply any specific origin, we believe that the Fire Clay tonstein represents an altered air-fall tuff that was deposited over a wide geographic area in an instant of geologic time. Thus, the Fire Clay tonstein represents a timeline useful for the correlation of stratigraphic sections in which it

Kunk, M. J., and Rice, C. L., 1994, High precision $^{40}Ar/^{39}Ar$ age spectrum dating of sanidine from the Middle Pennsylvanian Fire Clay tonstein of the Appalachian basin, *in* Rice, C. L., ed., Elements of Pennsylvanian Stratigraphy, Central Appalachian Basin: Boulder, Colorado, Geological Society of America Special Paper 294.

occurs, and, if precisely dated, is useful for refining the geologic time scale. The Fire Clay coal bed has many correlatives, all of which are characterized by the presence of the tonstein. The Fire Clay tonstein is most commonly encased in coal; in places, however, it occurs at the top or bottom of the coal bed, and is present where coal is absent (Huddle et al., 1963). All samples discussed herein were encased in coal.

The Fire Clay tonstein is one of about 10 tonsteins known to occur in Middle Pennsylvanian rocks of the central Appalachian basin (Burger and Damburger, 1985). It occurs halfway between the bases of the Kendrick and Magoffin Members of the Middle Pennsylvanian Breathitt Formation and has been placed in the upper part of the *Diadoloceras nermeiri–Bisatoceras micromphalus* Goniatite Zone as it is projected from the Ozarks to the Appalachian basin by Rice, Belkin et al. (this volume). On the basis of spores in coal, the tonstein has been projected to a stratigraphic position near the Westphalian B-C boundary in Western Europe. For a more detailed discussion of the biostratigraphic assignment, see Rice, Belkin et al. (this volume).

The primary purpose of this study was to obtain a precise numerical age for the time of deposition of the Fire Clay tonstein across the widest possible geographic area in which it has been preserved. We collected samples of the tonstein, separated the sanidine, and obtained $^{40}Ar/^{39}Ar$ dates. Similar studies of sanidines from tonsteins using $^{40}Ar/^{39}Ar$ age spectrum dating techniques in the Upper Carboniferous of Western Europe have been the basis for significant revisions in the geologic time scale (Hess and Lippolt, 1986). Previous $^{40}Ar/^{39}Ar$ age spectrum dating (Hess et al., 1988) of what is probably the Fire Clay tonstein collected in eastern Kentucky has suggested an age of 311 Ma for this unit.

The second objective of this study was to identify the stratigraphic position of the Fire Clay tonstein at the southwestern end of the Cumberland overthrust sheet (Fig. 1). Although the Fire Clay tonstein has been traced from West Virginia across Virginia and Kentucky to Tennessee, both on the Cumberland overthrust sheet and in the autochthonous portion of the basin northwest of the overthrust sheet (Wanless, 1946), there is an area that is stratigraphically and structurally isolated at the southwestern end of the Cumberland overthrust sheet where the identity of the tonstein has been questioned. Rice (1984) determined that the stratigraphic position of the Fire Clay tonstein in that area was in the Walnut Mountain coal bed, about 80 m higher in the Pennsylvanian section than had been previously indicated by Glenn (1925) and Englund (1968). We collected sample RI of the tonstein in Glenn's (1925) type Walnut Mountain coal bed on Walnut Mountain in Campbell county, Tennessee, for $^{40}Ar/^{39}Ar$ age spectrum dating and for a comparison of its phenocryst mineralogy with that of other samples of the Fire Clay tonstein dated in this study. Elsewhere in this volume, Rice, Belkin et al. (this volume) show that the chemistry of the primary igneous phenocrysts and the chemistry of their silicate-melt in-

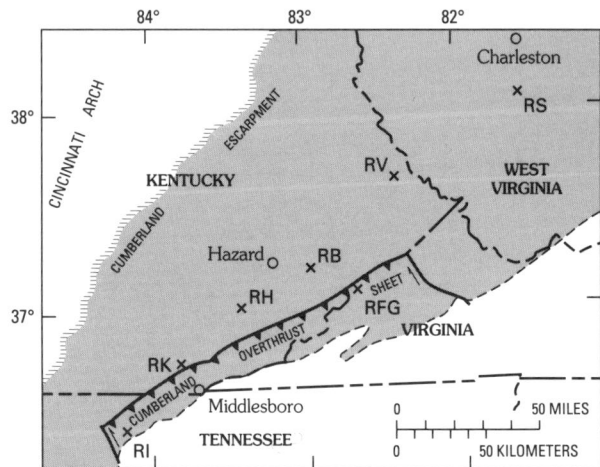

Figure 1. Map showing location sites of samples from the Fire Clay tonstein in the central Appalachian basin (shaded). The Fire Clay tonstein has not been identified outside the area depicted.

clusions are the same for all of the samples of the Fire Clay tonstein analyzed, including the sample from Glenn's (1925) Walnut Mountain coal bed, and suggest that all of the samples are the product of a single volcanic eruption of rhyolitic material. $^{40}Ar/^{39}Ar$ age spectrum dating of sample RI confirms this correlation.

The third objective of this study was to place a numerical constraint on the movement of the Pine Mountain thrust fault. Estimates of the emplacement age of the Cumberland overthrust sheet have all been indirect and are based on stratigraphic relations, some of which are poorly known. An approximation of the timing of its emplacement can be drawn from Secor et al. (1986), and references therein, who noted that the Alleghanian orogeny is bracketed by the deposition of the Permian-Carboniferous clastic wedge from southwestern Pennsylvania to Alabama. The base of the wedge occurs in the Late Mississippian, and its top is represented by the Dunkard Group of Late Pennsylvanian to Early Permian age in Ohio, Pennsylvania, and West Virginia. Secor et al. (1986) estimated the age of this interval to range from 327 ± 21 to 266 ± 17 Ma. Because the Fire Clay Tonstein is cut by the Pine Mountain thrust fault, its age will provide a maximum estimate for the time of thrusting and emplacement of the Cumberland overthrust sheet.

SAMPLE DESCRIPTION, COLLECTION, AND PREPARATION

The Fire Clay tonstein is composed largely of well-crystallized kaolinite, generally dark brownish gray in color, and has conchoidal fracture with a waxy luster. It commonly ranges from 10 to 20 cm in thickness. Where thickest, the Fire Clay tonstein is graded; the lower part contains the most phenocrysts, from 3 to 5% of the rock. In general, the southernmost and thickest Fire Clay tonstein samples were more easily processed because they contained more and larger primary igneous phenocrysts.

Seven samples of the Fire Clay tonstein for $^{40}Ar/^{39}Ar$ age spectrum dating were collected along a 300-km traverse in the central part of the Appalachian basin (Fig. 1; Table 1). The thickest, locally reported occurrences were sampled. The samples were disaggregated by standard grinding techniques, and most of the kaolinite was dispersed by using detergents and dimethyl sulfoxide (a solvent). The phenocryst population was then sieved and its sanidine component concentrated with the aid of ultrasonic cleaners, heavy liquids, and hand picking, to a purity of >99.9%. Sanidine grains used for argon analytical work ranged in size from 125 to 250 µm, except for sample RB, which contained less sanidine of predominantly smaller grain sizes, ranging from 74 to 125 µm.

Two samples, RS and RH, had residual, resistant kaolinite present as coatings on individual grains. These coatings were removed by placing the samples in a 7% solution of cold hydrofluoric acid in an ultrasonic cleaner for 3 min. This treatment was followed by rinsing the samples three times, again in an ultrasonic bath, in deionized, distilled water. $^{40}Ar/^{39}Ar$ age spectrum results of sanidine from the Fish Canyon Tuff, prepared using the same hydrofluoric acid leaching technique, demonstrate that it does not affect the numerical value of its plateau age or the overall age spectrum pattern when compared with age spectrum results of Fish Canyon sanidine prepared to the same level of purity by strictly mechanical means (M. J. Kunk, unpublished data, 1990).

Previous studies identifying sanidine (Seiders, 1965) and beta-form quartz (Robl and Bland, 1977) in the Fire Clay tonstein by microscopic examination of thin sections clearly indicated a volcanic origin but were not able to assess the possibility of contamination by grains of either detrital or authigenic origin. Newer, more rigorous studies of the phenocryst mineralogy of the Fire Clay tonstein from the same collections from which our sanidine samples were derived have recently been completed by Belkin and Rice (1989) and Rice, Belkin et al. (this volume). Their studies show that grains larger than 38 µm in samples of the Fire Clay tonstein are mostly primary igneous phenocrysts. The phenocrysts, as large as 1 mm in diameter, include quartz and sanidine, and traces of apatite, zircon, pyroxene, amphibole, monazite, and garnet, all of igneous origin; also present are sulfides and traces of iron and titanium oxides of secondary origin. Electron microprobe analyses of the sanidine and quartz grains from the samples show that they contain chemically indistinguishable silicate-melt inclusions, which strongly suggests that they are cogenetic phenocrysts of a rhyolitic magma (Belkin and Rice, 1989; Rice, Belkin et al., this volume). Current laminae were observed in only one sample studied (RV), and cathodoluminescence studies of its quartz grains suggest that only about 6% have a nonvolcanic origin. However, no discernible subpopulations of feldspars or authigenic K-feldspar overgrowths were detected in this or any other samples in this study.

$^{40}Ar/^{39}Ar$ ANALYTICAL TECHNIQUE

About 100 mg of sanidine from each of the seven tonstein samples was irradiated in the U.S. Geological Survey TRIGA reactor (Dalrymple et al., 1981) for 30 hr to convert a portion of the ^{39}K in the sample to ^{39}Ar. FCT-3 sanidine (Kunk et al., 1985) and MMhb-1 hornblende (Alexander et al., 1978) were irradiated along with the samples to monitor this conversion (McDougall and Harrison, 1988). The irradiated samples were heated in a low blank furnace similar in design to that of Staudacher et al. (1978) and were degassed in a stepwise manner; individual increments of gas were cleaned using SAES ST 707 and 101 getters, and a Ti getter prior to analysis in a VG MM 1200B mass spectrometer. The resultant argon isotopic data were reduced and analytical errors were calculated with the computer program ArAr* (Haugerud and Kunk, 1988) using the decay constant values recommended by Steiger and Jäeger (1977). An age of 519.4 ± 2.4 Ma was used for MMhb-1 (Alexander et al., 1978; Dalrymple et al., 1981). An age of 27.79 ± 0.07 Ma was used for FCT-3 sanidine (Kunk et al., 1985); this age and uncertainty are relative to the above quoted age for MMhb-1 hornblende. Although a more recent compilation of results (Samson and Alexander, 1987) has suggested an age of 520.4 ± 1.7 Ma, we have chosen to use the original age and analytical uncertainty of MMhb-1 because we believe that the uncertainty in the age stated for MMhb-1 by Samson and Alexander (1987) may be too small. The ages presented in this study can be easily recalculated for age values other than 519.4 ± 2.4 Ma for MMhb-1 and 27.79 ± 0.07 Ma for FCT-3, if so desired.

The uncertainty stated for the age of an individual step within an age spectrum does not include the uncertainty in the monitor-derived J-value, as it is a constant within a single sample. The uncertainties associated with individual plateau ages do include the analytical uncertainty in measuring the J-value. The uncertainty associated with the mean age of several plateaux that agree with each other within the limits of analytical precision includes the error in the J-value, except as noted in Table 3. The first type of uncertainty is a sample standard deviation of the mean of the measured ages that has been

TABLE 1. LOCATION AND ELEVATION OF SAMPLES USED IN THIS CHAPTER

Sample	Latitude N	Longitude W	Elevation m	ft
RS	38°03'00"	81°33'10"	287	940
RV	37°37'45"	82°22'07"	315	1,035
RB	37°13'06"	82°58'44"	369	1,210
RK	36°43'19"	83°49'22"	501	1,645
RH	37°00'03"	83°23'53"	341	1,120
RI	36°24'18"	84°10'48"	817	2,680
RFG	37°01'23"	82°41'11"	847	2,780

TABLE 2. ^{40}AR/^{39}AR AGE SPECTRUM DATA FOR THE SAMPLES USED IN THIS CHAPTER

T (C°)	^{39}Ar of Total (%)	^{40}Ar$_R$/^{39}Ar$_K$	Apparent K/Ca*	Radiogenic Yield (%)	^{39}Ar$_K$ (x 10^{-12} moles)†	Apparent Age and Precision§
RS Sanidine	**J-Value = 0.007862 ± 0.19%**		**Sample Wt. 0.1000 g**			
750	11.8	24.475	58.7	99.4	0.446	316.14 ± 0.69
950	8.0	24.102	53.3	99.8	1.953	311.72 ± 0.38
1100	15.2	24.044	35.9	99.9	3.709	311.03 ± 0.17
1150	13.1	24.049	37.4	99.9	3.196	311.09 ± 0.17
1175	9.6	24.070	37.0	99.9	2.347	311.34 ± 0.27
1200	11.2	24.069	35.6	99.9	2.731	311.33 ± 0.25
1225	10.7	24.107	35.3	99.9	2.603	311.79 ± 0.29
1250	7.3	24.130	37.0	99.7	1.789	312.05 ± 0.50
1275	12.0	24.223	37.0	99.7	2.917	313.15 ± 0.47
1300	9.2	24.243	34.6	99.7	2.185	313.40 ± 0.44
1350	2.1	24.381	52.2	99.7	0.517	315.04 ± 0.49
Total Gas		K/Ca	38.4		Age	312.0
Plateau Age (75% of gas in 950°–1250°C steps)						311.30 ± 0.55
RV Sanidine	**J-Value = 0.007823 ± 0.19%**		**Sample Wt. 0.1003 g**			
750	1.5	24.567	40.2	96.7	0.385	317.12 ± 0.37
950	4.1	24.068	47.4	99.8	1.071	311.20 ± 0.36
1100	10.4	23.963	35.5	99.9	2.751	309.96 ± 0.19
1150	9.7	23.975	36.9	99.9	2.557	310.11 ± 0.24
1175	8.0	23.993	35.5	99.9	2.110	310.32 ± 0.20
1200	8.7	23.999	33.6	99.9	2.287	310.38 ± 0.18
1225	8.2	23.995	33.3	99.9	2.161	310.34 ± 0.18
1250	14.6	24.009	35.8	99.9	3.850	310.51 ± 0.24
1275	8.6	24.066	44.1	99.9	2.277	311.18 ± 0.16
1300	6.1	24.110	52.7	99.8	1.160	311.71 ± 0.34
1350	12.1	24.129	36.0	99.8	3.193	311.94 ± 0.20
1450	8.1	24.189	33.8	99.8	2.139	312.64 ± 0.32
Total Gas		K/Ca	37.6		Age	311.0
Plateau Age (60% of gas in 1100°–1250°C steps)						310.27 ± 0.54
RB Sanidine	**J-Value = 0.007826 ± 0.19%**		**Sample wt. 0.0614 g**			
750	1.2	24.833	34.0	87.2	0.180	320.39 ± 1.76
950	8.7	24.043	48.2	99.7	1.271	311.02 ± 0.27
1100	22.4	23.947	33.6	99.9	3.265	309.88 ± 0.11
1150	16.4	23.973	34.2	99.9	2.393	310.18 ± 0.34
1175	12.8	24.045	47.1	99.8	1.868	311.04 ± 0.37
1200	10.4	24.069	47.3	99.8	1.518	311.33 ± 0.35
1225	6.6	24.115	51.4	99.8	9.601	311.88 ± 0.27
1250	7.6	24.125	49.0	99.8	1.113	312.00 ± 0.35
1275	8.0	24.204	48.0	99.7	1.170	312.93 ± 0.29
1300	4.5	24.217	47.0	99.7	6.572	313.09 ± 0.41
1350	1.2	24.368	40.0	99.4	1.730	314.87 ± 0.88
Total Gas		K/Ca	42.3		Age	311.2
Plateau Age (60.4% of gas in 950°–1175°C steps)						310.25 ± 0.55
RK Sanidine	**J-Value = 0.007826 ± 0.19**		**Sample wt. 0.1010 g**			
750	1.2	24.591	8.9	98.4	0.250	317.52 ± 0.75
950	6.2	24.030	59.2	99.7	1.282	310.86 ± 0.30
1000	5.7	23.995	61.6	99.7	1.189	310.46 ± 0.39
1050	5.6	23.996	54.8	99.7	1.162	310.46 ± 0.37
1100	7.0	24.014	35.4	99.5	1.464	310.68 ± 0.55
1150	9.9	23.981	32.4	99.6	2.068	310.28 ± 0.56
1200	12.5	23.997	34.4	99.8	2.600	310.48 ± 0.27
1250	40.8	24.060	31.5	99.8	8.498	311.23 ± 0.28
1275	4.8	24.174	54.1	99.7	1.008	312.58 ± 0.63
1300	6.3	24.216	33.7	99.7	1.332	313.08 ± 0.52
Total Gas		K/Ca	37.9		Age	311.2
Plateau Age (88% of gas in 950°–1250°C steps)						310.71 ± 0.55

TABLE 2. ^{40}AR/^{39}AR AGE SPECTRUM DATA FOR THE SAMPLES USED IN THIS CHAPTER
(continued)

T (C°)	^{39}Ar of Total (%)	^{40}Ar$_R$/^{39}Ar$_K$	Apparent K/Ca*	Radiogenic Yield (%)	^{39}Ar$_K$ (x 10^{-12} moles)†	Apparent Age and Precision§
RH Sanidine	**J-Value = 0.007826 ± 0.19%**		**Sample Wt. 0.1005 g**			
750	1.1	24.676	40.8	99.2	0.242	318.52 ± 1.22
950	6.6	24.089	59.9	99.8	1.511	311.52 ± 0.35
1000	4.4	24.040	57.3	99.7	1.008	310.98 ± 0.44
1050	4.6	24.037	53.9	99.7	1.056	310.95 ± 0.22
1100	8.2	24.061	57.8	99.8	1.877	311.24 ± 0.31
1150	10.6	24.006	34.2	99.8	2.419	310.58 ± 0.48
1175	7.6	24.066	58.7	99.7	1.740	311.30 ± 0.23
1200	8.8	24.080	57.7	99.8	2.000	311.46 ± 0.32
1225	8.3	24.031	36.9	99.6	1.895	310.88 ± 0.57
1250	17.8	24.095	35.9	99.7	3.980	311.64 ± 0.42
1275	8.2	24.147	38.1	99.7	1.867	312.25 ± 0.71
1300	12.6	24.026	36.1	99.9	2.874	310.82 ± 0.27
1350	1.5	24.313	62.0	99.6	3.455	314.23 ± 1.45
Total Gas		K/Ca	45.3		Age	311.4
Plateau Age (77% of gas in 950 - 1250°C steps)						311.20 ± 0.55
RI Sanidine	**J-Value = 0.007824**		**Sample Wt. 0.1025 g**			
750	1.0	24.487	57.3	99.1	0.245	316.21 ± 0.51
850	2.5	24.057	61.9	99.6	0.604	311.12 ± 0.35
950	4.6	24.014	65.4	99.7	1.142	310.60 ± 0.38
1000	5.3	23.984	68.2	99.7	1.314	310.24 ± 0.32
1025	4.2	23.986	62.2	99.7	1.026	310.27 ± 0.43
1050	4.4	24.030	64.0	99.7	1.082	310.79 ± 0.44
1075	5.1	24.011	65.6	99.7	1.249	310.56 ± 0.32
1100	7.7	24.089	64.8	99.7	1.887	311.49 ± 0.43
1125	6.2	23.987	30.6	99.7	1.531	310.28 ± 0.50
1150	6.5	23.986	41.3	99.6	1.589	310.27 ± 0.48
1175	6.0	24.034	64.7	99.7	1.472	310.84 ± 0.40
1200	6.8	24.086	63.1	99.7	1.679	311.46 ± 0.41
1250	14.9	24.060	35.1	99.7	3.672	311.15 ± 0.44
1300	21.4	24.099	36.3	99.8	5.255	311.62 ± 0.32
1350	3.4	24.286	60.4	99.7	0.833	313.83 ± 0.44
Total Gas		K/Ca	50.3		Age	311.2
Plateau Age (58% of gas in 1050°–1250°C steps)						310.87 ± 0.55
RFG Sanidine	**J-Value = 0.007825 ± 0.19%**		**Sample Wt. 0.1012 g**			
750	1.4	24.676	37.8	99.0	0.279	318.49 ± 1.09
950	6.1	24.068	52.4	99.7	1.198	311.28 ± 0.30
1100	14.3	24.020	34.5	99.7	2.829	310.71 ± 0.34
1150	9.8	24.066	56.0	99.8	1.928	311.25 ± 0.31
1175	7.6	24.061	51.4	99.8	1.507	311.20 ± 0.25
1200	6.9	24.069	35.1	99.9	1.371	311.30 ± 0.46
1225	6.4	24.097	53.7	99.8	1.266	311.62 ± 0.44
1250	16.2	24.066	35.9	99.7	3.199	311.25 ± 0.35
1275	31.2	24.166	25.8	99.9	6.160	311.86 ± 0.19
Total Gas		K/Ca	37.8		Age	311.5
Plateau Age (99% of gas in 950°–1275°C steps)						311.38 ± 0.55

*Apparent K/Ca ratios were calculated by using the equation given in Fleck et al., 1977.
†^{39}Ar concentrations were calculated using the measured sensitivity of the mass spectrometer and thus have a precision of about 5%.
§Comparisons between steps for the determination of the existence of an age plateau were done using the larger of the calculated uncertainty from the individual analyses, or the reproducibility limit of the mass spectrometer. The mass spectrometer reproducibility limit was determined by replicate measurements of FCT-3. During the period of time in which these samples were analyzed, the reproducibility limit was 0.15%.

quadratically combined with the uncertainty in the J-value. All uncertainties are stated at 1σ, and all comparisons between ages were made using the critical value test as described by Dalrymple and Lanphere (1969, p. 120), using this level of precision. The occurrence of age plateaux was determined using the definition of Fleck et al. (1977), as modified by Haugerud and Kunk (1988). A more rigorous discussion of the error propagation calculations, as well as age calculations, can be found in Haugerud and Kunk (1988).

ARGON RESULTS AND DISCUSSION

The age spectra of all seven samples (Fig. 2; Table 2) are in general very similar. All samples have age plateaux according to any widely accepted published definition of the term (e.g., see Dalrymple and Lanphere [1974] or Fleck et al. [1977]). These plateau ages range from 311.38 ± 0.55 to 310.25 ± 0.55 with 58 to 99% of the ^{39}Ar released, contained in four to nine contiguous steps. Isochron ages of the samples are identical within limits of analytical error for all samples, but are not presented here due to relatively low precision, caused by the high radiogenic yield of the samples. Most of the age spectra have somewhat older apparent ages in their lowest and highest temperature fractions that are statistically different in age from their respective plateau ages. One sample (RFG) has only an older low-temperature increment. These older ages give the spectra a slight U-shape (or L-shape for RFG) that is most pronounced in samples RS, RV, RB, and RH. This pattern in sanidine ^{40}Ar/^{39}Ar age spectrum data is not uncommon and its cause is not fully understood (Hess and Lippolt, 1986;

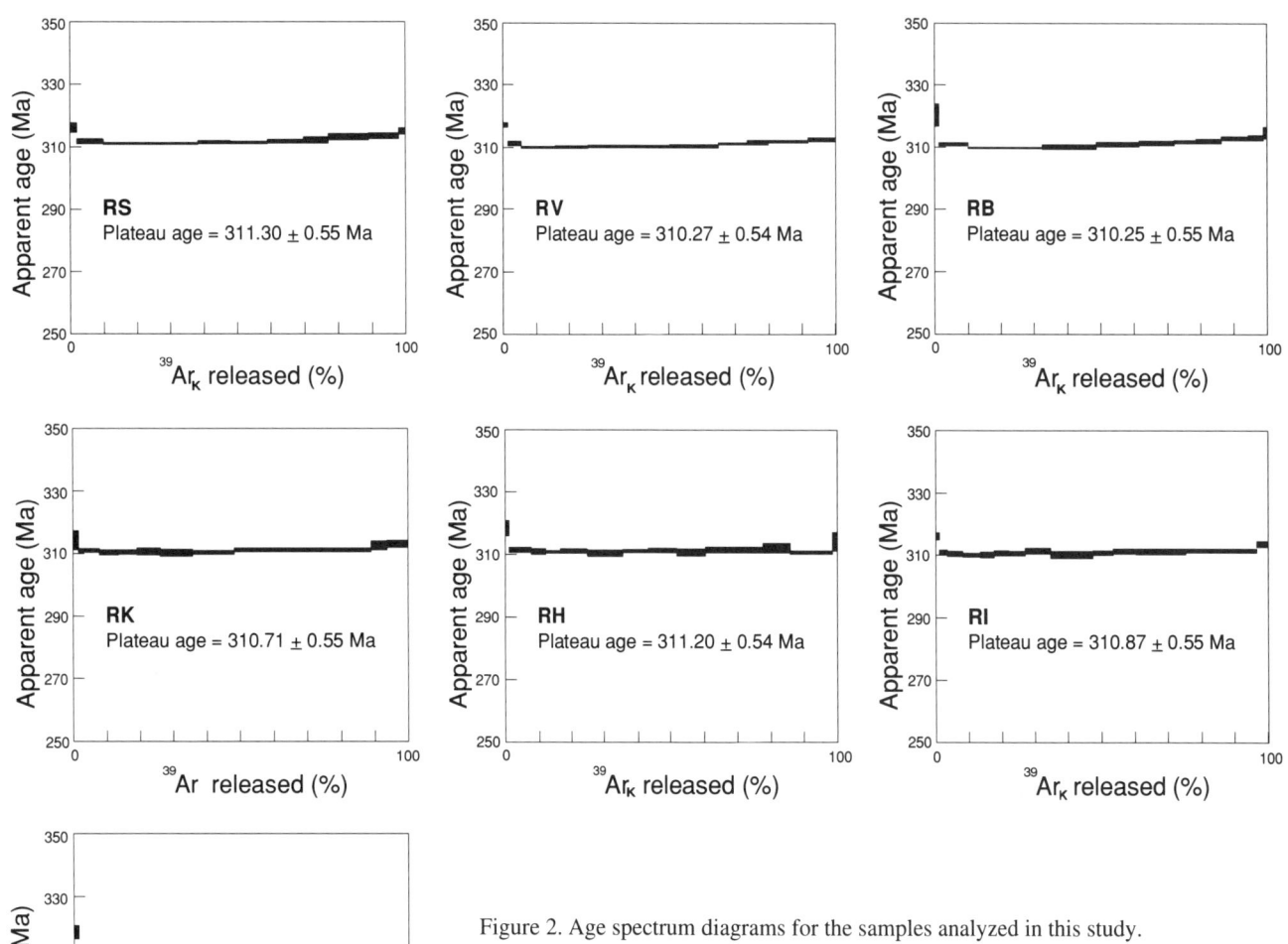

Figure 2. Age spectrum diagrams for the samples analyzed in this study.

TABLE 3. COMPARISON OF TOTAL GAS, MINIMUM, AND PLATEAU AGES OF SANIDINE FROM ALL SAMPLES*

Sample	Total Gas Age† (Ma)	Minimum Age§ (Ma)	Plateau Age** (Ma)
RS	312.0	311.03 ± 0.58	311.30 ± 0.55
RV	311.0	309.96 ± 0.59	310.27 ± 0.54
RB	311.2	309.88 ± 0.56	310.25 ± 0.55
RK	311.2	310.46 ± 0.67	310.71 ± 0.55
RH	311.4	310.58 ± 0.73	311.20 ± 0.55
RI	311.2	310.24 ± 0.64	310.87 ± 0.55
RFG	311.5	310.71 ± 0.65	311.38 ± 0.55
Mean ages**	311.4 ± 0.3	310.4 ± 0.4	310.9 ± 0.5

*Minimum ages are those increments from each age spectrum with the lowest apparent age. For purposes of comparison, the total gas, minimum, and plateau mean ages do not include the error in J-value because it is a constant for all three mean ages.
†No uncertainty is calculated for individual total gas ages.
§Stated uncertainty includes error in J-value.
**Stated uncertainty is the standard deviation of the mean of all seven samples, and does not include the error in J-value as it is a constant for this suite of samples.

McIntosh, 1989). We do not believe that this slight disturbance in the age spectra affects the accuracy of our data because the plateau ages of all the samples agree with each other, within the limits of analytical precision. In addition, a comparison of the mean plateau, total gas, and minimum ages (Table 3) strongly suggests that any disturbance in the age spectra is quite minor because none of these values can be statistically distinguished within the limits of the stated precision.

The agreement of the ^{40}Ar/^{39}Ar plateau ages among the seven sanidine samples, within the limits of analytical precision, is consistent with the conclusion of Belkin and Rice (1989) and Rice, Belkin et al. (this volume) that all of our samples are probably the product of one eruptive event. The best estimate of the age of that event, the eruption and deposition of the Fire Clay tonstein, is the mean of these seven plateau ages, 310.9 ± 0.8 Ma.

EFFECT ON NORTH AMERICAN AND EUROPEAN TIME SCALES

The time scale for the Carboniferous of North America has been based largely on indirect calculations, of processes such as rates of evolution and sedimentation, because of a lack of datable samples in stratigraphic sections with good biostratigraphic control. In contrast, the Carboniferous of Europe is better calibrated numerically because there are many more datable horizons with good biostratigraphic control. However, application of the numerical constraints from the Carboniferous of Europe to stratigraphic sections in the Carboniferous of North America has been difficult because of problems in correlating floral and faunal assemblages between the two areas.

The mean ^{40}Ar/^{39}Ar plateau age of the Fire Clay tonstein provides a precise radiometric age with which to improve the age calibration of this part of the Paleozoic in North America. When used together with the biostratigraphic correlation of Rice, Belkin et al. (this volume), our ^{40}Ar/^{39}Ar age data allow for a critical comparison with equivalent data from European sections.

Figure 3 shows the effect of the Fire Clay tonstein ^{40}Ar/^{39}Ar age spectrum data on the positioning of age boundaries in several published time scales for North America and Europe. The U.S. Geological Survey time scale for North America has no numerical estimates in this part of the scale. The Pennsylvanian is estimated to occupy the interval from 330 to 290–303 Ma because no reliable data from units within the Pennsylvanian of North America have previously been available. The Fire Clay tonstein ^{40}Ar/^{39}Ar age spectrum data thus provide the first high-precision age from within the Pennsylvanian in North America.

The time scale of Harland et al. (1990) has suggested an age of 311 Ma for the Westphalian A-B boundary and 303 Ma for the top of Westphalian D. By interpolation from that time scale, the age of the Westphalian B-C boundary can be estimated at about 307 to 308 Ma. Our data from the Fire Clay tonstein suggest that the age of the Westphalian B-C boundary should be somewhat older than their chart implies to accommodate the 310.9-Ma age from rocks that are thought to be biostratigraphically equivalent to the Westphalian B-C boundary. In addition, the age suggested by Harland et al. (1990) for the Westphalian A-B boundary is somewhat younger than our data would allow and should be revised downward.

Finally, a comparison with the work of Hess and Lippolt (1986) suggests excellent agreement with their estimate of the age of the Westphalian B-C boundary. Our age of 310.9 ± 0.8 Ma for what we believe to be the North American biostratigraphic equivalent of this boundary is virtually identical to their estimate of 311 ± 4 Ma.

It is obvious from our data that, if sanidine can be dated from tonsteins at other levels with similar biostratigraphic control within the Pennsylvanian, significant improvements in the numerical resolution of this part of the North American time scale are possible, and correlation with European sections can be improved.

CONCLUSIONS

The seven ^{40}Ar/^{39}Ar plateau ages presented in this study agree with each other within the limits of analytical precision. The mean of these results, 310.9 ± 0.8 Ma, represents a precise numerical estimate for the age of the Fire Clay tonstein. This age also represents a precise numerical estimate for the age of the upper part of the *Diadoloceras nermeiri–Bisatoceras micromphalus* Goniatite Zone of the Ozarks. The upper boundary of this goniatite zone has been placed at the top of the Trace Creek Shale Member of the Atoka Formation (usage of Zachry and Sutherland, 1984), and correlated by Rice, Belkin

U.S. Geological Survey Geologic Names Committee (1983)			Reconstructed from Harland et al. (1990)				Hess and Lippolt (1986), modified				This report			
System	Series	Age (Ma)	System	Series (U.S.A.)	W. Europe, succession and age (Ma)		W. Europe, stage, substage, and age (Ma)				System	Midcontinent, U.S.A.	Appalachian basin, U.S.A.	Age (Ma)
Permian		290 (290–303)*	Permian		Stephanian	290	Autunian			295	Permian			
								? C		300				
Pennsylvanian	Upper		Pennsylvanian	Virgilian			Stephanian	B		303	Pennsylvanian		Breathitt Formation	
				Missourian		303		A		306				
	Middle			Desmoinesian	Westphalian	D		D		309		Atoka Formation	Fire Clay tonstein	310.9
				Atokan		C B	311	Westphalian	C	311		Trace Creek Shale Member		
				Morrowan		A		? B		315		Morrow Formation		
	Lower	~330			Namurian			A		319				
						323		B+C						
Mississippian			Mississippian			333	Namurian	A		326	Mississippian			
				Visean	Dinantian									

*Range reflects uncertainty of isotopic and biostratigraphic age assignments.

Figure 3. Comparison of data presented in this study with published time scales.

et al. (this volume) to the upper part of Westphalian B in western Europe. Our mean ^{40}Ar/^{39}Ar plateau age of 310.9 ± 0.8 Ma is consistent with a numerical estimate of the age of the Westphalian B-C boundary of 311 ± 4 Ma by Hess and Lippolt (1986).

No difference in age was found between samples collected from the Cumberland overthrust sheet (RI and RFG) and those collected from the autochthonous portions of the Appalachian basin. This uniformity of age, together with the primary igneous phenocryst mineralogy and melt inclusion data of Belkin and Rice (1989) and Rice, Belkin et al. (this volume), strongly suggests that the tonstein collected from Glenn's (1925) Walnut Mountain coal is the Fire Clay tonstein, and that the Fire Clay tonstein can be reliably traced onto the Cumberland overthrust sheet.

Finally, our age of 310.9 ± 0.8 Ma for the Fire Clay tonstein can be used to place an older numerical limit on the timing of emplacement of the Cumberland Overthrust sheet and movement on the Pine Mountain thrust fault. Because the tonstein is cut by the Pine Mountain thrust fault, the emplacement age of the Cumberland overthrust sheet is required to be younger than the age of the Fire Clay tonstein, or less than 310.9 ± 0.8 Ma.

ACKNOWLEDGMENTS

We thank M. A. Lanphere, L. W. Snee, and R. A. Zartman, of the U.S. Geological Survey, and C. Swisher of the Institute of Human Origins at the University of California at Berkeley for thoughtful and constructive reviews of earlier versions of this manuscript.

REFERENCES CITED

Alexander, E. C., Jr., Michelson, G. M., and Lanphere, M. A., 1978, MMhb-1—A new ^{40}Ar/^{39}Ar dating standard, in Zartman, R. E., ed., Short papers of International Conference, Geochronology, Cosmochronology, Isotope Geology: U.S. Geological Survey Open-File Report 78-701, p. 6–8.

Ashley, G. H., 1928, Bituminous coal fields of Pennsylvania, pt. I: Pennsylvania Topographic and Geological Survey Bulletin M-6, ser. 4, 241 p.

Belkin, H. E., and Rice, C. L., 1989, A rhyolite ash origin for the Hazard No. 4 Flint Clay (Appalachian Basin)—Evidence from silicate melt inclusions [abs.]: Geological Society of America Abstracts with Programs, v. 21, p. A360.

Burger, K., and Damberger, H. H., 1985, Tonsteins in the coal fields of Western Europe and North America, in Cross, A. T., ed., Economic geology; Coal, oil, and gas: Congrès International de Stratigraphie et de Géologie du Carbonifère, 9th, Champaign-Urbana, Illinois, 1979, Compte Rendu, v. 4, p. 433–448.

Chesnut, D. R., 1985, Source of the volcanic ash deposit (flint clay) in a coal of the Appalachian basin: Congrès International de Stratigraphie et de Géologie du Carbonifère, 10th, Madrid, 1983: Compte Rendu, v. 1, p. 145–154.

Dalrymple, G. B., and Lanphere, M. A., 1969, Potassium-argon dating Principles, techniques and applications to geochronology: San Francisco, W. H. Freeman, 258 p.

Dalrymple, G. B., and Lanphere, M. A., 1974, Age spectra of some undisturbed terrestrial samples: Geochimica et Cosmochimica Acta, v. 38, p. 715–738.

Dalrymple, G. B., Alexander, E. C., Jr., Lanphere, M. A., and Kraker, G. P., 1981, Irradiation of samples for ^{40}Ar/^{39}Ar dating using the Geological Survey TRIGA reactor: U.S. Geological Survey Professional Paper 1176, 55 p.

Englund, K. J., 1968, Geology and coal resources of the Elk Valley area, Tennessee and Kentucky: U.S. Geological Survey Professional Paper 572, 59 p.

Fleck, R. J., Sutter, J. F., and Elliot, D. H., 1977, Interpretation of discordant ^{40}Ar/^{39}Ar age spectra of Mesozoic tholeiites from Antarctica: Geochimica et Cosmochimica Acta, v. 41, p. 15–32.

Geologic Names Committee, 1983, Geologic time chart: U.S Geological Survey Bulletin 1573-8, p. 1–4.

Glenn, L. C., 1925, The northern Tennessee coal field: Tennessee Division of Geology Bulletin 33-B, 478 p.

Harland, W. B., Armstrong, R. L., Cox, A. V., Craig, L. E., Smith, A. G., and Smith, D. G., 1990, A geologic time scale, 1989: Cambridge, England, Cambridge University Press, 263 p.

Haugerud, R. A. and Kunk, M. J., 1988, ArAr* A computer program for reduction of ^{40}Ar-^{39}Ar data: U.S. Geological Survey Open-File Report 88-261, 67 p.

Hess, J. C., and Lippolt, H. J., 1986, ^{40}Ar/^{39}Ar ages of tonstein and tuff sanidines—New calibration points for the improvement of the Upper Carboniferous time scale: Chemical Geology, v. 59, p. 143–154.

Hess, J. C., Lippolt, H. J., and Burger, K., 1988, New time-scale calibration points in the Upper Carboniferous from Kentucky, Donetz Basin, Poland, and West Germany [abs.]: Besançon, France, 6th International Conference on Fission-Track Dating.

Huddle, J.W., Lyons, E. J., Smith, H. L., and Ferm, J. C., 1963, Coal reserves of eastern Kentucky: U.S. Geological Survey Bulletin 1120, 247 p.

Kunk, M. J., Sutter, J. F., and Naeser, C. W., 1985, High-precision ^{40}Ar/^{39}Ar ages of sanidine, biotite, hornblende, and plagioclase from the Fish Canyon Tuff, San Juan volcanic field, south-central Colorado: Geological Society of America Abstracts with Programs, v. 17, p. 636.

McDougall, I., and Harrison, T. M., 1988, Geochronology and thermochronology by the ^{40}Ar/^{39}Ar method: New York, Oxford University Press, 212 p.

McIntosh, W. C., 1989, Ages and distribution of ignimbrites in the Mogollon-Datil volcanic field, southwest New Mexico—A stratigraphic framework using ^{40}Ar/^{39}Ar dating and paleomagnetism [Ph.D. thesis]: Socorro, New Mexico Institute of Mining and Technology, 291 p.

Rice, C. L., 1984, Stratigraphic framework and nomenclatural problems in the Pennsylvanian of the Cumberland overthrust sheet, Kentucky and Tennessee: Geological Society of America Bulletin, v. 95, p. 1475–1481.

Robl, T. L., and Bland, A. E., 1977, The distribution of aluminum in shales associated with the major economic coal seams of eastern Kentucky, in Rose, T. G., ed., Proceedings, Kentucky Coal Refuse Disposal and Utilization Seminar, 3rd, May 11–12, 1977, Lexington: Lexington, University of Kentucky, p. 97–101.

Samson, S. C., and Alexander, E. C., 1987, Calibration of the interlaboratory ^{40}Ar-^{39}Ar dating standard, MMhb-1: Chemical Geology, v. 66, p. 27–34.

Secor, D. T., Snoke, A. W., and Dallmeyer, R. D., 1986, Character of the Alleghanian orogeny in the southern Appalachians—Part III, Regional tectonic relations: Geological Society of America Bulletin, v. 97, p. 1345–1353.

Seiders, V. M., 1965, Volcanic origin of flint clay in the Fire Clay coal bed, Breathitt Formation, eastern Kentucky: U.S. Geological Survey Professional Paper 525-D, p. D52–D54.

Staudacher, Th., Jessberger, E. K., Dorflinger, D., and Kiko, J., 1978, A refined ultrahigh-vacuum furnace for rare gas analysis: Journal of Physical Earth: Science Instrumentation, v. 11, p. 781–784.

Steiger, R. H. and Jäeger, E., 1977, Subcommission on Geochronology—Convention on the use of decay constants in geo- and cosmochronology: Earth and Planetary Science Letters, v. 36, p. 359–362.

Wanless, H.R., 1946, Pennsylvanian geology of a part of the southern Appalachian coal field: Geological Society of America Memoir 13, 162 p.

Zachry, D. L., and Sutherland, P. K., 1984, Stratigraphy and depositional framework of the Atoka Formation (Pennsylvanian), Arkoma basin of Arkansas and Oklahoma, in Sutherland, P. K., and Manger, W. L., eds., The Atokan Series (Pennsylvanian) and its boundaries—A symposium: Oklahoma Geological Survey Bulletin 136, p. 9–17.

MANUSCRIPT ACCEPTED BY THE SOCIETY FEBRUARY 1, 1994

Glossary of Pennsylvanian stratigraphic names, central Appalachian basin

Charles L. Rice
U.S. Geological Survey, National Center, MS 926, Reston, Virginia 22092
John K. Hiett
Kentucky Department of Mines and Minerals, P.O. Box 14080, Lexington, Kentucky 40512
Elizabeth D. Koozmin
U.S. Geological Survey, National Center, MS 902, Reston, Virginia 22092

INTRODUCTION

The number of stratigraphic names and nomenclatural treatments of Pennsylvanian rock units in the central Appalachian basin is bewildering, even for those generally familiar with the stratigraphy of the region. Long before the adoption or general recognition of a uniform code of stratigraphic nomenclature (1933 and subsequent years; see North American Commission on Stratigraphic Nomenclature, 1983), most of the thicker beds of coal, limestone, shale, sandstone, and other strata had been given several names, each with a specific local application. The vertical recurrence of identical sedimentary facies and sequences of facies (or of cyclic or rhythmic sequences) in coal-bearing deposits made identification of individual units difficult across broad areas of the basin. Thus, over time, much of the Pennsylvanian stratigraphic framework for the Appalachian basin was constructed largely of sequences of informal units (mainly coal beds) whose positions in the rock column were not well determined or adequately described. The result was the use of the same name for different coal beds in different parts of the basin (in some cases for commercial reasons). Miscorrelation of key Pennsylvanian rock units is widespread, even in mapped areas of the Appalachian basin or areas thought to be stratigraphically well known. There have been, of course, many correlation charts drawn up by coal company geologists and engineers (mostly unpublished) and geologists from government (such as Rice and Smith, 1980) and academia (such as Wanless, 1939, 1946, 1975) that relate Pennsylvanian stratigraphic sections from one area to another. But many correlation charts, even in comprehensive geologic reports, do little more than relate lists of names.

Some states, such as Pennsylvania, Ohio, and West Virginia, attempted to establish standard sections for the Pennsylvanian System, so that each stratigraphic position was represented by a single unit name. Unfortunately, those attempts have commonly led to long-standing regional miscorrelations. But the names of many stratigraphic units, particularly coal-bed and sandstone units, have arisen from local usages, particularly by coal companies. The names have been taken from geographic locations within relatively small mining districts or within coherent drainage basins where transportation routes tended to produce a uniform nomenclature. Harold R. Wanless, professor of geology at the University of Illinois, was the first to recognize that the complexity of the Pennsylvanian stratigraphic section required a careful description of each name and its use. Because of the large number of names, he found it useful to list them systematically in glossaries. His purpose (Wanless, 1946, p. 135) was to establish "for each name the area in which the name was used, the source of the name, the larger stratigraphic unit of which it is a part, and its correlation with other named beds in other areas." As such, his glossaries (Wanless, 1939, 1946), which encompass much of the Pennsylvanian of the Eastern Interior and Appalachian basins, are among the best and most convenient sources of Pennsylvanian stratigraphic information.

Our goal is to provide a new, more comprehensive glossary for the central Appalachian basin (see Fig. 1 of the Introduction for this volume), which combines appropriate parts of Wanless's older glossaries and expands the list of Pennsylvanian nomenclature to include names from areas in the basin not covered by his two publications. In some cases, we have been able to improve Wanless's correlation of stratigraphic units because of new data developed by detailed geologic mapping in the coal fields of the Appalachian basin. An important source used in the compilation of our glossary was the joint SEAMS database of the University of Kentucky's Center for Applied Energy Research and the Kentucky Geological Survey. The SEAMS database contains all coal-bed names of both the eastern and western Kentucky coal fields as well as those names used in bordering areas of adjacent states. Resource studies require consistent usage of coal-bed names, and various government agencies need reliable coal-bed data for planning and other purposes. The SEAMS coal-bed database has been under development for many years and reflects the nomenclature used in publications arising from the joint U.S. Geological Survey/Kentucky Geological Survey mapping project (1960-1978) by which the entire state of Kentucky was mapped at a scale of 1:24,000. The database includes references to all coal-bed names used in the coal industry in Kentucky, such as those used on coal company mine maps. The database is designed to interface with other databases, such as the coal resource and coal quality databases for Kentucky; cross-checks between databases make possible the detection of errors in correlation. However, many names that appear in the SEAMS database have been omitted here because they represent individual benches or splits of coal beds that are otherwise described. Also included in the database but not in this glossary are references to the local names of some clay-bed deposits. The abbreviated output of the SEAMS database for eastern Kentucky and our edited versions of Wanless's two glossaries were merged to provide the main part of

---

Rice, C. L., Hiett, J. K., and Koozmin, E. D., 1994, Glossary of Pennsylvanian stratigraphic names, central Appalachian basin, *in* Rice, C. L., ed., Elements of Pennsylvanian Stratigraphy, Central Appalachian Basin: Boulder, Colorado, Geological Society of America Special Paper 294.

our glossary. A search of published geologic literature for the states of Ohio, Pennsylvania, Maryland, West Virginia, Virginia, Kentucky, and Tennessee was made in order to update the glossary and, where possible, to augment the description of units.

In this expanded glossary, we have listed both formal and informal lithic units. Only the more important of the informal lithic units are listed, including regionally extensive units such as coal beds, marine shales and limestones that contain marine fossils, and some sandstone beds; those units that have been shown to be discontinuous and easily miscorrelated are, in general, not included. However, many freshwater or nonmarine limestones of the Upper Pennsylvanian are included because they appear to be both continuous and, at least locally, of commercial value. For informal units, the rank term is not capitalized (as in North Fork shale or Skelley limestone), and the glossary entry designates them as "unranked." In many cases, no fieldwork has been done to determine whether these units could be considered properly defined formal units.

Formal rock units contain both formal members or beds and unranked informal rock units whose status is either uncertain or unstudied. The formal nomenclature for the Pennsylvanian of the central Appalachian basin varies from state to state as a result of different approaches by state geologic surveys and the U.S. Geological Survey. For example, in Ohio, Pennsylvanian rock units of formation rank as used in Pennsylvania have been raised to group rank. Thus, in Ohio, equivalents of formal members of the Allegheny Formation in Pennsylvania, such as the Vanport Limestone Member, are considered to be informal members of the Allegheny Group; in some reports, such as that by Sturgeon et al. (1958), the Vanport limestone was an informal member of one of 12 cyclothems into which the Allegheny Group was divided. According to the North American Stratigraphic Code (North American Commission on Stratigraphic Nomenclature, 1983), the cyclothem is not a formal stratigraphic unit. Another example is the various nomenclatural treatments of the basal quartzarenite of the Pennsylvanian System in Pennsylvania, Ohio, Maryland, and northern West Virginia. This unit has been called informally the Sharon sandstone or Sharon conglomerate (or conglomeratic sandstone) or identified formally as the Sharon Member or Sharon Sandstone (Conglomerate) Member of the Pottsville Formation or as a formation within the Pottsville Group and even as the Sharon Group. It is beyond the scope of the glossary to include or consider all of these usages. To avoid redefining a stratigraphic unit in the glossary, we commonly follow as closely as possible the treatment in our source; therefore, the stratigraphic ranks of units do differ among the unit descriptions. Our citation of what appears to be formal usage, such as Upper and Lower Mercer Limestone Members—two uses for the same name, clearly prohibited by the Code—is not intended as a statement of approval.

Coal beds commonly split into several distinct beds or "benches," requiring modification of the parent coal-bed name by such terms as Lower, Upper, and Little (meaning lower) or combinations of terms. The term "marker," as in Taggart Marker coal, indicates a coal (or a "leader" coal) below the main Taggart coal bed. The addition of "rider" or the capital letter A to a coal-bed name designates a coal bed above the main coal. In some cases, there is a clear genetic relation between the coal beds of such a coal "zone," but that relation should not be taken for granted. For example, the Stockton coal bed directly underlies the Kanawha black flint (a marine unit), and the Stockton "A" coal bed overlies the flint. The two coal beds appear to be separated by a major transgressive event and probably are not related. To keep the glossary to a convenient size, descriptions of many coal-bed names using the combining terms listed above are omitted because their position with respect to the main coal bed is readily apparent to the reader.

The glossary lists only a few of the cyclothems described in the literature, because they commonly require an unnecessary duplication of names. Many cyclothems take their names from the coal bed or the limestone that they include. Commonly, the cyclothem, its coal bed, and its limestone bed all have the same geographic name. Thus, the basic data of origin, age, and stratigraphic position of the cyclothem can be found listed under the geographic name.

Many of our references are not original sources but rather are general sources of stratigraphic information that contain references to original usages. We reference Wanless (1939, 1946) extensively because we use his wording in the stratigraphic descriptions. Commonly, we have shortened and slightly modified Wanless's descriptions and those of other workers without notation. Wanless's publications, particularly Geological Society of America Memoir 13 (1946), provide many references to stratigraphic columns and other source material that we do not include because of space limitations. Where we have made a minor but important modification to a description by an author cited in the glossary, the modification is added after the citation. Where the following modification is of major significance, we cite its source separately; in some cases, where two sources provide a more comprehensive description, we cite both together. The citation of a 7.5-min topographic quadrangle map indicates that it is the best source for locating the geographic names that have been used for many of the more recently named stratigraphic units. In some cases, the stratigraphic columns of detailed 1:24,000-scale (7.5-min) geologic maps also provide the best description of the stratigraphic position of the subject units.

GLOSSARY

"A" coal (GA). Name applied locally to a coal 60 ft (18 m) above No. 5 Durham coal at Durham in Walden Sandstone, Lookout Mountain, Walker County, Georgia; in Tennessee, was tentatively placed in Eastland Shale (Wanless, 1946); coal in strata now assigned to Whitwell Shale of Crab Orchard Mountains Group in Tennessee (Wilson et al., 1956).

"A" coal (KY). Name applied locally to a coal in Harlan County, Kentucky; in Breathitt Formation; probably equivalent to Imboden coal (Wanless, 1939).

"A" seam (KY). See Tiptop coal.

Addington coal (VA). Coal locally mined on Rocky Fork of Guest River near Addington, Wise County, Virginia; in lower part of Wise Formation between Rocky Fork (above) and Clintwood coals (Wanless, 1946); underlies Betsie Shale Member of Wise Formation.

Adele coal (KY). Coal named for small mine on State Road Fork near Adele, Morgan County, Kentucky (Englund, 1955); approximately 80 ft (24 m) above Magoffin Member of Breathitt Formation, in lower part of Hazard or Prater coal zone.

Aily coal (VA). Name applied to coal below McClure Sandstone Member of Norton Formation cropping out on Lick Creek of Russell Fork of Big Sandy River near Aily, Dickenson County, Virginia; between Kennedy (above) and Raven coals (Wanless, 1939; Diffenbach, 1989); locally known as Pebble seam in New River Formation in southern West Virginia.

Albright limestone (WV, PA, MD, OH). Unranked nonmarine limestone exposed at Albright, Preston County, West Virginia; in middle part of Glenshaw Formation of Conemaugh Group; probably equivalent to Bloomfield limestone of southeastern Ohio (Sturgeon et al., 1958); underlies Lower Bakerstown coal in Pennsylvania and incorrectly placed below Upper Bakerstown coal in Maryland by Swartz (1922) (Flint, 1965).

Allegheny Formation (Group) (PA, MD, OH, WV). Formation exposed in Allegheny River valley, western Pennsylvania; includes strata from top of Upper Freeport coal (above) to top of Homewood Sandstone Member of Pottsville Formation or base of Brookville coal in Pennsylvania; formerly called Lower Productive Measures (Wanless, 1939).

Alma coal (WV, KY). Coal named for locality on Tug Fork of Big Sandy River near Sprigg, Mingo County, southern West Virginia; in Kanawha Formation (Wanless, 1939); between No. 2 Gas (above) and Cannelton limestone of White (1885), equivalent to Powellton coal zone in Kanawha Valley, West Virginia; in Breathitt Formation, equivalent to Upper Elkhorn No. 1 and No. 2 coals, Pike County, Kentucky (Blake et al., this volume.; see also Rice et al., 1977).

Alma "A" coal (WV.). Thin coal 10 to 15 ft (3 to 4.6 m) above Alma coal, Mingo County, West Virginia; in Kanawha Formation (Wanless, 1939).

Almy coal (TN). Coal mined at Almy, Scott County, Tennessee; previously in Briceville Formation; correlated with Coal Creek coal but considered equivalent to Poplar Creek coal, which is older than Coal Creek (Wanless, 1946); coal now designated as top of Crooked Fork Group (Wilson et al., 1956, as revised by Hardeman et al., 1966).

Alvy Creek unit (KY) Informal subdivision of Breathitt Formation in eastern Kentucky extending from base of unnamed marine zone overlying improperly termed "Bee Rock Sandstone of Lee Formation" (above) to base of unnamed marine zone overlying improperly termed "Sewanee Sandstone Member of Lee Formation" (Chesnut, 1989); unit unranked and not otherwise described or defined; Bee Rock and Sewanee are incorrectly projected across Kentucky into West Virginia.

Amburgy coal (zone) (KY). Coal mined on Little Carr Fork (formerly Amburgy Branch) of Carr Fork of North Fork, Kentucky River, Knott County, Kentucky; underlies Kendrick Shale Member of Breathitt Formation between Whitesburg (above) and Elkhorn coals; equivalent to Williamson coal in West Virginia and Kentucky (Wanless, 1939); commonly in two beds.

Ames limestone (OH, PA, WV, KY). Widespread unranked marine limestone exposed in Ames Township, Athens County, Ohio; near middle of Conemaugh Formation between Skelley (above) and Portersville limestones (Wanless, 1939); formal Ames Limestone Member of Conemaugh Formation in states other than Ohio.

Anderson coal (OH). Coal formerly mined by Mr. Anderson near Love City, Guernsey County, Ohio; in lower part of Conemaugh Formation between Portersville (above) and Cambridge limestones (Wanless, 1939); equivalent to Bakerstown coal in West Virginia.

Anderson Formation (TN). *Name herein abandoned.* Formation was uppermost Pennsylvanian division in Tennessee, capping high ridges in Anderson County and adjacent counties; previously overlaid Scott Formation and included all strata from Pilot Knob sandstone (now called Pilot Mountain Sandstone Member) to top of section (Wanless, 1946); strata now assigned to upper part of Vowell Mountain Formation and overlying Cross Mountain Formation (Wilson et al., 1956, as revised by Hardeman et al., 1966).

Angel coal (TN). Coal locally mined in Cumberland Plateau, Cumberland and Bledsoe Counties, Tennessee (source of name unknown); in upper part of Gizzard Group below Sewanee Conglomerate; erroneously correlated with Battle Creek coal (Wanless, 1946); now assigned to Signal Point Shale, just below Sewanee Conglomerate, hence equivalent to Wilder coal (Wilson et al., 1956).

Ant coal (TN). Thin coal 60 ft (18 m) above Poplar Creek coal near Coalfield, Morgan County, Tennessee (source of name unknown); previously in Briceville Formation; correlated with "rider" coal above Williamsburg or Lily coals in Kentucky (Wanless, 1946); now assigned to lower part of Slatestone Formation (Wilson et al., 1956, as revised by Hardeman et al., 1966).

Anthony coal (OH, KY). Coal crops out on land of Samuel Anthony, in Coal Township, Jackson County, Ohio; in Pottsville Formation in Ohio immediately above Sciotoville fire clay (Wanless, 1939); in Breathitt Formation in Kentucky directly above Olive Hill flint clay.

Armes Gap Sandstone Member (TN). Sandstone exposed in Armes Gap north of Petros, Morgan County, Tennessee; unit commonly ranges in thickness from 20 to 40 ft (6 to 12 m) and is in middle part of Graves Gap Formation (Wilson et al., 1956, as revised by Hardeman et al., 1966).

Arnett limestone (KY). An unranked thin marine limestone exposed about 35 ft (11 m) above Kentucky Route 28 in gap 1.8 mi (2.9 km) east of village of Arnett on line between Breathitt and Owsley Counties, Kentucky; in Breathitt Formation just above Hazard coal or within Hazard coal zone (see Outerbridge, 1978).

Arnoldsburg cyclothem (OH). Name applied to strata in Ohio that includes Arnoldsburg nonmarine limestone whose type area is near Arnoldsburg, Calhoun County, West Virginia; in Monongahela Group between Lower Uniontown (above) and Benwood cyclothems (Sturgeon et al., 1958).

Artemus coal (KY). Coal exposed near Artemus, Knox County, Kentucky; in Breathitt Formation; a local name for Jellico coal (Rice and Smith, 1980, Table 2).

Auxier coal (KY). Type locality at Federal Mine No. 2 of Elswick Coal Company, at mouth of John Moore Branch of Russell Fork of Big Sandy River, near Elkhorn City, Pike County, Kentucky; in Breathitt Formation between Millard (above) and Elswick coals; probably equivalent to Hagy coal of Virginia (Wanless, 1939).

"B" coal (KY). Name applied locally to a coal in Harlan County, Kentucky; in Breathitt Formation; probably equivalent to Taggart Marker coal (Wanless, 1939).

Bacon Creek coal (KY). Coal mined on Bacon Creek near Corbin, Whitley County, Kentucky; in lower part of Breathitt Formation between Blue Gem (above) and Lily coals; also known as Little Blue Gem coal (Wanless, 1939).

Bakerstown coal (PA, WV). Coal exposed near Bakerstown, Allegheny County, Pennsylvanian; below middle of Conemaugh Formation, a little above Cambridge limestone in Pennsylvania and Pine Creek limestone in West Virginia (Wanless, 1939).

Bald Knob cannel coal (TN). Cannel coal exposed on Bald Knob, Frozen Head Mountain, and other high summits in Morgan and Anderson Counties, Tennessee; previously in Anderson Formation above Wild Cat coal (Wanless, 1946); now assigned to Cross Mountain Formation (Wilson et al., 1956, as revised by Hardeman et al., 1966).

Bald Rock Conglomerate Member (VA). Conglomerate named for Bald Rock, a prominent outcrop 4 mi (6 km) west of Dungannon,

Scott County, Virginia; middle of three prominent cliff-forming conglomerates in Lee Formation (Wanless, 1939; Miller, 1969).

Banner coal (zone) (VA, KY). Coal beds mined near Banner, eastern Wise County, Virginia; in Norton Formation (Virginia) or Breathitt Formation (Kentucky) between Splash Dam (above) and Kennedy coals (Nolde and Diffenbach, 1988).

Barner coal (KY). Coal mined on F. Barner property on Cannon Creek just west of Rocky Face uplift, Bell County, Kentucky; in Breathitt Formation; correlated with Turner coal; correlation uncertain because of folding associated with Rocky Face fault (Wanless, 1946); probably equivalent to Mason coal of Middlesboro area, Bell County, Kentucky (Rice and Ping, 1989).

Barrelville coal (MD). See Lower Freeport coal.

Barren Fork coal (KY). Coal mined on Barren Fork of Cumberland River, near Flat Rock, McCreary County, Kentucky; in Breathitt Formation where it intertongues with Corbin Sandstone Member (above) and Rockcastle Sandstone Member, both of Lee Formation (Wanless, 1939).

Barton coal (PA, MD, OH). Coal mined near Barton in Georges Creek valley, Allegany County, Maryland; in lower part of Casselman Formation above Barton nonmarine limestone (Flint, 1965); probably equivalent to Bakerstown coal; locally called "Four-foot" coal in Maryland (Swartz, 1922).

Battle Creek coal (TN). Coal mined many places on Battle Creek, Cumberland Plateau, Marion County, Tennessee; in Gizzard Group below Warren Point Sandstone; correlated with Etna coal on Raccoon Mountain, Marion County, Tennessee (Wanless, 1946; Wilson et al., 1956).

Bear Creek coal (TN). Coal mined on Bear Creek at Roberta, near Oneida, Scott County, Tennessee; previously in Briceville Formation; correlated with Coal Creek coal by Wanless (1946) but now considered equivalent to Poplar Creek coal; coal in strata now assigned to top of Crooked Fork Group (Wilson et al., 1956, as revised by Hardeman et al., 1966).

Bear Run coal (OH). Thin coal exposed on Bear Run, Bloom Township, Scioto County, Ohio; in Pottsville Formation, normally between Vandusen coal (above) and Massillon sandstone; roof is commonly cannel coal and black-band iron ore (Wanless, 1939).

Bearwallow coal (VA). Coal named for Bearwallow Ridge on county line between Buchanan and Tazewell Counties, Virginia; in Kanawha Formation between Big Fork (above) and Kennedy coals (Englund, 1981).

Bearwallow Sandstone Member (VA). Sandstone named for Bearwallow Ridge on county line between Buchanan and Tazewell Counties, Virginia; as much as 120 ft (37 m) thick in Kanawha Formation between Lower Banner (above) and Big Fork coals (Englund, 1981).

Beattyville coal (KY). Coal mined at Beattyville, Lee County, Kentucky; in Breathitt Formation (Wanless, 1939); between Gray Hawk (above) and Barren Fork coals (Rice and Smith, 1980).

Beaver Creek coal (KY). Coal named for Beaver Creek Mines, McCreary County, Kentucky; in Breathitt Formation below Rockcastle Conglomerate Member of Lee Formation above Hudson coal; now generally called No. 2 Stearns coal (Wanless, 1939, 1946).

Beckley coal (WV). Coal mined extensively at Beckley, Raleigh County, southern West Virginia; in lower part of New River Formation between Lower Raleigh (above) and Quinnimont sandstones; locally called War Creek coal (Wanless, 1939).

Beckley rider coal (WV). Thin coal 3 to 8 ft (1 to 2 m) above Beckley coal, Raleigh County, West Virginia; in New River Formation (Wanless, 1939).

Bedford coal (OH). Cannel coal mined in Bedford Township, Coshocton County, Ohio; in Pottsville Formation between Upper Mercer limestone (above) and Upper Mercer coal (Wanless, 1939).

Bee Rock Conglomerate (Sandstone) Member (VA, KY, TN). Conglomerate named for "Bee Rock" in Big Stone Gap, Wise County, Virginia; top member of Lee Formation in Virginia (Wanless, 1939); unit extended into Kentucky and Tennessee on Cumberland overthrust sheet (Rice and Newell, 1990), even though Englund (1968b) stated that Bee Rock is truncated west of Cumberland Gap by an unconformity at base of Naese Sandstone Member; ranges from 80 to 325 ft (24 to 99 m) in thickness.

Bee Tree shale (KY). Unranked blue-gray shale marked by large calcareous "cartwheel" concretions exposed on Bee Tree Branch of Smoot Creek, Letcher County, Kentucky; in Breathitt Formation between Amburgy (above) and Elkhorn coals (Wanless, 1946).

Beech Grove coal (TN). Coal exposed in mountain between Beech Fork of New River, Campbell County, and Beech Grove (formerly called Better Chance), Anderson County, Tennessee; previously in Scott Formation between Sharp (above) and Big Mary coals; formerly called Block coal; tentatively correlated with Smith 11-foot coal (Harlan County, Kentucky) and Wax coal (Lee County, Virginia) (Wanless, 1946); now assigned to Redoak Mountain Formation (Wilson et al., 1956, as revised by Hardeman et al., 1966).

Bennetts Fork coals (KY, TN). Two coals mined at Winona mine on Bennetts Fork, southern Bell County, Kentucky; in Breathitt Formation correlated with Hance and Matewan coals; coals underlie Betsie Shale Member (Rice and Maughan, 1978; Rice and Ping, 1989).

Bens Creek coal (WV). Coal exposed on Bens Creek, Stafford district, Mingo County, West Virginia; in lower part of Kanawha Formation (Wanless, 1939); in a zone of two coals with Matewan coal at top.

Bentley coal (VA). Coal in lower part of Wise Formation in Lee County, Virginia; probably equivalent to Blair coal (Giles, 1925).

Benwood cyclothem (OH). Name applied to strata in Ohio that include Benwood nonmarine limestone (see below); in Monongahela Group (Sturgeon et al., 1958).

Benwood limestone (PA, OH, WV). Unranked nonmarine limestone exposed near town of Benwood, Marshall County, West Virginia; overlies Sewickley coal and also locally called Sewickley Member of Pittsburgh Formation of Monongahela Group in Pennsylvania (Berryhill et al., 1971).

Betsie Shale Member (KY, WV, VA, TN). Extensive, generally fossiliferous marine shale as much as 140 ft (43 m) thick, commonly marked by thin basal bed having distinctive high gamma-ray signature; exposed above mine in Hance coal about 3,200 ft (975 m) north-northwest of Betsie Gap between Blacklick and Coal Stone Branches of Brownies Creek, Bell County, Kentucky; in Breathitt Formation in

Kentucky and Tennessee (in Cumberland overthrust sheet); in Wise Formation in Virginia; in Kanawha Formation in West Virginia, where it replaces Eagle limestone of White (1891); and in Slatestone Formation in Tennessee in areas other than Cumberland overthrust sheet; unit commonly misidentified as Cannelton limestone of White (1885) in Big Sandy River drainage area of Kentucky and West Virginia; (Rice et al., 1987; Blake et al., this volume).

Bevins coal (KY). Coal mined near store of J. M. Bevins on Bent Branch of Johns Creek, Pike County, Kentucky; in Breathitt Formation above Williamson coal; probably equivalent to Fire Clay coal (Wanless, 1939).

Big Fork coal (VA). Coal crops out in Big Fork Ridge, southeastern Buchanan County, Virginia; in lower part of Norton Formation between Lower Banner (above) and Kennedy coals (Wanless, 1939).

Big Mary coal (TN). Name applied to a coal formerly mined at Old Pioneer, Campbell County, Tennessee (source of name unknown); previously in Scott Formation between Beech Grove (above) and Windrock coals; characterized by marine fossils in roof shale; commonly but erroneously called Dean coal in Tennessee; commonly correlated with Fire Clay rider coal in Kentucky (Wanless, 1946); now assigned to lower part of Redoak Mountain Formation (Wilson et al., 1956, as revised by Hardeman et al., 1966).

Big Red Block ore (OH). Persistent iron ore mined in Scioto County from Jackson County line to Ohio River; in Pottsville Formation between Tionesta coal (above) and Upper Mercer limestone (Stout, 1916).

Big Vein coal (PA). See Pittsburgh coal.

Big Wheel coal (TN, KY). Coal named for Big Wheel Gap on county line between Scott and Campbell Counties, Ketchen 7.5-min quadrangle, Tennessee-Kentucky, above Braden Mountain coal; incorrectly placed in Hignite Formation by Englund (1968b) (see Rice, 1984b); in Breathitt Formation in Kentucky (McDowell et al., 1985); probably equivalent to Pewee coal, which is now assigned to uppermost part of Redoak Mountain Formation in Tennessee (Wilson et al., 1956, as revised by Hardeman et al., 1966).

Billy Goat coal (KY, TN). Name locally applied to coal also known as Yellow Creek or Stray vein (source of name unknown); in Breathitt Formation, Bell County, Kentucky, and previously at base of Jellico Formation, Claiborne County, Tennessee; correlated with Blue Gem and Rich Mountain coals in Kentucky and Tennessee (Wanless, 1946); now also assigned to Breathitt Formation on Cumberland overthrust sheet in Tennessee.

Bingham coal (zone) (KY). Coal mined on E. B. Bingham property, on Ferrell Creek, southern Pike County, Kentucky; in Breathitt Formation between Pond Creek (above) and Millard coals; equivalent to Matewan coal in West Virginia and Clintwood coal in Virginia (Wanless, 1939).

Birmingham shale (PA, OH, WV). Unranked marine or brackish-water shale cropping out at Birmingham, Allegheny County, Pennsylvania; near middle of Conemaugh Formation between Elk Lick coal (above) and Skelley limestone (Wanless, 1939).

Bishop coal (KY). Named for mine on Columbus Bishop's land on Road Run, Clay County, Kentucky (Russell, 1918); in Breathitt Formation; probably equivalent to coal in lower part of Whitesburg coal zone.

Black Band coal (WV). See Hernshaw coal.

Black Wax coal (TN). Thin coal exposed at Jellico and Newcomb, Campbell County, Tennessee (source of name unknown); previously in Briceville Formation between Blue Gem (above) and Dixie coals; correlated with Bacon Creek coal in Kentucky (Wanless, 1946); now assigned to Slatestone Formation (Wilson et al., 1956, as revised by Hardeman et al., 1966).

Blair coal (zone) (VA, KY). Coal formerly mined by Mr. Blair at mouth of Lick Branch of Indian Creek, northern Wise County, Virginia; in lower part of Wise Formation between Clintwood (above) and Lyons coals; in Breathitt Formation in Kentucky; probably equivalent to Bens Creek coal in West Virginia (Wanless, 1939).

Bloomfield limestone (OH). Unranked nonmarine limestone named for the village of Bloomfield, Highland Township, Muskingum County, Ohio; in middle part of Glenshaw Formation of Conemaugh Group; probably equivalent to Albright limestone.

Blue Gem coal (zone) (KY, TN). Trade name now also applied to a small creek, Blue Gem Branch, tributary to Cumberland River, 6 mi (10 km) north of Pineville, Bell County, Kentucky; in lower part of Breathitt Formation between Jellico (above) and Little Blue Gem coals (Wanless, 1939).

Blue Lick coals (PA). Coals mined along Blue Lick valley in Pine Hill area between Berlin and Meyersdale, Somerset County, Pennsylvania; in lower unnamed member of Pittsburgh Formation of Monongahela Group; locally miscorrelated with Redstone coal (Flint, 1965).

Bluff seam (TN). Name locally applied to coal in Marion County, Tennessee, west of Sequatchie Valley; in Gizzard Group below Warren Point Sandstone and above Battle Creek coal (Wanless, 1946; Wilson et al., 1956).

Blunt Run limestone (OH). New name for unranked marine equivalent of the Boggs ore; name is from type section on Blunt Run, Muskingum County, Ohio, according to M. C. Hansen of the Ohio Division of Geological Survey (oral communication, 1992); in Pottsville Group (Formation), see Boggs limestone and ore (below).

Boggs limestone and ore (OH). Unranked iron-ore deposits crop out on Boggs farm near South Webster, Scioto County, southern Ohio; in upper part of Pottsville Formation above Lower Mercer coal; farther north in Ohio, marine limestone and (locally) flint occur at this horizon (Wanless, 1939); Boggs marine limestone renamed Blunt Run limestone for stream of same name that contains type section in Muskingum County, Ohio, according to M. C. Hansen of the Ohio Division of Geological Survey (oral communication, 1992).

Bolivar clay (PA, OH). Fire clay produced near Bolivar, Westmoreland County, Pennsylvania; in upper part of Allegheny Formation beneath a thin coal a few feet below Upper Freeport coal (Wanless, 1939).

Bon Air coals (TN). Two coals mined at Bon Air, White County, Tennessee; in Gizzard Group below Warren Point Sandstone, although formerly placed in Whitwell Formation of overlying Crab Orchard Mountains Group by Wanless (1946) (Wilson et al., 1956).

Borden coal (MD). See Sewickley coal.

Bottom Creek unit (KY). Informal subdivision of Breathitt Formation between base of unnamed marine unit at top of improperly termed "Sewanee Sandstone Member of Lee Formation" (above) to top of improperly termed "Warren Point Sandstone Member of Lee Formation," where present (Chesnut, 1989); unit unranked and not otherwise described or defined; Sewanee and Warren Point are incorrectly used as members of Lee Formation and are incorrectly projected across eastern Kentucky.

Bowling coal (KY). Coal said to have been named by J. M. Hodge, probably after a prospect in Clay County, Kentucky; in Breathitt Formation between Bishop (above) and Moore coals; correlated with Amburgy coal (Wanless, 1946).

Braden Mountain coal (TN, KY). Coal named for Braden Mountain, Campbell and Scott Counties, Tennessee; incorrectly placed in Hignite Formation by Englund (1968b) (see McDowell et al., 1985); between Big Wheel (above) and Red Ash coals; now assigned to upper part of Redoak Mountain Formation in Tennessee (Wilson et al., 1956, as revised by Hardeman et al., 1966); in Breathitt Formation in Kentucky.

Breathitt Formation (KY, TN). Named for Breathitt County, Ky.; upper part of the Pennsylvanian in much of eastern Kentucky; includes all strata between Conemaugh Formation (above) and Lee Formation (Wanless, 1939); formation is as much as 3,120 ft (950 m) thick in eastern Harlan County, Kentucky; intertongues with sandstone members of Lee Formation (Rice et al., 1979); unit generally restricted to Cumberland overthrust sheet in Tennessee; elsewhere in Tennessee, Breathitt equivalent to Crooked Fork Group and Slatestone, Indian Bluff, Graves Gap, Redoak Mountain, Vowell Mountain, and Cross Mountain Formations (Wilson et al., 1956, as revised by Hardeman et al., 1966).

Briceville Formation (TN). *Name herein abandoned.* Exposed in lower slopes of mountains near Briceville, Anderson County, Tennessee; previously between Jellico Formation (above) and Lee Group, including 500 to 1,000 ft (150 to 300 m) of strata, predominantly shale; approximately equivalent to lower part of Breathitt Formation in Kentucky, lower part of Kanawha Formation in West Virginia, and part of Wise Formation in Virginia; questionably equivalent to Gladeville Sandstone and upper part of Norton Formation in Virginia (Wanless, 1946); strata now assigned to lower part of Slatestone Formation (Wilson et al., 1956, as revised by Hardeman et al., 1966).

Broas coal (zone) (KY). Coal formerly mined by R. M. Broas, near head of Middle Fork of Rockcastle Creek, southwestern Martin County, Kentucky; in Breathitt Formation; probably equivalent to Hindman or Hazard No. 9 coal (Wanless, 1939).

Brookville coal (PA, OH). Coal formerly mined at Brookville, Jefferson County, Pennsylvania; near base of Allegheny Formation (Wanless, 1939); as used in Ohio, probably a coal in Clarion coal zone of Pennsylvania (see Rice, Kosanke, and Henry, this volume).

Bruin coal (KY). Coal in roadcut exposure on Kentucky State Highway 7 at Bruin, Elliott County, Kentucky; in Breathitt Formation (Englund and DeLaney, 1966); overlies Grayson sandstone bed.

Brush Creek coal (PA, WV, OH). Thin coal just below Brush Creek Limestone Member (see below); in lower part of Conemaugh Formation (Wanless, 1939).

Brush Creek Limestone Member (PA, WV, OH, KY). Widespread marine limestone commonly in two beds (called Upper and Lower Brush Creek limestones by previous workers) 5 to 20 ft (1.5 to 6 m) apart, exposed on Brush Creek, Cranberry Township, Butler County, Pennsylvania; in lower part of Conemaugh Formation (Wanless, 1939).

Brushy Mountain coal (TN). See State coal.

Bryant Farm coal (KY). Local name for Barren Fork coal in parts of Laurel County, Kentucky; in Breathitt Formation (Stager, 1963).

Bryson Formation (TN, KY). *Name herein abandoned.* Named for Bryson Peak in Log Mountains, Claiborne County, Tennessee; was assigned as highest subdivision of Pennsylvanian in Cumberland Gap field overlying Hignite Formation; included strata above base of Red Springs coal; about 200 ft (61 m) thick in type area (Wanless, 1946; Rice and Maughan, 1978); formation boundaries were coal that is discontinuous and miscorrelated in different areas; strata now assigned to Breathitt Formation in Kentucky (McDowell et al., 1985) and upper part of Redoak Mountain Formation and overlying Vowell Mountain and Cross Mountain Formations in Tennessee (Wilson et al., 1956, as revised by Hardeman et al., 1966).

Buckeye Spring coal (KY). Name used for coal also called Sandstone Parting coal in southwestern Bell County, Kentucky (see Ashley and Glenn, 1906); in Breathitt Formation between Poplar Lick (above) and Mingo coals (Rice and Maughan, 1978).

Buffalo Creek coal (WV, KY). Coal crops out on Buffalo Creek, Lee district, Mingo County, West Virginia; in Kanawha Formation; equivalent to Winifrede coal of Kanawha Valley (in West Virginia) or Hazard coal (in Kentucky) (Wanless, 1939; Blake et al., this volume).

Buffalo Creek limestone (WV). *Name herein abandoned.* Marine limestone and shale in Kanawha Formation, Mingo County, West Virginia, equivalent to Winifrede Shale Member (Blake et al., this volume).

Buffalo Sandstone Member (PA, OH, WV, MD). Massive sandstone cropping out on Buffalo Creek, Buffalo Township, eastern Butler County, Pennsylvania; in lower part of Conemaugh Formation between Cambridge or Meyersdale limestone (above) and Brush Creek Limestone Member (Wanless, 1939).

Bull Creek coal (KY). Coal named for Bull Creek in southern part of Hima 7.5-min quadrangle, Knox County, Kentucky (Reeves, 1964); in Breathitt Formation; equivalent to a coal bed in Whitesburg coal zone.

Bumbee coal (TN). Thin coal mined on Bumbee Creek (tributary of Piney Creek) 7 mi (11 km) southwest of Spring City, Rhea County, Tennessee; erroneously placed in Duskin Creek Formation about 125 ft (38 m) above Rockcastle Sandstone (Nelson, 1925); now assigned to Rockcastle Conglomerate (Wilson et al., 1956).

Burns coal (KY). Coal mined by Dudley Burns on Barn Branch of Bullskin Creek, Clay County, Kentucky; in Breathitt Formation between Lower Howard (above) and Manchester coals; correlated with Bacon Creek coal in Kentucky (Wanless, 1946).

Burnt Mill Shale (TN). Shale exposed near Burnt Mill Bridge in Robbins and Oneida South 7.5-min quadrangles, Scott County, Tennessee; commonly ranges from 35 to 130 ft (11 to 40 m) in thickness and is in middle part of Crooked Fork Group (Wilson et al., 1956).

Burtons Ford coal (VA). Coal crops out at Burtons Ford on Clinch River, Wise County, Virginia; in Lee Formation below Bald Rock

Conglomerate Member; roof shale has flora indicating correlation with Pocahontas Formation (Wanless, 1939).

Butler Sandstone Member (PA, MD). Sandstone exposed near town of Butler, Butler County, Pennsylvania; in Allegheny Formation about 35 ft (11 m) below Upper Freeport coal (Keroher et al., 1966).

"C" coal (KY). Name applied locally to Darby coal in Harlan County, Kentucky; in Breathitt Formation; equivalent to Taggart coal; equivalent to Keokee coal of Wise Formation in Wise County, Virginia (Wanless, 1939).

Cadell coal (KY). Coal exposed and locally mined in Whitley County, Kentucky (source of name unknown); in Breathitt Formation above Jellico coal; correlated with Moss and Joyner coals in Tennessee (Wanless, 1946); equivalent to Upper Elkhorn No. 3 coal in Kentucky (Rice and Smith, 1980, Table 2).

Cambridge limestone (OH, PA, MD). Unranked widespread marine limestone exposed at Cambridge, Guernsey County, Ohio; in lower part of Conemaugh Formation between Portersville limestone (above) and Brush Creek Limestone Member; equivalent to Pine Creek limestone of Pennsylvania and West Virginia (Wanless, 1939).

Campbell Creek coal (WV, VA). Coal exposed at mouth of Campbell Creek, near Malden, Kanawha County, West Virginia; widely mined coal in Kanawha Formation; in many places is multiple-benched and includes upper Peerless coal and lower No. 2 Gas coal; locally miscorrelated with Pond Creek coal in southern West Virginia and Virginia (Wanless, 1939).

Campbell Creek limestone (WV). Named by White (1885) for Campbell Creek, Kanawha County, West Virginia; a slightly fossiliferous calcareous dark shale having limestone concretions and lenses, 30 to 40 ft (9 to 12 m) above Campbell Creek coal; in Kanawha Formation (Wanless, 1939); unit between Williamson (above) and Peerless coals in West Virginia but stratigraphically misplaced in many geologic reports concerning southern West Virginia and adjacent parts of Kentucky and Virginia as marine unit above Eagle, Pond Creek, or Lower Elkhorn coals; probably equivalent to Elkins Fork shale of Morse (1931) in Kentucky (Blake et al., this volume).

Cannel City coal (zone) (KY). Coal named for Cannel City, Morgan County, Kentucky; contains thin flint clay parting (tonstein) (Englund, 1955); underlies Kendrick Shale Member of Breathitt Formation; equivalent to Amburgy coal.

Cannelton limestone (WV). Formerly worked as a cement rock by Mr. Stockton, near Cannelton, Kanawha County, West Virginia; a slightly fossiliferous marine calcareous shale having limestone concretions (White, 1885); near middle of Kanawha Formation between Powellton (above) and Eagle "A" coals (Wanless, 1939); unit miscorrelated with Betsie Shale Member in southwestern part of West Virginia and adjacent parts of Kentucky in many geologic reports; equivalent to Crummies Member of Breathitt Formation in Kentucky, where it occurs above Pond Creek or Lower Elkhorn coals (Blake et al., this volume).

Carter coal (VA). Coal in lower part of Lee Formation (source of name unknown) in basal sandstone (quartz arenite) unit in Scott County, Virginia (Nolde and Diffenbach, 1988).

Caryville Sandstone Member (TN). Sandstone exposed above Caryville, Jacksboro 7.5-min quadrangle, Campbell County, Tennessee; unit as much as 60 ft (18 m) thick and is in middle part of Redoak Mountain Formation (Wilson et al., 1956, as revised by Hardeman et al., 1966).

Casselman Formation (PA, MD, WV). Exposed along Casselman River in Somerset County, Pennsylvania, and in Garrett County, Maryland; a sequence of strata including sandstone, siltstone, shale, limestone, and coal extending from base of Pittsburgh coal (above) to top of Ames Limestone Member of Glenshaw Formation; in upper part of Conemaugh Group; formation about 500 ft (152 m) thick (Flint, 1965).

Cassville Shale Member (WV, PA, OH). Shale named for Cassville, Monongahelia County, West Virginia; formerly at base of Washington Formation above Waynesburg coal; strata now assigned to lower member of Waynesburg Formation in Pennsylvania at or near base of Dunkard Group (Berryhill et al., 1971); contains flora of Zone 12 of Read and Mamay (1964); strata assigned Late Pennsylvanian or Early Permian age (Englund et al., 1979).

Castle coal (WV). Thin coal cropping out near Castle Post Office, Center district, Wyoming County, West Virginia; in upper part of New River Formation (Wanless, 1939); may be correlative with Sewell "B" coal.

Castle Rock coal (TN, GA). See Etna or Dade coals.

Catoosa coal (TN). Coal formerly mined at Catoosa on Emory River, Morgan County, Tennessee; also called Oakdale coal; incorrectly correlated as part of Duskin Creek Formation (Wanless, 1946); now assigned to Vandever Formation (Wilson et al., 1956).

Catron Formation (KY). *Name herein abandoned.* Exposed on Coon Branch of Catron Creek, Harlan County, Kentucky; included strata between top of Jesse Sandstone Member and base of Wallins Creek coal or between Hignite (above) and Mingo Formations (Wanless, 1946); formation boundaries are sandstones and coals that are discontinuous and miscorrelated in different areas; strata now assigned to Breathitt Formation (McDowell et al., 1985).

Cawood sandstone (KY). Unranked sandstone forming massive bluffs along Martins Fork of Cumberland River at Cawood, Harlan County, Kentucky (Ashley and Glenn, 1906); in Breathitt Formation, underlies Path Fork coal.

Cedar coal (WV, KY). Coal formerly mined near Cedar, Magnolia district, Mingo County, West Virginia; in Kanawha Formation, about 40 ft (12 m) above marine shale called Eagle by Krebs and Teets (1916); probably equivalent to Dorchester coal in Virginia or Millard coal in Kentucky (Wanless, 1939).

Cedar Grove coal (zone) (WV, KY). Coals mined at Cedar Grove, Cabin Creek district, Kanawha County, West Virginia; in upper part of Kanawha Formation between Fire Clay coal (above) and Dingess Shale Member; locally miscorrelated with Thacker or Island Creek coal in southern West Virginia; equivalent to Whitesburg coal zone in Breathitt Formation in Kentucky (Blake et al., this volume).

Chadwell Member (VA, KY, TN). Exposed southeast of Chadwell Gap on Cumberland Mountain, Lee County, Virginia; pebbly sandstone in lower part of Lee Formation as much as 350 ft (107 m) thick; previously identified as "sandstone member A" (Englund, 1964a); originally assigned to Upper Mississippian, but preliminary analyses of microspores suggest Chadwell and enclosing strata may be in basal part of Pennsylvanian of that area (C. F. Eble, oral communication, 1993).

Charleston Sandstone (WV). Very massive sandstone near Charleston, Kanawha County, West Virginia, formed where several sandstones join to make a single sandstone bluff, intervening coals and shales being absent; between Conemaugh (above) and Kanawha Formations (Wanless, 1939).

Chenoa coal (zone) (KY). Coal mined near Chenoa, Bell County, Kentucky; in Breathitt Formation between Bennetts Fork (above) and Splitseam coals (Rice and Maughan, 1978).

Chicken Ridge Sandstone Member (VA). Sandstone exposed on Chicken Ridge, northeastern part of Jewell Ridge 7.5-min quadrangle, Buchanan County, Virginia; as much as 50 ft (15 m) thick in Kanawha Formation between Bearwallow (above) and Kennedy coals (Englund, 1981).

Chilton coal (WV). Coal mined at Chilton on Davis Creek, Loudon district, Kanawha County, West Virginia; unimportant coal in upper part of Kanawha Formation between Winifrede Shale Member (above) and Fire Clay coal; commonly miscorrelated with Fire Clay coal in southwestern West Virginia and adjacent parts of Kentucky (Blake et al., this volume).

Chilton "A" coal (WV). Minor coal about halfway between Winifrede (above) and Chilton coals; in upper part of Kanawha Formation a little above Winifrede Shale Member (Wanless, 1939; Blake et al., this volume).

Chilton rider coal (WV). Thin coal 20 to 40 ft (6 to 12 m) above Chilton coal; in Kanawha Formation, occurs at or near base of Winifrede Shale Member (Wanless, 1939).

Christmas coal (TN). Coal formerly mined near Christmas siding in DeArmond Gap of Little Emory River, Morgan County, Tennessee; previously placed at base of Briceville Formation by Glenn (1925) and previously placed in Duskin Creek Formation by Wanless (1946); now assigned to Dorton Shale in Crooked Fork Group (Wilson et al., 1956).

Clarion clay (PA, OH). Underclay of Clarion coal, lower part of Allegheny Formation; refractory clay used in ceramic industries (Wanless, 1939).

Clarion coal (PA, OH, WV, MD). Coal zone in western Pennsylvania in Allegheny Formation; underlies Vanport Limestone Member of that area and probably miscorrelated in other areas such as central or southern Ohio where Vanport is absent (Rice, Kosanke, and Henry, this volume).

Clarion Formation (PA). Name derived from Clarion, Clarion County, Pennsylvania; lower of three formations of Allegheny Group of Somerset County, Pennsylvania; extends from base of underclay below Lower Kittanning coal (above) to base of Brookville underclay; about 130 ft (40 m) thick (Flint, 1965; Keroher et al., 1966).

Clarksburg limestone (WV, OH). Unranked brackish-water limestone exposed along Elk Creek and West Fork River near Clarksburg, West Virginia; in Conemaugh Group between Clarksburg coal or underclay (above) and Morgantown sandstone (Sturgeon et al., 1958); commonly in two beds, Upper and Lower Clarksburg limestones, in Pennsylvania and Maryland (Flint, 1965).

Clarysville coals (PA, MD). Several coals in 35-ft (11-m) interval exposed in Hoffman drainage tunnel at Clarysville, Allegany County, Md. (Swartz, 1922); in middle of Casselman Formation, generally associated with Upper and Lower Clarksburg nonmarine limestones (Flint, 1965).

Clear Fork coal (KY). Coal mined along Clear Fork of Cumberland River at top of Yellow Creek Sandstone Member of Breathitt Formation, Bell County, Kentucky (Rice and Ping, 1989).

Cliff coal (GA). Name applied to coal a little below Warren Point Sandstone of Gizzard Group of Raccoon, Sand, and Lookout Mountains, Georgia; equivalent to Etna coal (Wanless, 1946).

Clifty coals (TN). Coals formerly mined extensively at Clifty and Eastland, White County, Tennessee; previously in lower part of Eastland Formation, Lee Group; tentatively correlated with Durham coals on Lookout Mountain (Wanless, 1946); now assigned to Whitwell Shale of Crab Orchard Mountains Group (Wilson et al., 1956).

Clintwood coal (zone) (VA, KY). Coal mined near Clintwood, Dickenson County, Virginia; in lower part of Wise Formation in Virginia and Breathitt Formation in Kentucky; equivalent to Matewan coal in West Virginia and Kentucky (Wanless, 1939); in many areas, this coal directly underlies Betsie Shale Member (Rice et al., 1987).

Clod coal (KY). See Mudseam coal.

Clover Fork Sandstone Member (VA, KY). Named for Clover Fork of Cumberland River in northeastern corner of Keokee 7.5-min quadrangle, Harlan County, Kentucky; as much as 78 ft (24 m) thick; in Breathitt Formation (Kentucky) or in Wise Formation (Virginia) between Taggart Marker (above) and Wilson coals (Miller, 1969).

Coalburg coal (WV). Coal mined high in Kanawha River bluff above Coalburg, Kanawha County, West Virginia; in Kanawha Formation between Stockton (above) and Little Coalburg coals (Wanless, 1939).

Coalburg "A" coal (WV). Name locally used in southern West Virginia for a thin coal 15 to 20 ft (5 to 6 m) above Coalburg coal; in Kanawha Formation (Wanless, 1939).

Coal Creek coal (TN). Extensively mined near Coal Creek (now called Lake City), Anderson County, Tennessee; originally called Wheeler coal; previously in upper part of Briceville Formation between Blue Gem (above) and Poplar Creek coals; correlated with Kent coal (LaFollette–Walnut Mountain district, Tennessee) (Wanless, 1946); correlated with Swamp Angel coal (Jellico district, Tennessee and Kentucky) by Englund (1968b), but Tennessee Division of Geology correlates Swamp Angel with older Poplar Creek coal; correlated with Clintwood coal of Virginia and Kentucky (Rice et al., 1987); now assigned to Slatestone Formation (Wilson et al., 1956, as revised by Hardeman et al., 1966).

Coalfield Sandstone (TN). Sandstone underlies town of Coalfield, Petros 7.5-min quadrangle, Morgan County, Tennessee; formation ranges from 0 to 80 ft (0 to 24 m) thick and is in middle part of Crooked Fork Group (Wilson et al., 1956; thickness revised by Hardeman et al., 1966).

Coal Gap coal (TN). Coal exposed 220 ft (67 m) above Coal Gap at head of Beech Fork, Anderson County, Tennessee; previously in Anderson Formation between Wild Cat (above) and Grassy Spring coal (Wanless, 1946); now assigned to lower part of Cross Mountain Formation above its Low Gap Sandstone Member (Wilson et al., 1956, as revised by Hardeman et al., 1966).

Coalton coal (KY). See Princess No. 7 coal.

Coke Oven coal (TN). See Richland coal.

Cold Gap coal (TN). See Wild Cat coal.

Collier coal (KY, VA). Coal mined on Collier farm on Right Fork of Franks Creek, Letcher County, Kentucky; in Breathitt Formation between Taggart (above) and Harlan coals or in Wise Formation between Taggart (above) and Wilson coals; may be equivalent to Standiford coal in Virginia (Wanless, 1946).

Colony coal (KY). Coal mined near Upper Colony Church, Laurel County, Kentucky (Hatch, 1963a); in Breathitt Formation; equivalent to Lily coal.

Columbiana Member (OH, PA). Marine shale exposed along Brookwood Hollow in Section 29, Perry Township, Columbiana County, Ohio; in Allegheny Formation between Middle and Lower Kittanning coals; unit replaces Hamden Limestone and is described as approximately 5.5 ft (1.7 m) of fossiliferous black and gray shale having nodular limestone concretions (Sturgeon and DeLong, 1964); extends into Pennsylvania and is probably equivalent to Obryan Member in southern Ohio and northeastern Kentucky (see Rice, Kosanke, and Henry, this volume).

Colvin coal (KY). Named for Colvin Branch (now called Buck Branch) of Licking River, Magoffin County, Kentucky (Adkison, 1957); in Breathitt Formation; equivalent to coal in Haddix coal zone.

Conemaugh Formation or Group (PA, OH, WV, MD, KY). Named for Conemaugh River, western Pennsylvania; one of four major divisions of typical Pennsylvanian; equivalent to older Lower Barren Measures; includes beds from base of Pittsburgh coal (above) to top of Upper Freeport coal (Wanless, 1939); where Ames Limestone Member is present, Conemaugh is locally raised in rank to group and has been divided at top of Ames into Casselman Formation (above) and Glenshaw Formation (Flint, 1965); locally in Ohio, Conemaugh Series (of old usage) was divided into 15 cyclothems, from top to bottom: Upper and Lower Little Pittsburgh, Little Clarksburg, Elk Lick, Duquesne, Gaysport, Ames, Harlem, Upper Bakerstown, Anderson, Wigus, Upper and Lower Brush Creek, Mason, and Mahoning cyclothems; all names derived from a named coal or limestone included in cyclothem (Sturgeon et al., 1958).

Connoquenessing Sandstone Member (PA, MD, OH, WV). Sandstone exposed along bed of Connoquenessing Creek, Lawrence County, Pennsylvania; in Pottsville Formation; where most complete, about 175 ft (53 m) thick and consists of subdivisions incorrectly called Upper and Lower Connoquenessing Sandstones of Pottsville Group, which are separated by Quakertown coal and shales (Dutcher et al., 1959); stratigraphic position of these sandstones in areas other than type area is not clear (see Slucher and Rice, this volume).

Contrary coal (KY). Coal named for exposures in Contrary Creek, northeastern Heidelburg 7.5-min quadrangle, Lee County, Kentucky; in Breathitt Formation; probably equivalent to coal in upper part of Stearns coal zone (Black, 1977).

Copland coal (zone) (KY). Coal beds directly underlying Magoffin Member of Breathitt Formation (Morse, 1931); source of name unknown.

Copper Spur coal (KY). Coal about 135 ft (41 m) above Hignite coal and below Magoffin Member of Breathitt Formation in Log Mountains area, Bell County, Kentucky; source of name unknown; equivalent to Hamlin coal (Rice and Maughan, 1978).

Corbin Sandstone Member (Tongue) (KY). Massive, locally conglomeratic sandstone exposed near Corbin, Whitley County, Kentucky; uppermost member of Lee Formation along western border of eastern Kentucky coal field (Wanless, 1939); sandstone ranges from 0 to 250 ft (76 m) thick; equivalent to Wartburg Sandstone of Tennessee (Wilson et al., 1956).

Corley Hollow coal (KY). Coal mined near western edge of Livingston 7.5-min quadrangle, Rockcastle County, Kentucky; in Breathitt Formation; source of name unknown; equivalent to Halsey Rough coal (Brown and Osolnik, 1974).

Cornett coal (KY, VA). Local name for Morris coal in Breathitt Formation, eastern Harlan County, Kentucky, between High Splint (above) and Limestone coals (Froelich and Stone, 1973); near top of Wise Formation in Wise County, Virginia.

Cornett coal (VA). See Dorchester coal.

Council Sandstone Member (VA). Sandstone exposed in a bluff and roadcut on Virginia State Highway 80 directly beneath Council Elementary School, Council, Buchanan County, Virginia; in lower part of Norton Formation (Miller and Meissner, 1977); locally extended into middle part of New River Formation in Virginia.

Cove Creek coal (VA). Coal exposed on Cove Creek, Scott County, Virginia; in Lee Formation between Stock Creek coal (above) and top of Middlesboro Member of Lee Formation; equivalent to Tacus and Starns coals (Henika, 1988).

Cow Cliff coal (KY). Local name for Haddix coal in Strong Branch, southern Breathitt County, Kentucky, arising from natural shelters, frequented by cattle, developed below an overlying and overhanging resistant sandstone on steep valley sides; in Breathitt Formation.

Crab Orchard Mountains Group (TN). Group exposed along railroad track that cuts through Crab Orchard Mountains, Cumberland County, Tennessee, and includes (in descending order): Rockcastle Conglomerate, Vandever Formation, Newton Sandstone, Whitwell Shale, and Sewanee Conglomerate; overlies Gizzard Group; unit ranges from 200 to 950 ft (61 to 290 m) thick and is about 640 ft (195 m) thick at type locality (Wilson et al., 1956; thickness revised by Hardeman et al., 1966).

Craig coal (TN). Coal named by Mr. Petree from an outcrop on Black Diamond Coal Company property near Beech Fork, Anderson County, Tennessee; previously in Scott Shale between Windrock (above) and Upper Pioneer coals or may be Upper Pioneer coal unusually close to Windrock coal at this locality (Wanless, 1946); now assigned to upper part of Graves Gap Formation (Wilson et al., 1956, as revised by Hardeman et al., 1966).

Cranes Creek coal (KY). Coal mined on Sugar Hollow Branch of Cranes Creek, near Colmar, Bell County, Kentucky; previously in Hance Formation a little below Cawood sandstone (Wanless, 1946); now assigned to Breathitt Formation; coal is about the level of Mason coal (see Ashley and Glenn, 1906; Rice and Ping, 1989).

Creech coal (zone) (KY). Coal prospected by Robert Creech in Jackson Mountain, near Puckett Creek, eastern Bell County, Kentucky; in

Breathitt Formation between Kendrick Shale Member (above) and Kellioka coal; correlated with Low Splint coal in Virginia, Amburgy coal in Kentucky, and Jordan coal in Tennessee (Wanless, 1946).

Crooked Fork Group (TN). Exposed southeast of Wartburg and along Crooked Fork, Morgan County, Tennessee, and includes all strata between top of Poplar Creek coal (above) and top of Rockcastle Conglomerate of Crab Orchard Mountains Group; unit ranges from 320 to 455 ft (98 to 139 m) thick and is 360 ft (110 m) thick at type section (Wilson et al., 1956; thickness revised by Hardeman et al., 1966).

Cross Mountain Formation (TN). Named for Cross Mountain, Anderson and Campbell Counties, Tennessee; uppermost Pennsylvanian formation in Tennessee; includes all strata above top of Frozen Head Sandstone Member of underlying Vowell Mountain Formation; at type section, formation is 554 ft (169 m) thick (Wilson et al., 1956, as revised by Hardeman et al., 1966).

Crossville Sandstone (TN). Exposed in stone quarries between Crab Orchard and Crossville, Cumberland County, Tennessee (Wanless, 1946); formation ranges from 30 to 70 ft (9 to 21 m) thick and is in lower part of Crooked Fork Group (Wilson et al., 1956; thickness revised by Hardeman et al., 1966).

Crummies Member (KY). Exposed on U.S. Highway 421 on Cranks Ridge about 1,500 ft (450 m) east of village of Crummies, Harlan County, Kentucky; fossiliferous marine shale, generally characterized by large limestone concretions, in Breathitt Formation; as much as 140 ft (43 m) thick, commonly only 20 ft (6 m) thick; overlies Path Fork, Lower Elkhorn, or Pond Creek coals and commonly misidentified as Campbell Creek limestone of White (1885) in Kentucky and adjacent parts of West Virginia and Virginia; correlates with Cannelton limestone of White (1885) in Kanawha River area of West Virginia (Rice et al., 1993; Blake et al., this volume).

Cumberland Gap coal (KY, VA). Coal locally found in Dark Ridge Member of Lee Formation in area of Cumberland Gap, in Bell County, Kentucky, and Lee County, Virginia (Englund, 1964b).

"D" coal (KY). Name applied to a coal mined on Looney Ridge and nearby areas north of Benham, Harlan County, Kentucky; in Breathitt Formation between Creech (above) and Darby coal zones (Froelich and Stone, 1973).

Dade coal (GA, TN). Coal formerly mined extensively near Cole City and Castle Rock, Dade County, Georgia; in Raccoon Mountain Formation of Gizzard Group between Etna or Castle Rock (above) and Rattlesnake coals (Wanless, 1946; Wilson et al., 1956).

Danleyton coal (KY). Coal exposed near Danleyton, southern Greenup County, Kentucky; in Breathitt Formation below Lower Stinson coal (Wanless, 1939).

Darby coal (VA, KY). Name locally applied to Taggart, Kellioka, or Keokee bed, near Darbyville in The Pocket, in Lee County, Virginia; in Wise Formation in Virginia and Breathitt Formation in parts of Harlan County, Kentucky, between Low Splint (above) and Taggart Marker coals (Wanless, 1946).

Dark Ridge Member (KY, VA, TN). Exposed on southern end of Dark Ridge on northern side of Cumberland Gap, Bell County, Kentucky; thin-bedded sandstone and shale as much as 300 ft (91 m) thick, in Lee Formation between Middlesboro Member (above) and White Rocks Sandstone Member or Chadwell Member; locally mapped as "sandstone and shale member B" (Englund, 1964a).

Davis coal (WV). See Upper Freeport coal.

Dean coal (KY). Coal crops out on Dean Branch of Greasy Creek, Bell County (north of Pine Mountain fault), Kentucky; in Breathitt Formation; equivalent to Fire Clay coal (Wanless, 1939).

Dingess coal (WV). Coal named for Dingess tunnel, Harvey district, Mingo County, West Virginia; in Kanawha Formation directly above Dingess Shale Member; correlates with lower part of Cedar Grove coal zone of Kanawha County, West Virginia, and lower part of Whitesburg coal zone in Kentucky.

Dingess Shale Member (WV). Exposed south of Dingess and southeast of Williamson, Mingo County, West Virginia; marine shale containing limestone concretions in Kanawha Formation between Cedar Grove (above) and Williamson coals, equivalent to Kendrick Shale Member of Breathitt Formation in Kentucky (Blake et al., this volume).

Dirty-Six coal (VA). Name given a coal between Middle and Lower Seaboard coals, Tazewell County, Virginia; in New River or Lee Formations (Englund, 1981).

Dirty-Six coal (PA). See Morantown coal.

Dismal Sandstone Member (VA, WV). Sandstone exposed on Dismal Creek, Buchanan County, Virginia; locally more than 100 ft (30 m) thick in New River Formation between Aily (above) and Jewell or Raven coals (Englund, 1981).

Dixie coal (TN, KY). Coal locally worked near Newcomb, Campbell County, Tennessee (source of name not known); previously in Briceville Formation between Black Wax (above) and Swamp Angel coals (Wanless, 1946); previously in Hance Formation (Englund, 1968b); considered equivalent to Coal Creek coal in Tennessee, now assigned to Slatestone Formation (Wilson et al., 1956, as revised by Hardeman et al., 1966) and to Breathitt Formation in Kentucky (McDowell et al., 1985).

Domas coal (KY). Local name for Raven coal in upper part of Lee Formation near Elkhorn City, Pike County, Kentucky (Alvord and Miller, 1972).

Dorchester coal (VA). Coal mined at Dorchester, 1 mi (1.6 km) west of Norton, Wise County, Virginia; lowest coal bed in Wise Formation, resting directly on Gladeville Sandstone; formerly called Glamorgan coal (Wanless, 1939); locally known as Marcee or Cornett coal in Lee County, Virginia (Brown et al., 1952).

Dorothy coal (WV). See Winifrede coal.

Dorothy limestone and shale (WV). Unranked marine unit exposed near Dorothy, Raleigh County, West Virginia; in Kanawha Formation between Glenalum Tunnel (above) and Gilbert coals (B. M. Blake, Jr., written communication, 1994).

Dorr Run shale (OH). Thin unranked shale containing marine to brackish-water fauna exposed along Dorr Run in Athens, Hocking, and Perry Counties, Ohio; in Allegheny Formation directly above Lower Freeport coal (Sturgeon and Merrill, 1949).

Dorton Shale (TN). Shale exposed near Dorton, Dorton 7.5-min quadrangle, Cumberland County, Tennessee; formation commonly 50 to 80 ft (15-24 m) thick and locally as much as 150 ft (46 m) thick;

occurs near base of Crooked Fork Group (Wilson et al., 1956; thickness revised by Hardeman et al., 1966).

Douglas coal (WV). Coal mined on Shortpole Branch, near Douglas Station, Sandy River district, McDowell County, West Virginia; in Kanawha Formation (Wanless, 1939; Englund et al, 1979); widely known and mined as the Red Ash coal.

Douglas shale (WV). Unranked black, sandy shale containing brackish-water to marine fauna above Lower Douglas coal, southern West Virginia; lowest marine horizon in Kanawha Formation (Wanless, 1939).

Duncan coal (VA). Coal named for a locality, probably a mine, on Coalpit Branch of Stony Creek, Scott County, Virginia; in upper part of Lee Formation (Wanless, 1939).

Duquesne coal (OH, PA). Coal named for Duquesne district of Pittsburgh, Pennsylvania; in middle of Conemaugh Formation between Skelley (above) and Gaysport marine limestones (Sturgeon et al., 1958).

Durham coals (TN). See Clifty coal or No. 5 Durham coal.

Duskin Creek Formation (TN). *Name herein abandoned.* Exposed on Duskin Creek, southwest of Spring City, Rhea County, Tennessee; strata in Walden Ridge syncline, overlying Rockcastle Sandstone (Wanless, 1946); unit probably represents a sequence of shale and siltstone in the Rockcastle Sandstone.

Dwale shale (KY). Unranked marine shale named by Morse (1931) and exposed at Dwale, Floyd County, Kentucky; in lower part of Breathitt Formation, forming roof of Prestonsburg or Upper Elkhorn No. 3 coal (Wanless, 1939).

"E" coal (KY). Name applied locally to upper coal of Creech coal zone in Harlan County, Kentucky; in Breathitt Formation; also locally called Low Splint coal (Froelich and Stone, 1973).

Eagle coal (WV, KY, VA). Coal mined at Eagle on Kanawha River, Kanawha district, Fayette County, West Virginia; widely mined coal in lower part of Kanawha Formation above Betsie Shale Member and below Cannelton limestone of White (1885); generally miscorrelated with coal in Millard coal zone in Kentucky and adjacent parts of Virginia; equivalent to Lower Elkhorn (Pond Creek) coal zone in Kentucky (Blake et al., this volume); widely miscorrelated with a coal (Blair?) below the Clintwood coal in southwestern Virginia.

Eagle "A" coal (WV). Coal exposed in southern Mingo County, West Virginia, so named because it is 30 to 50 ft (9 to 15 m) above the miscorrelated Eagle coal; *not present at type exposure of Eagle coal* but reserved as a name for coal splits above Eagle coal in Kanawha River valley; in lower part of Kanawha Formation (Blake et al., this volume).

Eagle limestone and shale (WV). Unranked marine limestone and shale named by White (1891), exposed near Eagle, Fayette County, West Virginia; in lower part of Kanawha Formation in West Virginia about 100 ft (30 m) below Eagle coal; name mistakenly applied to older marine unit above Hagy coal zone in Kentucky and Virginia and the Dorothy limestone and shale in southern West Virginia; now called Betsie Shale Member (Rice et al., 1987).

East Lynn sandstone (WV, KY). Massive unranked sandstone exposed at East Lynn, Stonewall district, Wayne County, West Virginia; in lower part of Allegheny Formation; unit 50 to 100 ft (15 to 30 m) thick and locally conglomeratic (Wanless, 1939).

Eastland Formation (Shale) (TN). Exposed near Eastland, an abandoned mining town in White County, Tennessee; about 75 ft (23 m) of shale containing Clifty coals; below Newton Sandstone (Wanless, 1946); *name abandoned* because unit was determined to be same as Whitwell Shale of Crab Orchard Mountains Group (Wilson et al., 1956).

Eckman sandstone (VA, WV). See No. 5 Pocahontas coal.

Egan coal (VA). See Stock Creek coal.

Elk fire clay (WV). Fire clay exposed along Elk River near Charleston, Kanawha County, West Virginia, where it has been used in making fire brick; in Allegheny Formation above No. 6 Block coal (Wanless, 1939; Englund et al., 1979).

Elk Gap coal (TN, KY). Coal named for Elk Gap just north of Pioneer, Campbell County, Tennessee; previously in Mingo Formation in Kentucky and Tennessee (Englund, 1968b); equivalent to coal in Upper Elkhorn No. 3 coal zone in Breathitt Formation in Kentucky; in Tennessee, now assigned to lower part of Indian Bluff Formation above Jellico coal (Wilson et al., 1956, as revised by Hardeman et al., 1966); in Kentucky now assigned to Breathitt Formation.

Elkhorn coal zone (KY). See Lower and Upper Elkhorn coals.

Elkins Fork shale (KY). Unranked marine shale named by Morse (1931) and exposed near Elkins Fork school, northern Pike County, Kentucky; in Breathitt Formation (Wanless, 1939); probably equivalent to marine Seth limestone of Krebs and Teets (1915) and Campbell Creek limestone of White (1885) in West Virginia (Blake et al., this volume).

Elk Lick coal (OH, WV). Coal exposed near Elk Lick, Somerset County, Pennsylvania; in upper part of Conemaugh Group between Clarksburg nonmarine limestone (above) and Skelley marine limestone (Sturgeon et al., 1958).

Ellerslie coal (MD). See Lower Kittanning coal.

Elswick coal (KY). Coal mined in Federal Mine No. 1 of Elswick Coal Company, on Russell Fork of Big Sandy River, 0.5 mi (0.8 km) northwest of Elkhorn City, Pike County, Kentucky; in Breathitt Formation below Auxier coal; probably equivalent to Upper Banner coal in Virginia (Wanless, 1939).

Etna (or Aetna) coal (TN, GA). Coal long mined in Etna Mountain, an isolated segment of Raccoon Mountain, near Whiteside, Marion County, Tennessee; also called Cliff or Castle Rock coal; correlated with Battle Creek coal of Cumberland Plateau (Marion County, Tennessee) (Wanless, 1946); assigned to Raccoon Mountain Formation of Gizzard Group between Warren Point Sandstone (above) and Dade coal (Wilson et al., 1956).

Ewing limestone (OH, PA, WV). Unranked nonmarine limestone exposed at Ewing site, Trimble Township, Athens County, Ohio (Sturgeon et al., 1958); below Upper Bakerstown or Barton coal in Glenshaw Formation of Conemaugh Group in Pennsylvania; probably miscorrelated with Lavansville limestone of Pennsylvania in Maryland (Flint, 1965).

Federal Hill coal (PA, MD). Thin coal about 75 ft (23 m) above Harlem coal exposed in railroad cut through northern end of Federal Hill 1.5 mi (2.4 km) southwest of Barrelville, Allegany County, Maryland (Swartz, 1922); in lower part of Casselman Formation of Conemaugh Group (Flint, 1965).

Fentress Formation (Shale) (TN). Name proposed for strata of Lee Formation below Rockcastle Sandstone in Fentress and Overton Counties, Tennessee, where Warren Point Sandstone, Sewanee Conglomerate, and Newton Sandstone are thin or absent and all of lower part of Lee is predominantly shale; correlated with Gizzard Group through Vandever Formation of southern Tennessee; as much as 340 ft (104 m) thick (Wanless, 1946; Wilson et al., 1956).

Ferriferous ore (PA, OH, KY). Band of iron-ore overlying unranked Ferriferous (Vanport) limestone; in Allegheny Formation in western Pennsylvania; name applied also to iron ore above Obryan Member of Allegheny Formation in southern Ohio and of Breathitt Formation in northeastern Kentucky (Wanless, 1939; Rice, Kosanke, and Henry, this volume); also called Hanging Rock ore.

Fire Clay coal (KY, WV). Coal commonly containing a distinctive parting (4 to 6 in. [10 to 15 cm] thick) of flint fire clay of volcanic origin below middle of coal; in Breathitt Formation between Hamlin (above) and Whitesburg coals; also called Dean, Hyden, and No. 4 coals; locally miscorrelated with younger Chilton coal in West Virginia (Blake et al., this volume); equivalent to Windrock coal of Tennessee, Wallins Creek and Hazard No. 4 coals in Kentucky, and Phillips coal in Virginia.

Fire Clay rider coal (zone) (KY). Coal or zone of coals commonly 20 to 50 ft (6 to 15 m) above Fire Clay coal; in Breathitt Formation; probably equivalent to Chilton coal in Kanawha River valley in West Virginia and Big Mary coal in Tennessee; coal commonly overlain by marine to brackish-water shale (Blake et al., this volume; Wanless, 1946).

Fire Creek coal (WV). Coal mined on Fire Creek, near New River Canyon, Sewell Mountain district, Fayette County, West Virginia; in lower part of New River Formation between Quinnimont sandstone (above) and Pineville Sandstone Member (Wanless, 1939).

Fishpot coal (PA, WV, MD, OH). Persistent coal in Monongahela Group between Sewickley or Meigs Creek coal (above) and Redstone coal; equivalent to Lower Sewickley coal (Sturgeon et al., 1958); should not be confused with unnamed coal in Lower Sewickley cyclothem.

Fishpot Member (PA). Exposed at mouth of Fishpot Run in southern Washington County, Pennsylvania; mainly siltstone and mudstone between Sewickley (above) and Redstone Members of Pittsburgh Formation (Monongahela Group) in Pennsylvania (Berryhill et al., 1971); equivalent in Ohio to Fishpot cyclothem (Monongahela Group) which includes sandstone, redbeds, nonmarine limestone, underclay, and Fishpot coal (Sturgeon et al., 1958).

Flag coal (KY). Cannel coal having appearance of flagstone where weathered, near mouth of Troublesome Creek, Perry County, Kentucky; in upper part of Breathitt Formation between Francis (above) and Hazard coals (Wanless, 1939).

Flag rider coal (KY). Thin coal 20 to 40 ft (6 to 12 m) above Flag coal; in Breathitt Formation (Wanless, 1939).

Flattop Mountain Sandstone Member (VA, WV). Massive sandstone, capping Flattop Mountain, 2 mi (3.2 km) northwest of Pocahontas, Virginia, at boundary between McDowell and Mercer Counties, West Virginia, and Tazewell County, Virginia; uppermost member of Pocahontas Formation (Wanless, 1939).

Flatwoods coal (KY). Coal named for exposures at Flatwoods, southern Pike County, Kentucky; in Breathitt Formation; miscorrelated by Hunt et al. (1937) to be above Taylor coal but equivalent to Fire Clay coal (Rice and Smith, 1980).

Flint Ridge coal (OH). Coal exposed on Flint Ridge, Hopewell Township, Licking County, Ohio; in Pottsville Formation between Middle Mercer (above) and Lower Mercer coals (Wanless, 1939).

Flint Ridge flint (KY). Flint exposed on Flint Ridge, at head of Leatherwood Branch of South Quicksand Creek, Breathitt County, Kentucky; in unranked, partially silicified limestone as much as 30 ft (9 m) thick; contains marine fossils; in upper part of Breathitt Formation about 40 ft (12 m) above top coal of Skyline coal zone; here correlated with Putnam Hill Limestone Member of Allegheny Formation in Ohio and Kanawha black flint of White (1891) in West Virginia and with Kilgore Flint Member in eastern Kentucky (Wanless, 1939; Rice, Kosanke, and Henry, this volume).

Fodderstack Sandstone Member (TN). Sandstone named for Little Fodderstack Mountain, Petros 7.5-min quadrangle, Anderson and Morgan Counties, Tennessee, where it forms a bench on which a trail encircles the mountain; unit at least 12 ft (4 m) thick and in upper part of Redoak Mountain Formation (Wilson et al., 1956, as revised by Hardeman et al., 1966).

Forked coal (KY). Divided or "forked" coal at head of Left Fork of Wolf Creek, Whitley County, Kentucky; in Breathitt Formation; equivalent to Moss and Cadell coals (Wanless, 1946).

Fossil limestone (KY). Name applied to a marine limestone in Breathitt Formation in eastern Kentucky; probably equivalent to Magoffin Member at most places and to Stoney Fork Member at a few (Wanless, 1946).

Four Corners unit (KY). Informal subdivision of Breathitt Formation in eastern Kentucky between base of Stoney Fork Member (above) and base of Magoffin Member (Chesnut, 1989); unit unranked and not otherwise described or defined.

Francis coal (zone) (KY). Coal mined by George Francis near gap between Yellow Creek of Carr Fork and Kelly Fork of Lots Creek, in southwestern Knott County, Kentucky; in upper part of Breathitt Formation between Hindman (above) and Flag coals (Wanless, 1939).

Franklin coal (PA, MD). Coal strip-mined at St. Paul, west of Salisbury, Somerset County, Pennsylvania; in Casselman Formation of Conemaugh Group about 110 ft (34 m) below Pittsburgh coal (Flint, 1965).

Freeburn coal (KY). Coal name used locally in Pike County, Kentucky; equivalent to Lower Elkhorn and Pond Creek coals in Breathitt Formation (Hunt et al., 1937).

Freeport Formation (PA). Uppermost of three formations making up Allegheny Group of Somerset County, Pennsylvania; named for Freeport, Armstrong County, Pennsylvania; extends from top of Upper Freeport coal (above) to top of lower bench of Upper Kittanning coal group; unit is about 100 ft (30 m) thick (Flint, 1965).

Friendsville shale (MD, PA). Unranked marine shale exposed at Friendsville, Garrett County, Maryland; in middle of Conemaugh Formation (Swartz, 1922); equivalent to Woods Run limestone in Pennsylvania and Maryland and Portersville limestone in Ohio (Flint, 1965).

Frozenhead coal (TN). Coal formerly mined on Frozen Head Mountain, near Petros, Morgan County, Tennessee; also called Sharp coal; previously in Scott Formation (Wanless, 1946); now assigned to middle part of Redoak Mountain Formation (Wilson et al., 1956, as revised by Hardeman et al., 1966).

Frozen Head Sandstone Member (TN). Sandstone named by Glenn (1925) from Frozen Head Mountain, Petros 7.5-min quadrangle, Morgan County, Tennessee; unit as much as 60 ft (18 m) thick; previously in Anderson Formation (Wanless, 1946); now assigned as top member of Vowell Mountain Formation (Wilson et al., 1956, as revised by Hardeman et al., 1966).

Frozen Sandstone Member (KY). Sandstone exposed on Little Frozen Creek, Breathitt County, Kentucky; in lower part of Breathitt Formation between Vires (above) and Van Cleve coals (Hansen and Johnston, 1963).

Fugate coal (KY). Name used for a coal in Breathitt Formation between Hindman (above) and Flag rider coals in Magoffin County, Kentucky; name said to be derived from report on coals of Troublesome Creek, Perry County, Kentucky; however, the name Fugate not used in that report; probably correlates with Francis coal, a name more widely used in Kentucky (Wanless, 1939).

Gallitzin coal (PA, MD). Probably named for Gallitzin, Cambria County, Pennsylvania; in lower part of Glenshaw Formation of Conemaugh Group; probably equivalent to Humbert coal (Flint, 1965).

Gaysport limestone (OH). Unranked marine limestone exposed in vicinity of Gaysport, Bluerock Township, Muskingum County, Ohio; in middle of Conemaugh Group between Duquesne coal (above) and Ames limestone (Sturgeon et al., 1958).

Gilbert coal (WV, KY). Coal exposed at Gilbert, Stafford district, southeastern Mingo County, West Virginia (Wanless, 1939); in lower part of Kanawha Formation between Lower War Eagle(?) (above) and Douglas coals (Blake et al., this volume).

Gilbert "A" coal (WV). Thin coal 30 to 50 ft (9 to 15 m) above Gilbert coal, southern West Virginia; in lower part of Kanawha Formation above Gilbert shale (Wanless, 1939).

Gilbert shale (WV). Unranked marine shale forming roof of Gilbert coal, southern West Virginia; in Kanawha Formation (Wanless, 1939).

Gin Creek coal (VA). Coal exposed on southern slope of Little Black Mountain on Gin Creek north of Darbyville, Lee County, Virginia; formerly called No. 8 coal; in Wise Formation between Wax (above) and Phillips coals; correlated with Fire Clay rider coal (Wanless, 1946).

Gizzard Group (TN). Name applied to lowest division of Pennsylvanian in southern Tennessee for Little Fiery Gizzard Creek 2 mi (3.2 km) south of Tracy City, Marion County, Tennessee; between Sewanee Conglomerate (above) and Pennington Shale (Mississippian); ranges in thickness from 0 to 520 ft (159 m) (Wanless, 1946; Wilson et al., 1956).

Gladeville Sandstone (VA). Massive sandstone exposed at Wise (formerly called Gladeville), Wise County, Virginia; between Wise (above) and Norton Formations (Wanless, 1939); ranges in thickness from 0 to 51 ft (16 m); difficult to identify beyond its type area (Miller, 1969).

Glamorgan coal (zone) (KY, VA). Coal mined at Glamorgan, Wise County, Virginia (Hinds, 1918); two coal beds above marine unit (mistakenly called Eagle shale by Alvord and Miller [1972] but probably Oceana limestone of Hennen and Gawthrop [1915]) and below Millard coal zone in Breathitt Formation; may represent coals in lower part of Millard coal zone (Rice and Smith, 1980).

Glenalum Tunnel coal (WV). Coal exposed at eastern portal of N & W Railroad tunnel east of Glenalum Station, Stafford district, southern Mingo County, West Virginia; in lower part of Kanawha Formation between Lower War Eagle coal (above) and Gilbert shale (Wanless, 1939).

Glenmary coal (TN). Coal mined at Glenmary, Scott County, Tennessee; previously in Briceville Formation (Wanless, 1946); correlated with Poplar Creek coal; now assigned to top of Crooked Fork Group (Wilson et al., 1956).

Glenmary Shale (TN). Shale named for Glenmary, Robbins 7.5-min quadrangle, Scott County, Tennessee; in upper part of Crooked Fork Group and commonly 50 to 150 ft (15 to 46 m) thick (Wilson et al., 1956; thickness revised by Hardeman et al., 1966).

Glenshaw Formation (PA, OH). Exposed at Glenshaw in valley of Pine Creek north of Pittsburgh, Allegheny County, Pennsylvania; formation averages about 375 ft (114 m) in thickness; name proposed for lower part of Conemaugh Group where Ames Limestone Member is present and forms top of formation; extends down to top of Upper Freeport coal (Flint, 1965).

Goodrich coal (TN). Coal mined near Dayton, Rhea County, Tennessee (source of name not known); in Gizzard Group about 30 ft (9 m) below Nelson coal (Wanless, 1946; Wilson et al., 1956).

Goodwill coal (WV, VA). Coal found in strip-mine exposures on ridgetop about 0.75 mi (1.2 km) west of community of Goodwill, Mercer County, West Virginia; in Pocahontas Formation between Flattop Mountain Sandstone Member (above) and No. 7 Pocahontas coal (Englund, 1968a).

Grassy coal (zone) (KY). Coal named for small mine near town of Grassy Creek, Morgan County, Kentucky (Englund, 1955); in Breathitt Formation; probably equivalent to Lower Elkhorn coal bed.

Grassy Spring coal (TN). Coal exposed at a grassy spring on trail up western end of Vowell Mountain, Campbell County, Tennessee; previously in Anderson Formation between Coal Gap (above) and Rock Spring coals (Wanless, 1946); now assigned to lower part of Cross Mountain Formation (Wilson et al., 1956, as revised by Hardeman et al., 1966).

Graves Gap Formation (TN). Formation named for Graves Gap on Cross Mountain, Anderson and Campbell Counties, Tennessee, and includes all strata between top of Windrock coal (above) to top of Pioneer Sandstone Member of Indian Bluff Formation; underlies Redoak Mountain Formation; unit ranges from 275 to 420 ft (84 to 128 m) thick and is 365 ft (111 m) thick at type section (Wilson et al., 1956, as revised by Hardeman et al., 1966).

Gray Hawk coal (KY). Coal named in Sturgeon 7.5-min quadrangle in southern Lee County and eastern Jackson County, Kentucky

(Weir, 1978), for a coal in lower part of Breathitt Formation below tongue of Corbin Sandstone Member of Lee Formation; approximately equivalent to Splash Dam coal.

Grayson sandstone bed (KY). Name given to a thin ganisterlike sandstone directly underlying Wolf Creek (or Bruin) coal in lower part of Breathitt Formation near Grayson, Carter County, Kentucky (Whittington and Ferm, 1967); sandstone thickens to west and southwest to as much as 70 ft (21 m) and was there reassigned to Lee Formation by DeLaney and Englund (1973).

Greasy Creek coal (VA). Coal exposed on Greasy Creek, north of Dog Fork anticline, western Tazewell County, Virginia; in upper part of Lee Formation; may be equivalent to Sewell "B" coal of Kanawha Formation in West Virginia (Wanless, 1939).

Grundy unit (KY). Informal member of Breathitt Formation in eastern Kentucky extending from base of Betsie Shale Member (above) to base of unnamed marine zone above improperly termed "Bee Rock Sandstone Member of Lee Formation" (Chesnut, 1989); unit unranked and not otherwise described or defined; Bee Rock incorrectly projected across Kentucky into West Virginia.

Guinea Fowl ore (OH). Name given by miners to thin iron ore formerly worked at Scioto Furnace, Bloom Township, Scioto County, Ohio; in Pottsville Formation between Huckleberry (above) and Anthony coals (Bownocker and Dean, 1929).

Gun Creek coal (KY). Coal exposed on Gun Creek, southeastern Magoffin County, Kentucky; in Breathitt Formation between Whitesburg (above) and Tom Cooper coals; locally contains a thin flint fire clay parting (tonstein) resembling that in Fire Clay coal; directly underlies Kendrick Shale Member (Wanless, 1939; Rice, 1969).

Guyandot sandstone (WV). Massive unranked sandstone cropping out on Laurel Fork of Guyandot River, 1 mi (1.6 km) southwest of McGraw, Slab Fork district, Wyoming County, West Virginia; in upper part of New River Formation between Castle (above) and Sewell "B" coals (Wanless, 1939).

Haddix coal (zone) (KY). Coal mined high in bluff of North Fork of Kentucky River above Haddix, Breathitt County, Kentucky; in Breathitt Formation between Hazard (above) and Copland coals (Wanless, 1939).

Hagy coal (zone) (VA, KY). Coal named for Hagy mine, on Trace Fork of Prater Creek, Dickenson County, Virginia; in Norton Formation in Virginia between Norton (above) and Splash Dam coals (Wanless, 1939); in Breathitt Formation in Kentucky between Millard (above) and Splash Dam coals.

Halsey Rough coal (KY). Coal mined on Halsey Rough, Billows 7.5-min quadrangle, Pulaski County, Kentucky (Hatch, 1963b); in Breathitt Formation, approximately equivalent to Beattyville coal.

Hamden ore and limestone (OH). Iron ore is named for Hamden Furnace, Clinton Township, Vinton County, Ohio; in lower part of Allegheny Formation, above Lower Kittanning coal; in northern Ohio, marine limestone occurs at this horizon (Wanless, 1939); unit renamed Columbiana Member by Sturgeon and DeLong (1964).

Hamlin coal (zone) (KY). Coal probably named for outcrops on Hays Branch, near Lost Creek Post Office, Perry County, Kentucky; in Breathitt Formation between Copland (above) and Fire Clay coals (Wanless, 1939).

Hance coal (zone) (KY). Two coals 11 ft (3 m) apart on Betsie Branch, Brownies Creek, and at various places along Hance Creek, Bell County, Kentucky (Wanless, 1946); in Breathitt Formation; correlated with Bennetts Fork and Clintwood coals (Rice and Ping, 1989).

Hance Formation (KY, TN, VA). *Name herein abandoned.* Formation exposed in Hance Ridge, eastern Bell County, Kentucky; included strata below Lower Hance coal (of overlying Mingo Formation) to top of Naese Sandstone Member of Lee Formation; formation was about 600 ft (183 m) thick (Wanless, 1946); formation redefined by Englund et al. (1963b) to extend to base of Harlan coal for a thickness of about 1,200 ft (366 m); formation boundaries were sandstone and coal that are discontinuous and miscorrelated in different areas; strata now assigned to Breathitt Formation in Kentucky (McDowell et al., 1985) and to Crooked Fork Group and overlying Slatestone Formation in Tennessee.

Harlan coal (zone) (KY). Extensively mined coal at Harlan, Harlan County, Kentucky; in Breathitt Formation; probably equivalent to Lower Standiford coal, Wise County, Virginia, and Alma coal in West Virginia (Wanless, 1939).

Harlan Sandstone (Formation) (VA, KY). Uppermost Pennsylvanian strata of Virginia, named for Harlan County, Kentucky, where it caps Big Black and Little Black Mountains; includes massive sandstones, shales, and coals; above High Splint coal of Wise Formation; formation as much as 780 ft (239 m) thick (Wanless, 1939); extended into border area of Kentucky by Miller (1969), where it is equivalent to uppermost part of Breathitt Formation.

Harlem coal (OH, PA, MD). Coal probably named for Harlem Springs, Lee Township, Carroll County, Ohio; in Conemaugh Group between Ames limestone (above) and Ewing limestone (Sturgeon et al., 1958).

Harrison ore (OH). Iron ore, locally fossiliferous, crops out on Niner Ridge, Harrison Township, Scioto County, southern Ohio (Wanless, 1939); of Early Pennsylvanian age, may contain fossils reworked from other marine horizons, including underlying Mississippian strata; occurs locally at base of Pennsylvanian strata in Ohio; in Pottsville Formation below Sharon Member.

Hartridge shale (WV). Unranked brackish to marginal marine and plant-bearing shale cropping out at Hartridge, Middle Fork district, Randolph County, Virginia; in New River Formation directly above Sewell coal (Wanless, 1939).

Harvey conglomerate (WV). Unranked conglomeratic sandstone exposed at Bolt Post Office (formerly Harvey) on Marsh Fork, Trap Hill district, Raleigh County, West Virginia; in New River Formation between Lower Iaeger shale and Castle coal (Wanless, 1939).

Hatfield coal (TN). Coal mined at head of Hatfield Creek, Jellico West 7.5-min quadrangle, Campbell County, Tennessee, 15 to 35 ft (5 to 11 m) above Big Mary coal (Englund, 1968b); possibly equivalent to Beech Grove coal in Redoak Mountain Formation (Wilson et al., 1956, as revised by Hardeman et al., 1966).

Hazard coal (zone) (KY). Coal mined high in ridges above Hazard, Perry County, Kentucky; in Breathitt Formation between Flag (above) and Haddix coals; coal zone includes Hazard No. 5 and No. 6 (Wanless, 1946).

Hazard No. 4 coal (KY). Local name for Fire Clay coal in area of Hazard, Perry County, Kentucky; in Breathitt Formation (Huddle et al., 1963).

Hazard No. 7 coal (zone) (KY). Coal mined in area of Perry County, Kentucky, equivalent to Oakley coal (Huddle et al., 1963); also in part equivalent to Peach Orchard coal zone; in Breathitt Formation.

Hazard No. 8 coal (KY). Coal mined in area of Perry County, Kentucky, equivalent to Francis, Fugate, and Flag coals; in Breathitt Formation (Huddle et al., 1963).

Hazard No. 9 coal (zone) (KY). Coals equivalent to Hindman coal, which underlies Stoney Fork Member (formerly Lost Creek limestone) of Breathitt Formation (Huddle et al., 1963).

Hazard No. 10 coal (zone) (KY). Name of coal beds on Potato Knob, Leslie County, Kentucky; in Breathitt Formation, locally as much as 100 ft (30 m) above Hindman coal (Rice, 1975).

Helenwood coal (TN). Coal mined at Helenwood, Scott County, Tennessee; previously in Briceville Formation; one of several names applied in Scott County, Tennessee, to coal incorrectly correlated with Coal Creek coal (Wanless, 1946); equivalent to Poplar Creek coal; now assigned to Slatestone Formation (Wilson et al., 1956, as revised by Hardeman et al., 1966).

Helton coal (KY). Coal mined by R. L. Helton near head of Spruce Pine Creek of Middle Fork of Kentucky River, southern Leslie County, Kentucky; in Breathitt Formation above Hindman coal; highest coal mined in Kentucky River valley north of Pine Mountain fault (Wanless, 1939).

Hensley Member (KY, VA, TN). Exposed in Hensley Flats on northwestern side of Cumberland Gap, Bell County, Kentucky; unit consists mostly of shale, siltstone, and thin-bedded sandstone as much as 450 ft (137 m) thick; in Lee Formation between Bee Rock Sandstone Member (above) and Middlesboro Member; previously mapped as "sandstone and shale member D" (Englund, 1964a).

Herbert Sandstone (TN). Massive, locally conglomeratic sandstone exposed on Glade Creek near Herbert, Bledsoe County, Tennessee; strata previously assigned to Lower Bon Air sandstone (Wanless, 1946); *name abandoned* and strata now assigned to Newton Sandstone of Crab Orchard Mountains Group (Wilson et al., 1956).

Hernshaw coal (WV). Coal mined on Lens Creek near Hernshaw, Loudon district, Kanawha County, West Virginia; in Kanawha Formation between Fire Clay coal (above) and Dingess Shale Member; correlates with Cedar Grove coal; locally known as Black Band coal and locally confused with Fire Clay coal (Englund et al., 1979) but is 25 to 35 ft (8 to 11 m) below Fire Clay at Hernshaw, West Virginia (Wanless, 1939; see also, Blake et al., this volume); equivalent to Whitesburg coal in Kentucky

High Splint coal (VA, KY). Splint coal exposed high in Black Mountain in Big Stone Gap field; about 20 ft (6 m) below top of Wise Formation in Virginia (Wanless, 1939); in upper part of Breathitt Formation in Kentucky.

Hignite coal (zone) (KY). Coal beds mined in Hignite Creek, Bell County, Kentucky; in Breathitt Formation between Magoffin Member (above) and Stray coal zone (Rice and Maughan, 1978); probably equivalent to coals in upper part of Fire Clay coal zone.

Hignite Formation (KY, TN). *Name herein abandoned.* Formation named for Hignite Creek in Log Mountains east of Middlesboro, Bell County, Kentucky; included strata extending from top of Red Springs coal (above) to base of Lower Hignite coal or between Bryson (above) and Catron Formations; formation was as much as 560 ft (170 m) thick (Wanless, 1946); formation boundaries were sandstone and coal, which are discontinuous and miscorrelated in different areas; strata now assigned to Breathitt Formation (McDowell et al., 1985); name extended into Tennessee by Englund (1968b) for strata now included in Redoak Mountain Formation (Wilson et al., 1956, as revised by Hardeman et al., 1966).

Hindman coal (KY). Coal reported to be 10 ft (3 m) thick near hilltop above Right Fork of Troublesome Creek, 3 mi (5 km) southeast of Hindman, Knott County, Kentucky; in Breathitt Formation locally underlies Stoney Fork Member and is above Francis coal (Wanless, 1939).

Hitchins clay bed (KY). Unranked ceramic clay bed named for Hitchins, Carter County, Kentucky; in Breathitt Formation below Princess No. 6 coal (Whittington and Ferm, 1967).

Hoffman coals (MD, PA). Group of as many as three coals exposed in Hoffman drainage tunnel near Clarysville, Allegany County, Maryland; in Conemaugh Formation between Lonaconing (above) and Upper Clarysville coals (Swartz, 1922).

Homewood Formation (Sandstone Member) (PA, OH, WV, KY). Massive sandstone exposed at Homewood Station, Big Beaver Township, northern Beaver County, Pennsylvania; uppermost member of Pottsville Group (Formation); below Brookville coal (Wanless, 1939).

Honeycomb coal (PA, MD). See Lower Bakerstown coal.

Hooper coal (TN). Coal formerly mined at Hooper mines near Winslow siding on Little Emory River, Morgan County, Tennessee; previously in lower part of Briceville Formation between Poplar Creek (above) and Christmas coals; tentatively correlated with Rex coal (LaFollette region, Tennessee) (Wanless, 1946); now assigned to Burnt Mill Shale of Crooked Fork Group, stratigraphically *above* Rex coal (Wilson et al., 1956).

Hopewell coal (KY). Exposed 0.5 mi (0.8 km) south of Hopewell, near southern boundary of Greenup County, Kentucky, along Kentucky Route 1; in Breathitt Formation lies 5 to 10 ft (1.5 to 3 m) above Bruin coal (Whittington and Ferm, 1965).

Horse Creek coal (KY). Coal exposed along Horse Creek, Clay and Knox Counties, Kentucky; in Breathitt Formation, equivalent to Manchester coal (Reeves, 1964).

Howard coal (KY). Coal mined by James Howard on Mine Fork, Magoffin County, Kentucky; in Breathitt Formation between Lacey Creek (above) and Wheelersburg coals; equivalent to Lower Elkhorn coal in Kentucky or Eagle coal in West Virginia (Wanless, 1939).

Howard coal (KY). Coal formerly mined by E. C. Howard on Rocky Branch of Goose Creek, 7 mi (11 km) south of Manchester, Clay County, Kentucky; in Breathitt Formation between Whitesburg (above) and Manchester coals (Wanless, 1939); locally in two beds.

Huckleberry coal (KY, OH). Several different uses in Kentucky and Ohio; thin coal above Olive Hill flint clay in Breathitt Formation in northeastern Kentucky; essentially equivalent to Anthony coal in Pottsville Formation in southern Ohio; mined coal in Breathitt Formation at Jellico coal horizon on Huckleberry Branch of Sexton Creek, Clay County, Kentucky; name also used in Pottsville Formation in southern Ohio for coal between Quakertown (above) and Anthony coals.

Hudson coal (KY). Named for Hudson mine on southern bank of Cumberland River between mouths of Laurel and Rockcastle Rivers, McCreary County, Kentucky; in Breathitt Formation, lowest commercial coal in western border of eastern Kentucky coal field; below Beaver Creek coal (Wanless, 1939).

Hughes Ferry coal (WV). Coal crops out at old site of Hughes Ferry on Gauley River, Nicholas County, West Virginia; in upper part of New River Formation; now generally known as Iaeger coal in West Virginia (Wanless, 1939).

Humbert coal (PA). Coal mined near village of Humbert, 4 mi (6.4 km) northeast of Confluence, Somerset County, Pennsylvania; in Glenshaw Formation of Conemaugh Group between Brush Creek (above) and Mahoning coals (Flint, 1965).

Hunnewell coal (KY). See Lower Stinson coal.

Hyden coal (KY). Coal mined near Hyden, Leslie County, Kentucky; in Breathitt Formation; equivalent to Fire Clay coal (Wanless, 1939).

Hyden unit (KY). Informal subdivision of Breathitt Formation between base of Magoffin Member (above) and base of Kendrick Shale Member (Chesnut, 1989); unit unranked and not otherwise described or defined.

Iaeger coal (WV). Coal mined at Iaeger, on Tug Fork of Big Sandy River, Sandy River district, McDowell County, West Virginia; previously called Hughes Ferry coal; in upper part of New River Formation (Wanless, 1939).

Iaeger "A" coal (WV). Thin coal 40 to 70 ft (12 to 21 m) above Iaeger coal, McDowell County, West Virginia; in New River Formation (Wanless, 1939).

Iaeger "B" coal (WV). Thin coal about 100 ft (30 m) above Iaeger coal, McDowell County, West Virginia; in New River Formation (Wanless, 1939).

Imboden coal (VA, KY). Coal mined at Imboden, on Pigeon Creek, western Wise County, Virginia; in Wise Formation between Kelly (above) and Rocky Fork coals; equivalent to Eagle coal in West Virginia (Wanless, 1939); also in border areas of Kentucky.

Index coal (KY). Exposed along U.S. Highway 40 a few feet below gap in ridge between Index and West Liberty, Morgan County, Kentucky (Adkison, 1957); in Breathitt Formation; equivalent to coal in lower part of Hazard coal zone.

Indian Bluff Formation (TN). Formation named for Indian Bluff, Lake City 7.5-min quadrangle, Anderson County, Tennessee, and includes all strata from top of Pioneer Sandstone Member (above) to top of Jellico coal of Slatestone Formation; unit ranges from 150 to 415 ft (46 to 126 m) thick and is 455 ft (139 m) thick at type section on Cross Mountain (Wilson et al., 1956, as revised by Hardeman et al., 1966).

Indian Fork coal (TN). Thin coal exposed along northern side of Indian Fork, Fork Mountain 7.5-min quadrangle, Morgan County, Tennessee; in upper part of Indian Bluff Formation (Wilson et al., 1956, as revised by Hardeman et al., 1966).

Indian Fork Sandstone Member (TN). Sandstone exposed on northern side of Indian Fork, Fork Mountain 7.5-min quadrangle, Morgan County, Tennessee; in upper part of Indian Bluff Formation; as much as 60 ft (18 m) thick (Wilson et al., 1956, as revised by Hardeman et al., 1966).

Island Creek coal (WV). Local name for coal mined along Island Creek in Logan County, West Virginia; in Kanawha Formation; equivalent to Thacker coal.

Isoline coal (TN). Coal mined at Isoline, northern Cumberland County, Tennessee; in upper part of Vandever Shale (Formation), below Rockcastle Sandstone; probably equivalent to Morgan Springs coal (Wanless, 1946).

Jack Rock coal (VA). Coal named by Giles (1925) to describe a particularly tough coal, probably a splint coal; in Wise Formation; probably equivalent to Kelly coal in Lee County, Virginia, locally called No. 3 coal; should not be confused with Fire Clay or Hazard No. 4 coal of eastern Kentucky, which is younger (see Miller and Roen, 1973).

Jackson coal (TN). See Lower Sewanee coal.

Jackson Shaft coal (OH). Coal formerly mined by shaft at Jackson, Jackson County, Ohio; in Pottsville Formation; commonly called Sharon coal (Wanless, 1939).

Jawbone coal (VA, WV). Coal mined in Jawbone Hollow, a branch of Bull Run, southeast of Banner, Wise County, Virginia; in lower part of Norton Formation between Raven (above) and Tiller coals in Virginia (Wanless, 1939); in upper part of New River Formation; correlates with Iaeger coal in McDowell County, West Virginia.

Jellico coal (zone) (KY, TN). Principal coal mined in highlands of Whitley County, Kentucky, north of Jellico, Campbell County, Tennessee; in Breathitt Formation in Kentucky or at top of Slatestone Formation in Tennessee, above Blue Gem coal; correlated with Alma coal in West Virginia and Harlan coal in Kentucky (Wanless, 1939; Wilson et al., 1956, as revised by Hardeman et al., 1966).

Jellico Formation (TN). *Name herein abandoned.* Formation name proposed as substitute for Wartburg Sandstone after discovery that latter name had been erroneously applied to strata overlying Briceville Formation; named because Jellico coal is its most valuable economic resource; included strata that extended from top of its Pioneer Sandstone Member (above) to base of Blue Gem coal (Wanless, 1946); strata now assigned to Slatestone and Indian Bluff Formations (Wilson et al., 1956, as revised by Hardeman et al., 1966).

Jesse Sandstone Member (KY). Massive slightly conglomeratic sandstone exposed at head of Jesse Creek, Wallins Creek district, Harlan County, Kentucky; previously top member of Catron Formation (Wanless, 1946); now assigned to Breathitt Formation between Limestone (above) and Pardee coals (McDowell et al., 1985).

Jewell coal (VA). Named for mines on Jewell Branch of Dismal Creek, Buchanan County, Virginia; in New River Formation between Iaeger "B" (above) and Jawbone coals (Englund, 1981).

Johnstown limestone (PA). Unranked nonmarine limestone exposed at Johnstown, Cambria County, Pennsylvania; in Allegheny Formation between Upper and Middle Kittanning coals or in part of Kittanning Formation as described by Flint (1965).

Jordan coal (TN, KY). Coal named for John Jordan, who formerly owned land where the Gem mine at Peabody, Campbell County, Tennessee, was located; previously lowest coal of Scott Formation (Wanless, 1946); correlated with lower coal in Amburgy coal zone and Creech coal zone in Kentucky, and Low Splint zone in Virginia; now assigned as basal coal of Graves Gap Formation (Wilson et al., 1956, as revised by Hardeman et al., 1966).

Joyner coal (TN). Coal mined at Joyner mine of Oliver Springs Coal Company in Brushy Mountain, 4 mi (6 km) northwest of Oliver Springs, Morgan County, Tennessee; previously in Jellico Formation between Pioneer Sandstone Member (above) and Jellico coal (Wanless, 1946); now assigned to lower part of Indian Bluff Formation overlying its Seeber Flats Sandstone Member (Wilson et al., 1956, as revised by Hardeman et al., 1966).

Kanawha black flint (WV). Marine chert named by White (1891), exposed along Kanawha River bluffs between Charleston and Kanawha Falls, Kanawha County, West Virginia; at top of Kanawha Formation between Stockton "A" (above) and Stockton rider coals (Wanless, 1939); here correlated in Kentucky with Kilgore Flint Member of Breathitt Formation and Flint Ridge flint of Morse (1931).

Kanawha Formation (WV, VA). Named for exposures along Kanawha River between Gauley Bridge and Charleston, West Virginia; includes strata from top of Kanawha black flint of White (1891) (above) to base of Lower Douglas coal of Douglas Station, McDowell County, West Virginia (Wanless, 1939).

Kelley coal (TN). Coal formerly mined at Kelley mines, near Whiteside, Marion County, southern Tennessee, in Etna Mountain, a hill rising above surface of Raccoon Mountain plateau; lowest of three coals above Sewanee Conglomerate, equivalent to Richland coal of Rhea County, Tennessee, and elsewhere; in Whitwell Shale of Crab Orchard Mountains Group just above Sewanee Conglomerate (Wilson et al., 1956).

Kellioka coal (KY). Coal mined at Kellioka on Poor Fork in Harlan County, Kentucky; in Breathitt Formation between Creech (above) and Harlan coals (Wanless, 1946); correlated with Taggart Marker coal.

Kelly coal (VA). Coal extensively mined in western Wise County, Virginia; probably named for Kelly View on Powell River, northeast of Appalachia, Virginia; in Wise Formation between Lower Standiford (above) and Imboden coals (Wanless, 1939).

Kendrick Shale Member (KY, TN, VA). Marine shale exposed at Dr. Kendrick's homestead at headwaters of Cow Creek, Floyd County, Kentucky; in Breathitt Formation, above Amburgy coal in Kentucky; above Lower Pioneer coal in Graves Gap Formation in Tennessee; above Low Splint coal in Wise Formation in Virginia; unit ranges from 0 to 69 ft (21 m) but is commonly about 30 ft (9 m) thick; equivalent to Dingess Shale Member of Kanawha Formation in West Virginia (Wanless, 1939; Rice, 1980).

Kennedy coal (VA). Formerly known as Widow Kennedy coal, probably mined near Dante, Russell County, Virginia; extensively mined in Norton Formation between Lower Banner (above) and Aily coals; equivalent to Douglas coal in West Virginia (Wanless, 1939).

Kenova limestone (WV). Marine limestone exposed along Norfork and Southern Railway south of Kenova, Credo district, Wayne County, West Virginia; about 25 ft (7.5 m) above base of Glenshaw Formation of Conemaugh Group and about 55 ft (16.8 m) below base of Lower Brush Creek limestone (Glen Merrill, written communication, 1993).

Kent coal (TN). Coal named for Kent mine, near Peabody, Campbell County, Tennessee; previously in Briceville Formation between Rich Mountain coal zone (above) and Murray coal (Wanless, 1946); equivalent to Coal Creek coal (west of Jacksboro fault in Tennessee); now assigned to lower part of Slatestone Formation (Wilson et al., 1956, as revised by Hardeman et al., 1966).

Keokee coal (VA, KY). Widely mined coal at Keokee, Lee County, Virginia; in Wise Formation between Low Splint (above) and Kirk coals; equivalent to Taggart coal in Wise County, Virginia, and Harlan County, Kentucky (Wanless, 1939).

Keokee Sandstone Member (VA). Sandstone exposed in ledges 1 mi (1.6 km) east of Keokee School in town of Keokee on Virginia State Highway 68, Lee County, Virginia; unit consistently more than 30 ft (9 m) thick; in Wise Formation about 150 to 200 ft (46 to 61 m) above Robbins Chapel Sandstone Member (Miller, 1969).

Keystone coal (WV). Thin coal exposed at railroad station at Keystone, Browns Creek district, McDowell County, West Virginia; in Pocahontas Formation, above North Fork shale (Wanless, 1939).

Kilgore Flint Member (KY). Silicified siltstone of limited extent exposed near Kilgore, Boyd County, Kentucky; chert is about 12 ft (3.7 m) thick and contains marine fossils; locally overlies Princess No. 5 coal in northeastern Kentucky; in Breathitt Formation; correlates with Flint Ridge flint of Morse (1931) in Kentucky, Putnam Hill Limestone Member of Allegheny Formation in Ohio, and Kanawha black flint in West Virginia (Rice, Kosanke, and Henry, this volume).

Kirk coal (VA). Coal prospected on W. T. Kirk farm near head of Ely Creek, Lee County, Virginia; in Wise Formation between Taggart (above) and Harlan coals (Giles, 1925).

Kittanning Formation (PA). Middle formation of Allegheny Group in Somerset County, Pennsylvania; extends from top of lower bench of Upper Kittanning coal zone (above) to base of underclay below upper coal of Lower Kittanning coal zone; unit about 100 ft (30 m) thick (Flint, 1965).

Klondike coal (KY, TN). Coal named for Klondike mine, head of Bennetts Fork, Claiborne County, Tennessee; in Breathitt Formation at base of Kendrick Shale Member between Lower Hignite (above) and Poplar Lick coals (Wanless, 1939).

Knob coal (zone) (KY). Named for coals occurring at top of hills (Knobs) north of Buckhorn Creek, Noble 7.5-min quadrangle, Breathitt County, Kentucky; in Breathitt Formation between Flint Ridge flint of Morse (1931) (above) and Hindman coal (Williamson and Adkison, 1953); probably equivalent to Skyline coal zone

Koontz coal (PA). See Waynesburg coal.

Lacey Creek coal (KY). Coal mined on Lacey Creek, northern Magoffin County, Kentucky; in Breathitt Formation between Tom Cooper (above) and Howard coals (Wanless, 1939).

Landgraff coal (WV). Thin coal exposed at Landgraff, on Elkhorn Creek, Browns Creek district, McDowell County, West Virginia; in lower part of Pocahontas Formation (Wanless, 1939).

Lantana coal (TN). Coal formerly mined near Lantana Post Office, Cumberland County, Tennessee, where coal was locally as thick as 18 ft (5.5 m), probably because of tectonic thickening; at or near base of Vandever Shale of Crab Orchard Mountains Group (Wanless, 1946; Wilson et al., 1956).

Laurel coal (KY). Coal mined in ridge above Laurel Fork, Lenox 7.5-min quadrangle, Morgan County, Kentucky (Johnston, 1962); in Breathitt Formation; probably equivalent to Princess No. 5A or 5B coals.

Lavansville limestone (PA). Exposed near Lavansville, Somerset County, Pennsylvania; unranked nonmarine limestone lying directly below Ames coal and limestone in upper part of Glenshaw Formation of Conemaugh Group; probably misidentified in Maryland as Ewing limestone by Swartz (1922) (Flint, 1965).

Lawrence clay (OH). Refractory clay exposed at Lawrence Furnace, Lawrence County, Ohio; in lower part of Allegheny Formation (Wanless, 1939); between No. 5 coal (above) and Obryan Member.

Lawrence coal (OH). Thin coal in Lawrence County, southern Ohio, immediately above Lawrence clay; in Allegheny Formation (Wanless, 1939).

Lea coal (KY). Coal probably equivalent to coal in Hazard coal zone in Breathitt Formation (Rice and Smith, 1980, Table 2).

Leatherwood coal (KY). Coal mined near town of Leatherwood, southern Perry County, Kentucky (Prostka and Seiders, 1968); equivalent to coal in lower part of Hazard coal zone in Breathitt Formation.

Lee coal (zone) (KY, VA). Coals named by Miller (1910); Lee No. 1 equivalent to Hudson coal; Lee No. 2 equivalent to Beaver Creek coal; Lee No. 3 equivalent to Barren Fork coal; in Breathitt Formation where it intertongues with Lee Formation on western side of eastern Kentucky coal field. "Lee" is a name used locally to designate a coal directly overlying upper tongue of Middlesboro Member of Lee Formation in Virginia.

Lee Formation (KY, VA). Formation named for Lee County, Virginia; lowest division of the Pennsylvanian along border between Kentucky and Virginia; characterized by thick, pebbly quartzarenite units that are linear bodies as much as 700 ft (213 m) thick, about 40 to 50 mi (64 to 80 km) wide, which pinch out to southeast in Virginia and northwest in Kentucky; members overlap such that they are older toward southeast and younger toward northwest; on western side of basin in Kentucky, includes (from youngest to oldest) Grayson sandstone bed, Corbin and Rockcastle Sandstone Members, and Livingston Conglomerate Member; on eastern side of basin in Kentucky and Virginia, includes Naese and Bee Rock Sandstone Members, Hensley, Middlesboro, and Dark Ridge Members, White Rocks Sandstone Member, Chadwell Member, and Mississippian Pinnacle Overlook Member; intertongues with Breathitt Formation in Kentucky (Rice, 1984a); in Virginia, type section placed at Big Stone Gap, Wise County, Virginia, where the formation is about 1,400 ft (425 m) thick and includes Bee Rock Sandstone Member at top and all underlying Pennsylvanian strata; formation intertongues northeastward in Virginia with strata of Norton Formation, New River Formation, and underlying Pocahontas Formation (Miller, 1965, 1969); *usage of Lee Formation or Group is restricted from Tennessee except on Cumberland overthrust sheet,* where equivalent strata are assigned to (from youngest to oldest) Crooked Fork Group, Crab Orchard Mountains Group, and Gizzard Group of Wilson et al. (1956).

Leetonia limestone (OH). Unranked nonmarine limestone exposed at Leetonia, Columbiana County, Ohio; in Allegheny Group in clay or shale unit below Middle Kittanning coal; previously called Salem limestone (DeLong and White, 1963).

Lemon coal (PA). See Upper Freeport coal.

Lemoyne coal (TN). Coal formerly mined at LeMoyne mine east of Winfield in Scott County, Tennessee; correlated with Poplar Creek coal by Wanless (1946) but with Coal Creek coal by others; now assigned to top of Crooked Fork Group above Wartburg Sandstone (Wilson et al., 1956).

Lenox coal (KY). Coal exposed near Lenox, Morgan County, Kentucky; overlain by shale containing marine fossils in upper part of Hazard coal zone (Johnston, 1962); in Breathitt Formation; probably equivalent to Princess No. 3 coal.

Leonard coal (KY). Coal mined at Leonard Post Office on Childs Creek of Clover Fork of Cumberland River, Harlan County, Kentucky; in Breathitt Formation; probably equivalent to "B" or Taggart Marker coal (Wanless, 1939).

Lewiston coal (WV). See Stockton coal.

Lick Fork coal (KY, TN). Coal named for mines along Lick Fork of Little Elk Creek, Pioneer 7.5-min quadrangle, western Campbell County, Tennessee (Englund, 1968b); in Kentucky in Breathitt Formation, equivalent to coal in Upper Elkhorn No. 3 coal zone; in Tennessee, coal is in strata assigned to Graves Gap Formation (Wilson et al., 1956, as revised by Hardeman et al., 1966).

Lily coal (KY). Coal mined near Lily Station, south of London, Laurel County, Kentucky; in lower part of Breathitt Formation below Bacon Creek coal; correlated with Williamsburg and Manchester coals (Wanless, 1939)

Limekiln limestone (KY). Unranked marine limestone on Limekiln Knob, Lenox 7.5-min quadrangle, Morgan County, Kentucky (Johnston, 1962); in Breathitt Formation; probably equivalent to Kilgore Flint Member or Putnam Hill Limestone Member of Ohio.

Limestone coal (KY). Name used for coal commonly about 30 ft (9 m) below Magoffin Member on Cumberland overthrust sheet in Kentucky; in Breathitt Formation between Hazard (above) and Hamlin coals; equivalent to Copland coal in Kentucky, Pardee coal in Virginia, and Chilton coal in West Virginia (Wanless, 1939).

Little Alma coal (WV). Thin coal 40 to 60 ft (12 to 18 m) below Alma coal, Mingo County, West Virginia; in Kanawha Formation (Wanless, 1939).

Little Blue Gem coal (KY). Local name for coal 70 to 100 ft (21 to 30 m) below Blue Gem coal; in Breathitt Formation of southeastern Kentucky; equivalent to Black Wax coal in Tennessee (Rice and Smith, 1980).

Little Brushy coal (TN). Local name for Poplar Creek coal in Little Brushy Mountain southeast of Wartburg, Morgan County, Tennessee; previously in Briceville Formation (Wanless, 1946); now assigned to top of Crooked Fork Group above Wartburg Sandstone (Wilson et al., 1956).

Little Caney coal (KY). Coal exposed along Little Caney River, Elliott County, Kentucky; equivalent to Tom Cooper or Upper Elkhorn No. 3 coal bed; in Breathitt Formation (Englund and DeLaney, 1966).

Little Cedar coal (WV, KY). Thin coal in Mingo County, West Virginia; in lower part of Kanawha Formation in West Virginia between Cedar (above) and Lower War Eagle coals (Wanless, 1939); correlated with Hagy coal in Kentucky.

Little Chilton coal (WV). Thin coal between Chilton (above) and Fire Clay coal in southern West Virginia; *not present at Chilton, Kanawha County, West Virginia;* in Kanawha Formation (Wanless, 1939; Blake et al., this volume).

Little Clarksburg coal (OH, WV, PA, MD). Thin coal exposed in vicinity of Clarksburg, Harrison County, West Virginia; in upper part of Conemaugh Group between Lower Pittsburgh (above) and Clarksburg nonmarine limestones (Sturgeon et al., 1958).

Little Coalburg coal (WV). Thin coal 15 to 20 ft (5 to 6 m) below Coalburg coal in southern West Virginia; it may be a split of Coalburg coal; upper part of Kanawha Formation (Wanless, 1939).

Little Eagle coal (WV, KY). Thin coal 20 to 30 ft (6 to 9 m) below Eagle coal at Eagle, Fayette County, West Virginia; in lower part of Kanawha Formation (West Virginia) between Eagle coal (above) and Betsie Shale Member (previously Eagle shale); may be split of Eagle coal (Wanless, 1939); locally miscorrelated with Blair coal in Breathitt Formation in Kentucky by Rice and Smith (1980).

Little Fire Clay coal (KY). Name locally applied to a thin coal a little below Fire Clay or Dean coal in eastern Kentucky; locally contains thin flint clay parting (tonstein); in Breathitt Formation; equivalent to uppermost coal of Whitesburg coal zone (Wanless, 1946); miscorrelated with Little Chilton coal in West Virginia.

Little Fire Creek coal (zone) (WV). Thin coal (or as many as three distinct mappable benches) 20 to 40 ft (6 to 12 m) below Fire Creek coal on Fire Creek, at Grandview, Raleigh County, West Virginia; lower part of New River Formation (Wanless, 1939).

Little Frozenhead coal (TN). Name locally applied by Mr. Petree to coal elsewhere called Rock Spring coal; previously in lower part of Anderson Formation (Wanless, 1946); now assigned to upper part of Vowell Mountain Formation (Wilson et al., 1956, as revised by Hardeman et al., 1966).

Little No. 5 Block coal (WV). Coal between No. 5 Block (above) and Stockton "A" coals; in Charleston Sandstone or Allegheny Formation of West Virginia.

Little Pittsburgh coals (MD, PA, OH). As many as three coals (Little, Second Little, and Third Little Pittsburgh coal) found between Pittsburgh coal (above) and Lower Pittsburgh limestone; in upper part of Conemaugh Formation (Swartz, 1922).

Little Pittsburgh limestone (MD, PA, OH). Unranked nonmarine limestone in Conemaugh Formation below Little Pittsburgh coal (Swartz, 1922).

Little Raleigh coal (WV). Thin coal exposed at Raleigh (now called Beckley), Raleigh County, West Virginia; in New River Formation between upper and lower parts of Raleigh Sandstone Member; locally in a zone of as many as three coals (Wanless, 1939).

Little Raleigh "A" coal (WV). Thin coal 10 to 25 ft (3 to 8 m) above Little Raleigh coal, in Raleigh County, West Virginia; in New River Formation (Wanless, 1939).

Livingston coal (KY). Exposed near community of Livingston, Rockcastle County, Kentucky; local name for Lee No. 1 coal in Breathitt Formation (Brown and Osolnik, 1974).

Livingston Conglomerate Member (Tongue) (KY). Conglomeratic sandstone filling an elongate channel at base of Pennsylvanian on Roundstone Creek, Rockcastle County, Kentucky; member is as much as 120 ft (37 m) thick; lentil of Lee Formation, as much as 150 ft (46 m) below Rockcastle Conglomerate Member of Lee Formation (Wanless, 1939; Rice, 1984a).

Log Mountain coal (TN). Local name applied to Jellico coal in LaFollette–Walnut Mountain field, Campbell County, Tennessee (Wanless, 1946); assigned to Breathitt Formation on Cumberland overthrust sheet (Rice and Newell, 1990) but equivalents elsewhere in Tennessee are assigned to top part of Slatestone Formation (Wilson et al., 1956, as revised by Hardeman et al., 1966).

Lonaconing coal (MD, PA). Coal exposed near Lonaconing, Allegany County, Maryland; in Conemaugh Formation between Franklin (above) and Hoffman coals (Swartz, 1922).

Lookout Formation (Sandstone) (Georgia). Named for Lookout Mountain, Georgia; equivalent to Sewanee Conglomerate of Crab Orchard Mountains Group and underlying Gizzard Group in adjacent areas of Tennessee (Keroher et al., 1966).

Looney coal (KY). Coal mined on Looney Ridge of Black Mountain, north of Benham, Harlan County, Kentucky; in Breathitt Formation; probably equivalent to Limestone coal (Froelich and Stone, 1973).

Lost Creek limestone (KY). Unranked marine limestone exposed on hilltops about 1 mi (1.6 km) southwest of village of Lost Creek, on North Fork of Kentucky River, Breathitt County, Kentucky (Wanless, 1939); in Breathitt Formation, directly above Hindman coal; strata now assigned to Stoney Fork Member of Breathitt Formation (Ping and Rice, 1979); here correlated with Boggs (Blunt Run) limestone in Ohio.

Lowellville limestone (OH). Exposed south of Lowellville, Mahoning County, Ohio, along Grindstone Run; in Pottsville Group between Quakertown (above) and Sharon coals; equivalent to Poverty Run limestone (Slucher and Rice, this volume).

Lower Bakerstown coal (PA, MD). Coal named for Bakerstown, Allegheny County, Pennsylvania; in Glenshaw Formation of Conemaugh Group between Ewing (above) and Cambridge limestones; equivalent to Thomas or Honeycomb coal of Maryland and Anderson coal of Ohio (Flint, 1965).

Lower Banner coal (VA, KY). Lower of two coals mined at Banner, on Toms Creek, eastern Wise County, Virginia; in Norton Formation between Upper Banner and Kennedy coals (Wanless, 1939); in Breathitt Formation in border areas in Kentucky.

Lower Barren Measures (Appalachian coal field). Name formerly applied to strata now assigned to Conemaugh Formation between Monongahela Formation (Upper Productive Measures) and Allegheny Formation (Lower Productive Measures) (Wanless, 1939).

Lower Bolling coal (VA). Former name of lower of two coals mined by Bolling family in southwestern part of Pound 7.5-min quadrangle, Wise County, Virginia; in Wise Formation; now called Lower Campbell Creek coal (Wanless, 1939).

Lower Bon Air coal (TN). See No. 1 Bon Air coal.

Lower Brush Creek limestone (PA, OH). Lower of two unranked marine limestones in east-central and southern Ohio; near same interval as Brush Creek limestone, which is exposed in Cranberry Township, Butler County, Pennsylvania; in lower part of Conemaugh Group (Sturgeon et al., 1958).

Lower Campbell Creek coal (WV). Locally minable coal in southern West Virginia, in Kanawha Formation (Wanless, 1939); coal probably a lower split of Eagle or Pond Creek coals where name misapplied in Tug Fork area of West Virginia.

Lower Cedar Grove coal (WV). Name misused for a commercial coal in southern West Virginia not related to Cedar Grove coal at Cedar Grove, Kanawha County, West Virginia; in Kanawha Formation, equivalent to No. 2 Gas coal of Kanawha Valley; probably equivalent to Upper Elkhorn No. 3 coal of adjacent parts of Kentucky (Blake et al., this volume; Rice et al., 1977).

Lower Chilton sandstone (WV). Massive unranked sandstone in Mingo County, West Virginia, between Fire Clay coal (above) and Hernshaw coal; *not present at Chilton, Kanawha County, West Virginia;* in Kanawha Formation.

Lower Clarion coal (KY). Local name for coal commonly called Lower Richardson in some areas of Martin and Johnson Counties. Kentucky; in Breathitt Formation (Outerbridge, 1964).

Lower Clarksburg limestone (PA, MD). See Clarksburg limestone.

Lower Coalburg sandstone (WV). Massive unranked sandstone typically exposed in Logan and Mingo Counties, West Virginia; in Kanawha Formation between Little Coalburg (above) and Winifrede (Buffalo Creek) coals (Wanless, 1939).

Lower Dotson sandstone (WV). Lower of two massive unranked sandstones exposed at Wyoming (formerly Dotson) station, Sandy River district, McDowell County, West Virginia; in lower part of Kanawha Formation between Douglas (above) and Lower Douglas coals; formerly erroneously correlated with Lower Nuttall sandstone of New River Formation (Wanless, 1939).

Lower Douglas coal (WV). Thin coal about 100 ft (30 m) below Douglas coal at Douglas Station near mouth of Shortpole Creek, on Tug Fork, Sandy River district, McDowell County, West Virginia; lowermost unit in Kanawha Formation (Wanless, 1939).

Lower Elkhorn coal (KY). Lowermost coal of Elkhorn coal zone named for Elkhorn Creek, Pike County, Kentucky (Alvord, 1971); in Breathitt Formation; equivalent to Pond Creek and Eagle coals.

Lower Freeport coal (PA, OH, WV, MD). Lower of two coals exposed at Freeport, on Allegheny River, South Buffalo Township, Armstrong County, western Pennsylvania; in upper part of Allegheny Formation between Upper Freeport (above) and Upper Kittanning coals (Wanless, 1939); locally called Barrelville coal in Maryland (Swartz, 1922).

Lower Freeport limestone (PA, OH). Unranked nonmarine limestone 5 to 20 ft (1.5 to 6 m) below Lower Freeport coal; in upper part of Allegheny Group in Ohio (Sturgeon et al., 1958); in middle part of Freeport Formation of Allegheny Group in Pennsylvania (Flint, 1965).

Lower Guyandot sandstone (WV). Massive unranked sandstone, 20 to 40 ft (6 to 12 m) below Guyandot sandstone; in New River Formation between Sewell "A" (above) and Sewell coals (Wanless, 1939).

Lower Hignite coal (KY, TN). Extensively mined on Hignite Creek, Log Mountains, Bell County, Kentucky; previously assigned as basal unit of Hignite Formation (Ashley and Glenn, 1906) but now assigned to Breathitt Formation (McDowell et al., 1985).

Lower Horsepen coal (VA). Lower of three coals mined south of Horsepen Post Office, Tazewell County, Virginia; in Lee Formation between War Creek (above) and No. 9 Pocahontas coals; equivalent to Little Fire Creek coal in West Virginia (Wanless, 1939).

Lower Howard coal (KY). See Howard coal.

Lower Iaeger sandstone (WV). Massive unranked sandstone exposed at Iaeger, McDowell County, West Virginia; in upper part of New River Formation between Lower Iaeger coal (above) and Lower Iaeger shale (Wanless, 1939).

Lower Iaeger shale (WV). Unranked dark gray shale bearing plant fossils exposed 0.25 mi (0.4 km) northeast of Iaeger, McDowell County, West Virginia; in New River Formation between Lower Iaeger sandstone (above) and Harvey conglomerate (Wanless, 1939).

Lower Kittanning clay (OH). Refractory clay below Lower Kittanning coal of Stout (1916) in southern Ohio (may be Middle Kittanning coal in Pennsylvania) (see Rice, Kosanke, and Henry, this volume); in lower part of Allegheny Formation above Lawrence coal (southern Ohio).

Lower Kittanning coal (PA, OH, WV). Formerly called Kittanning coal, exposed near Kittanning, Armstrong County, western Pennsylvania; in lower part of Allegheny Group between Hamden limestone (Columbiana Member) (above) and Vanport limestone in Ohio and adjacent parts of Pennsylvania (Bownocker and Dean, 1929); mapped into adjacent parts of West Virginia from Pennsylvania; equivalent to Ellerslie coal in Maryland (Swartz, 1922).

Lower Mahoning sandstone (PA, OH, WV). Lower of two massive unranked sandstones at base of Conemaugh Formation; type locality not known but probably along Mahoning Creek, Armstrong County, western Pennsylvania; between Mahoning (above) and Upper Freeport coals; locally coalesces with Upper Mahoning sandstone (Wanless, 1939).

Lower Mercer coal (PA, OH, WV). Thin coal 10 to 20 ft (3 to 6 m) below Lower Mercer limestone, in Mercer County, western Pennsylvania; in Ohio, coal commonly called No. 3 and underlies Boggs limestone or ore; in upper part of Pottsville Formation (Wanless, 1939); mapped into adjacent parts of West Virginia from Pennsylvania.

Lower Mercer limestone (PA, OH). Widespread unranked marine limestone cropping out near Mercer, Mercer County, in western Pennsylvania, where it is lower of two marine limestones in upper part of Pottsville Formation; overlies Middle Mercer coal in Ohio (Wanless, 1939).

Lower Mercer ore (PA, OH). Thin band of iron ore directly on or a few feet above Lower Mercer limestone; in upper part of Pottsville Formation (Wanless, 1939).

Lower Nuttall sandstone (WV). Massive unranked sandstone exposed in upper walls of New River Canyon above Nuttall Station, Fayette County, West Virginia; in upper part of New River Formation between Iaeger "B" (above) and Iaeger "A" coals (Wanless, 1939).

Lower Petros Sandstone Member (TN). Sandstone exposed at town of Petros, Morgan County, Tennessee; in middle part of Slatestone Formation below Petros coal; the thicker of two benches of sandstone as much as 60 ft (18 m) thick at type section (Wilson et al., 1956, as revised by Hardeman et al., 1966).

Lower Pioneer coal (TN). Lower of two coals formerly mined at Pioneer Gap (Old Pioneer), Campbell County, Tennessee; previously in Scott Formation between Upper Pioneer and Jordan coals; correlated with Amburgy coal in Kentucky (Wanless, 1946); now assigned to Graves Gap Formation (Wilson et al., 1956, as revised by Hardeman et al., 1966).

Lower Pine Bald coal (TN). Coal in middle of Vowell Mountain Formation in Cross Mountain section, Anderson County, Tennessee; misidentified in some reports as Petree coal, which is in lower part of Vowell Mountain Formation (Wilson et al., 1956, as revised by Hardeman et al., 1966).

Lower Pittsburgh limestone (PA, WV, OH). Unranked nonmarine limestone 1 to 5 ft (0.3-1.5 m) thick below Little Pittsburgh coal, exposed at Pittsburgh, Allegheny County, Pennsylvania; in upper part of Conemaugh Formation (Wanless, 1939); equivalent to Summerfield limestone *(name abandoned)* in Ohio (Sturgeon et al., 1958).

Lower Pocahontas sandstone (VA, WV). Massive unranked sandstone exposed near Pocahontas, Tazewell County, Virginia; in Pocahontas Formation between No. 3 Pocahontas (above) and No. 2 Pocahontas coals (Wanless, 1939).

Lower Productive Measures (northern Appalachian coal field). Name formerly applied to Allegheny Formation between Lower Barren Measures (Conemaugh Formation) (above) and Millstone Grit (Pottsville Formation) (Wanless, 1939).

Lower Raleigh sandstone (WV). Massive unranked sandstone exposed in New River Canyon, in Raleigh County, West Virginia; in New River Formation between Little Raleigh (above) and Beckley coals (Wanless, 1939).

Lower Richardson coal (KY). Lowermost of two coals in Richardson coal zone in Big Sandy River area, eastern Kentucky; in upper part of Breathitt Formation above Broas coal (Huddle and Englund, 1966); equivalent to coals in Princess No. 5 and Skyline coal zones.

Lower Seaboard coal (VA). Lower of three coals mined at Seaboard, on Big Creek, western Tazewell County, Virginia; in Lee Formation between Middle Seaboard (above) and Upper Horsepen coals; equivalent to Sewell coal in West Virginia (Wanless, 1939).

Lower Sewanee coal (TN). Lower of two coals formerly mined east of Sewanee, Franklin County, Tennessee; also called Jackson coal; in lower part of Whitwell Shale (Formation); correlated with Kelley coal (on Raccoon Mountain, Marion County, Tennessee) and Richland coal (on Walden Ridge near Morgantown, Rhea County, Tennessee) (Wanless, 1946).

Lower Sewickley coal (PA, WV, MD, OH). Coal named for Sewickley, Allegheny County, Pennsylvania; in Monongahela Formation between Sewickley (above) and Fishpot coals in Ohio (Sturgeon et al., 1958); locally called Tyson coal in Maryland (Swartz, 1922).

Lower St. Charles coal (VA). Exposed near railroad station at St. Charles, Lee County, Virginia; in Wise Formation between Upper St. Charles and Pinhook coals; also called No. 2 coal in The Pocket coal field (Wanless, 1946).

Lower Standiford coal (VA). Lower of two coals locally mined on Standiford farm, near head of South Fork of Pound River, Wise County, Virginia; in Wise Formation between Upper Standiford and Kelly coals; equivalent to Harlan coal in Kentucky (Wanless, 1939).

Lower Stinson coal (KY). Coal mined on Lower Stinson Creek, northern Carter County, Kentucky; in Breathitt Formation between Upper Stinson and Danleyton coals; locally known as Hunnewell cannel coal; tentatively correlated with Brookville coal in lower part of Allegheny Formation in Ohio (Wanless, 1939).

Lower War Eagle coal (WV). Lower of three important coals mined at War Eagle, Stafford district, Mingo County, West Virginia; in Kanawha Formation; about 60 ft (18 m) above a marine fossiliferous shale tentatively correlated with Dorothy limestone and shale of Krebs and Teets (1916).

Lower Wilder coal (TN). Coal mined along South Fork, Scott County, Tennessee; source of name unknown; occurs near base of Signal Point Shale in Fentress and Cumberland Counties, Tennessee (Wilson et al., 1956).

Low Gap Sandstone Member (TN). Sandstone exposed in Low Gap on crest of Redoak Mountain between Graves Gap and Grassy Gap along line between Anderson and Campbell Counties, southwest of Lake City, Tennessee; in Cross Mountain Formation; sandstone has a maximum thickness of 70 ft (21 m) (Wilson et al., 1956, as revised by Hardeman et al., 1966).

Low Splint coal (VA, KY). Splint coal occurring low in Black Mountains in Big Stone Gap field, Wise County, Virginia; in Wise Formation in Virginia between Phillips (above) and Taggart coals; equivalent to Amburgy coal in Kentucky and Williamson coal in West Virginia (Wanless, 1939); contains a 0.5-in. (1.3-cm) flint clay (tonstein) parting; in Wise County, Virginia, Low Splint occurs in as many as five splits, designated (from youngest to oldest) A, B, C, D, and E (Nolde et al., 1988); also used as a local name for coal probably in Haddix coal zone west of Middlesboro, Bell County, Kentucky (Rice and Maughan, 1978).

Luke coal (MD). See Middle Kittanning coal.

Lyons coal (VA). Coal exposed near Lyons Post Office on Big Ridge, southwestern Dickenson County, Virginia; in Wise Formation between Blair (above) and Dorchester coals (Wanless, 1939).

Mabie fire clay (WV). Fire clay (both flint and semiflint clay) mined near the community of Mabie, Randolph County, West Virginia; near base of Allegheny Formation (Englund and Goett, 1968); probably equivalent to Ruffner fire clay of Kanawha County, West Virginia.

Magoffin Member (KY, TN, VA). Marine shale commonly having large concretions of limestone exposed at head of Sycamore Branch of Oakley Creek, southern Magoffin County, Kentucky; in Kentucky, in Breathitt Formation between Haddix (above) and Fire Clay coals; in Virginia, in Wise Formation between Reynolds Sandstone Member (above) and Pardee coal; in Tennessee, in Redoak Mountain Formation between Fodderstack Sandstone Member (above) and Sharp coal; equivalent to Winifrede Shale Member of Kanawha Formation in West Virginia; member ranges from 0 to 120 ft (37 m) but averages about 40 ft (12 m) thick (Wanless, 1939; Outerbridge, 1976; Blake et al., this volume).

Mahoning coal (PA, OH, WV). Exposed in Mahoning Creek(s) in Indiana and Jefferson Counties, Pennsylvania; in lower part of Conemaugh Formation between Upper Mahoning sandstone (above) and Thornton clay or Lower Mahoning sandstone (Wanless, 1939; Bownocker and Dean, 1929).

Mahoning limestone (PA). Unranked limestone bed exposed in runs draining into Mahoning River 5 to 6 mi (8 to 10 km) above its junction with Shenango River, Lawrence County, Pennsylvania, between Tionesta sandstone (above) and Tionesta coal (Rogers, 1858); in Pottsville Formation, equivalent to Upper Mercer limestone; name locally used for freshwater limestone between Upper and Lower Mahoning sandstones near base of Conemaugh Group (Bownocker and Dean, 1929).

Mahoning Sandstone Member (PA, OH, MD). Named for Mahoning Creek(s) in Indiana and Jefferson Counties, Pennsylvania; in Conemaugh Formation (Maryland) or Glenshaw Formation (Pennsylvania, Ohio); commonly subdivided into lower and upper parts by Mahoning coal (Flint, 1965).

"Mahoning" sandstone member (WV). Conglomeratic sandstone coming out of bed of Kanawha River near Spring Hill in Kanawha County, West Virginia; named and assigned to Conemaugh Formation by Krebs and Teets (1914); reassigned incorrectly(?) to top of Charleston Sandstone by Englund et al. (1979); intertongues with red shale of overlying Conemaugh Group.

Manchester coal (KY). Coal extensively mined on Goose Creek and tributaries near Manchester, Clay County, Kentucky (Wanless, 1939); in Breathitt Formation between Betsie Shale Member (above) and Corbin Sandstone Tongue of Lee Formation; equivalent to Lily, Swamp Angel, and Clintwood coals.

Marcee coal (VA). See Dorchester coal.

Marcum Hollow Sandstone Member (VA). Sandstone exposed in Marcum Hollow just north of town of Keokee, along Virginia State Highway 624, Lee County, Virginia; as much as 100 ft (30 m) thick; in Wise Formation between Low Splint (above) and Taggart coals (Miller, 1969).

Marion coal (TN). Name applied by A. W. Evans to coal about 60 ft (18 m) above Red Ash coal near Newcomb, Campbell County, Tennessee, north of Pine Mountain (source of name not known); previously in Scott Formation; probably equivalent to Pewee coal (Wanless, 1946); coal probably in strata now assigned to upper part of Redoak Mountain Formation (Wilson et al., 1956, as revised by Hardeman et al., 1966).

Mason coal (zone) (KY). Coal exposed along Bear Creek in Log Mountains, Bell County, Kentucky; correlated with Mingo (Jellico) coal (formerly of Mingo Formation) of Bennetts and Stoney Forks in Log Mountains; the name Mason is also used in Harlan and Bell Counties, Kentucky, for the first extensively mined coal (formerly in Hance Formation) below Harlan coal, also correlative of Mingo coal, and therefore older than Mason in Log Mountains (Wanless, 1946); older Mason and younger Harlan or Mingo coals now assigned to Breathitt Formation (Rice and Ping, 1989).

Mason coal (OH). Coal mined at Mason on Elk River northeast of Charleston, Kanawha County, West Virginia; named Mason and placed above Mahoning sandstone of lower part of Conemaugh Formation; later studies showed this coal to be in Allegheny Formation, roughly equivalent to Lower Kittanning; thus, name Mason is used in Ohio for coal in Conemaugh Formation between Brush Creek coal (above) and Upper Mahoning sandstone, on the basis of this erroneous correlation (Wanless, 1939); in Athens County, Ohio, Sturgeon et al. (1958) named the Mason cyclothem, so-named because it included this coal.

Massillon sandstone (OH). Unranked massive, locally conglomeratic sandstone exposed near Massillon, Stark County, northeastern Ohio, but defined by Stout (1927) in southern Ohio; in Pottsville Formation generally between Bear Run (above) and Quakertown coals; may be equivalent to upper part of Connoquenessing Sandstone Member of Pottsville Formation in Pennsylvania (Wanless, 1939); stratigraphic position not clear (see Slucher and Rice, this volume).

Matewan coal (WV, KY). Coal exposed at mouth of Mate Creek at Matewan, Magnolia district, Mingo County, West Virginia; in Kanawha Formation (Wanless, 1939); coal directly below Betsie Shale Member; equivalent to Clintwood coal in Virginia.

Matewan sandstone (WV). Massive unranked sandstone formerly quarried at Matewan, Mingo County, West Virginia; in Kanawha Formation below Matewan coal (Wanless, 1939).

Maynadier coal (MD). See Upper Bakerstown coal.

McArthur limestone (OH). *Name herein abandoned.* Calcareous marine shale exposed northwest of McArthur, Vinton County, Ohio; correlated with Putnam Hill Limestone Member of Allegheny Formation by Stout (1927) (see also Rice, Kosanke, and Henry, this volume).

McClure Sandstone Member (VA). Massive, locally conglomeratic sandstone exposed along McClure River, central Dickenson County, Virginia; in lower part of Norton Formation below Kennedy coal; equivalent to Upper Nuttall sandstone, top member of New River Formation in West Virginia and in places in Virginia (Wanless, 1939); member 20 to 90 ft (6 to 27 m) thick and feldspathic in type area (Diffenbach, 1989); locally correlated with Bee Rock Conglomerate Member of Lee Formation.

McGuire coal (KY). Coal in Bell and Knox Counties, Kentucky, 30 to 100 ft (9 to 30 m) above Dean coal; in Breathitt Formation; coal commonly canneloid and correlated with Fire Clay rider coal in Kentucky and Big Mary coal in Tennessee (Wanless, 1946).

Meatscaffold coal (KY). Local name for coal in Haddix coal zone; source of name unknown; in Breathitt Formation, eastern Kentucky (Rice and Smith, 1980, Table 2).

Meigs Creek coal (OH). Coal exposed on Meigs Creek, Morgan County, Ohio; in middle part of Monongahela Group between Benwood (above) and Fishpot coals; equivalent to Sewickley, Mapletown, or No. 9 (Ohio) coal (Sturgeon et al., 1958).

Mercer Shale Member (PA). Thin shale named for Mercer, Mercer County, Pennsylvania; in Pottsville Formation between Homewood (above) and Connoquenessing Sandstone Members; contains plant flora of Zone 8 of Read and Mamay (1964) (see also, Keroher et al., 1966).

Merwin coal (TN). Coal prospected by H. J. Merwin in Massengale Mountain, Campbell County, Tennessee; previously in Scott Formation; more generally known as Pewee and rarely as "X" coal (Wanless, 1946); now assigned to upper part of Redoak Mountain Formation (Wilson et al., 1956, as revised by Hardeman et al., 1966).

Meyersdale limestone (MD, PA). Unranked marine limestone exposed in railroad cut east of Meyersdale, Somerset County, Pennsylvania; in Glenshaw Formation of Conemaugh Group; probably equivalent to Cambridge limestone (Flint, 1965).

Middle Horsepen coal (VA). Middle of three coals mined south of Horsepen Post Office, Tazewell County, Virginia; in upper part of Lee Formation; equivalent to Little Raleigh coal in West Virginia (Wanless, 1939).

Middle Kittanning coal (PA, OH). Middle of three coals exposed along Allegheny River at Kittanning, Armstrong County, Pennsylvania; mined coal near middle of Allegheny Formation (Wanless, 1939); underlies Washingtonville marine shale; equivalent to No. 6 (Ohio) coal; equivalent to Luke coal of Maryland (Swartz, 1922).

Middle Mercer coal (PA, OH). Thin coal below Lower Mercer limestone; in Pottsville Formation of southern Ohio (Stout, 1927).

Middlesboro Member (KY, VA, TN). Pebbly to conglomeratic sandstone exposed just east of Middlesboro and north of Cumberland Gap, Bell County, Kentucky; locally more than 700 ft (213 m) thick; in Lee Formation between Hensley Member (above) and Dark Ridge Member; strata previously assigned to "sandstone member C" (Englund, 1964a); also in New River Formation (Englund, 1981).

Middle Seaboard coal (VA). Middle of three coals formerly mined at Seaboard, on Big Creek, western Tazewell County, Virginia; in Lee Formation between Greasy Creek (above) and Lower Seaboard coals; equivalent to Sewell "A" coal in West Virginia (Wanless, 1939).

Mill Creek coal (TN). Lowest coal exposed along Mill Creek near base of old Etna (or Aetna) incline, Marion County, Tennessee; in Gizzard Group below Red Ash coal (Wanless, 1946).

Millard coal (zone) (KY). Coal mined on Russell Fork of Big Sandy River at Millard, Pike County, Kentucky; in Breathitt Formation between Bingham (above) and Hagy coals (Wanless, 1939).

Millers Creek coal (KY). Coal named for Miller Creek, Johnson County, Kentucky; in Breathitt Formation; equivalent to Van Lear or Upper Elkhorn No. 3 coal (Outerbridge, 1964).

Mills coal (KY). Coal mined by Finley Mills on Toms Branch of Salt Gum Fork of Middle Fork of Stinking Creek, Knox County, Kentucky; in Breathitt Formation between Fire Clay coal (above) and Upper Elkhorn No. 3 coal; may be equivalent to Amburgy coal (Wanless, 1946).

Milner coal (VA). Coal named for Milner mine, on Mountain Fork of Stony Creek, Scott County, Virginia; in Lee Formation between Duncan (above) and Starns coal (Wanless, 1939).

Mine Fork coal (KY). Coal exposed along Big Mine Fork of Paint Creek, western Johnson County, Kentucky (Outerbridge, 1967); in Breathitt Formation below Corbin Sandstone Tongue of Lee Formation; approximately equivalent to Gray Hawk coal.

Mingo coal (KY, TN). Coal named for Mingo No. 1 mine on Bennetts Fork, Claiborne County, Tennessee; in Breathitt Formation between Sandstone Parting coal (above) and Bennetts Fork coal (Wanless, 1939); equivalent to Harlan and Jellico coals.

Mingo Formation (KY, TN). *Name herein abandoned.* Named for Mingo Mountain, Claiborne County, Tennessee; included strata from base of Poplar Lick coal (above) to base of Bennetts Fork coal or between Catron (above) and Hance Formations; unit about 950 ft (290 m) thick in type area (Wanless, 1946); formation boundaries are discontinuous coals that are miscorrelated in different areas; now assigned to Breathitt Formation (McDowell et al., 1985).

Monongahela Formation (Group) (PA, OH, WV, MD, KY). Exposed along Monongahela River, western Pennsylvania; includes strata from top of Waynesburg coal (above) to base of Pittsburgh coal; formerly called Upper Productive Measures (Wanless, 1939); locally raised in rank to group and divided into Uniontown (upper) and Pittsburgh Formations (Berryhill et al., 1971); formation ranges from 200 to 500 ft (61 to 152 m) thick.

Montell coal (MD). See Upper Kittanning coal.

Moore Branch coal (KY). Coal named for mines along Moore Branch of Little Fork of Little Sandy River, southern Carter County, Kentucky (Brown, 1977); in Breathitt Formation; probably equivalent to Fire Clay rider coal bed.

Moore coal (KY) Probably a coal in Upper Elkhorn coal zone; source of name unknown; in Breathitt Formation; see Bowling coal.

Morantown coal (PA). Thick, low-quality coal named for village of Morantown, in Somerset County, Pennsylvania, north of Frostburg, Maryland; at top of Casselman Formation of Conemaugh Group, about 5 to 15 ft (1.5 to 4.6 m) below Pittsburgh coal; also locally known as "Dirty Six" coal (not to be confused with Dirty Six coal of New River Formation) (Flint, 1965).

Morgan Springs coal (TN). Mined by G. W. Morgan at Morgan Springs, Bledsoe County, Tennessee; at or near top of Vandever Formation, below Rockcastle Sandstone; probably equivalent to Isoline coal (Wanless, 1946).

Morgantown sandstone (WV, PA, OH). Unranked massive sandstone exposed at Morgantown, Monongalia County, West Virginia; in Conemaugh Formation between Clarksburg redbeds (above) and Elk Lick coal (Wanless, 1939).

Morris coal (zone) (VA, KY). Coals high in Black Mountains near Big Stone Gap, probably named for Morris Gap in Little Black Mountain between Lee County, Virginia, and Harlan County, Kentucky; in Wise Formation in Virginia or Breathitt Formation in Kentucky between High Splint (above) and Pardee coals (Wanless, 1939).

Moss coal (KY). Coal mined by M. J. Moss on Four-Mile Creek east of Pineville, Bell County, Kentucky; in Breathitt Formation between Upper Elkhorn No. 3 (above) and Straight Creek coals (Wanless, 1939).

Mount Savage coals (MD). Two coals named for Mount Savage, Garrett County, Maryland; in Allegheny Formation; lower coal described as overlying Homewood sandstone and equivalent to Brookville coal; upper coal overlies the Mount Savage fire clay (Swartz, 1922); probably equivalent to No. 5 coal of southern Ohio or Princess No. 6 of Kentucky

Mount Savage fire clay (MD). Clay mined on Big Savage Mountain, Garrett County, Maryland; consists of both plastic and flint clay, 5 to 10 ft (1.5 to 3 m) thick; near base of Allegheny Formation between Upper and Lower Mount Savage coals (Swartz, 1922); probably equivalent to Ruffner fire clay of West Virginia.

Mudseam coal (KY). Named for coal containing shale partings in Elliott County, Kentucky (Englund and DeLaney, 1966); in Breathitt Formation; equivalent to Clod or Princess No. 3 coals.

Mudslip coal (KY). Local name for Little Blue Gem coal; in Breathitt Formation, eastern Kentucky (Rice and Smith, 1980, Table 2).

Murray coal (TN). Thin coal exposed in LaFollette region, Campbell County, Tennessee (source of name not stated); previously in Briceville Formation between Kent (above) and Rex coals; correlated with Poplar Creek coal of Tennessee (Wanless, 1946); now assigned to upper part of Crooked Fork Group (Wilson et al., 1956).

Naese coal (KY). Coal locally directly overlying Naese Sandstone Member of Lee Formation in Bell County, Kentucky; in Breathitt Formation (Rice and Ping, 1989).

Naese Sandstone Member (KY). Sandstone exposed in Naese Cliff and The Seven Sisters along Cumberland River about 8 mi (13 km) upstream from Pineville, Bell County, Kentucky; designated as top member of Lee Formation where present on Cumberland overthrust sheet; locally mapped as member of overlying Breathitt Formation; unit as much as 250 ft (76 m) thick in type area (Wanless, 1946; McDowell et al., 1985).

Nelson coal (TN). Coal formerly worked at Nelson mines in Walden Ridge near Dayton, Rhea County, Tennessee; in Raccoon Mountain Formation between Warren Point Sandstone (above) and Goodrich coal; may be equivalent to Etna and Battle Creek coals (Wanless, 1946).

Nemo coal (TN). Coal named for community of Nemo where it was mined, southwest of Wartburg, Morgan County, Tennessee; in Rockcastle Conglomerate (Wilson et al., 1956).

Newcomb Sandstone Member (TN). Sandstone named for Newcomb, Campbell County, Tennessee, where it underlies widely mined Jellico coal; near top of Slatestone Formation; commonly 20 ft (6 m) thick and ranging from 0 to 60 ft (18 m) (Wilson et al., 1956, as revised by Hardeman et al., 1966); correlated by Englund (1968b) with Sand Gap Sandstone Member of Slatestone Formation.

Newland coal (OH). Coal mined on Newland farm near McArthur, Vinton County, Ohio; at base of Allegheny Formation below Putnam Hill Limestone Member; equivalent to No. 4 or miscorrelated Brookville coal of central Ohio (Stout, 1927).

New Livingston coal (KY). Coal in lower part of Breathitt Formation in Rockcastle County, Kentucky; correlated with No. 2 Lee coal (Brown and Osolnik, 1974).

New River Formation (WV, VA). Formation exposed in New River Canyon, Raleigh and Fayette Counties, West Virginia; includes strata from top of Upper Nuttall sandstone (above) to base of No. 8 Pocahontas coal; ranges in thickness from 800 ft (244 m) along New River in Fayette County to about 1,000 ft (305 m) in easternmost outcrops (Wanless, 1939; Englund et al., 1979).

Newton Sandstone (TN). Sandstone named for Newton, Cumberland County, Tennessee; in lower part of Crab Orchard Mountains Group; locally conglomeratic; unit as thick as 200 ft (61 m), about 110 ft (34 m) thick in its type area (Wilson et al., 1956, as revised by Hardeman et al., 1966).

Nickell coal (KY). Coal named for Nickell mines at head of Stone Coal Fork of Caney Creek, Morgan County, Kentucky; in Breathitt Formation (Englund, 1955); probably equivalent to Hazard No. 7 coal.

No. 1 coal (OH). Commonly called Sharon coal, which is above Sharon conglomerate; in Pottsville Formation, Jackson County, southern Ohio (Stout, 1916); name used in other areas of Ohio.

No. 1 Bon Air coal (TN). Lower of two coals mined at Bon Air, White County, Tennessee; also known as Lower Bon Air; in Raccoon Mountain Formation of Gizzard Group (Wanless, 1946).

No. 1 Pocahontas coal (VA, WV). Thin coal exposed at Pocahontas, Tazewell County, Virginia; in Pocahontas Formation between Vivian (above) and Landgraff sandstones (Wanless, 1939).

No. 1 Stearns coal (KY). Lower of two coals mined near Stearns, McCreary County, Kentucky; also called Hudson coal; in lower part of Breathitt Formation (Wanless, 1946).

No. 2 coal (OH), Coal also commonly but erroneously called Quakertown coal; in Pottsville Formation in Jackson County, Ohio, 80 to 100 ft (24 to 30 m) above No. 1 coal (Stout, 1916); locally overlain by fossiliferous marine shale; name used in other areas of Ohio.

No. 2 coal (VA). Name applied to coal also known as Lower St. Charles coal in The Pocket coal field, Lee County, Virginia; in Wise Formation between Upper St. Charles or No. 2A coal and Pinhook coal (Wanless, 1946).

No. 2 Bon Air coal (TN). Upper coal at Bon Air, White County, Tennessee, 25 ft (8 m) above No. 1 coal; also known as Upper Bon Air; in Gizzard Group (Wanless, 1946; Wilson et al., 1956).

No. 2 Gas coal (WV). Name long used in Kanawha River Valley, Kanawha County, West Virginia, for lower split of Campbell Creek coal; in Kanawha Formation (Wanless, 1939; Blake et al., this volume).

No. 2 Lee coal (KY). Coal in lower part of Breathitt Formation along Cumberland Escarpment in eastern Kentucky named by Miller (1910); equivalent to New Livingston or Wolf Creek coals.

No. 2 Pocahontas coal (VA, WV). Thin coal exposed at Pocahontas, Tazewell County, Virginia; in Pocahontas Formation between Lower Pocahontas (above) and Vivian sandstones (Wanless, 1939).

No. 2 Stearns coal (KY). Upper of two coals mined near Stearns, McCreary County, Kentucky; about 40 ft (12 m) above No. 1 Stearns coal; in lower part of Breathitt Formation; also called Beaver Creek coal (Wanless, 1946).

No. 2A coal (VA). Name applied to a coal also known as Upper St. Charles coal in Lee County, Virginia; in Wise Formation (Wanless, 1946).

No. 3 coal (OH). Coal in Scioto County, Ohio, equivalent to Lower Mercer coal in Pottsville Formation (Stout, 1916); name used in other areas of Ohio.

No. 3 Pocahontas coal (VA, WV). Principal coal mined at Pocahontas, Tazewell County, Virginia; in Pocahontas Formation between Upper and Lower Pocahontas sandstones (Wanless, 1939).

No. 4 coal (OH). Coal at base of Allegheny Formation in Ohio; equivalent to Newland coal below Putnam Hill Limestone Member (Stout, 1927); commonly miscorrelated with Brookville coal of Pennsylvania.

No. 4 Durham coal (GA). Name used in Durham coal field, Walker County, Georgia, for lower of two mined coals; also called Tatum coal; about 160 ft (49 m) below No. 5 or main Durham coal; previously in Walden Formation (Wanless, 1946); now assigned to Crab Orchard Mountains Group (Wilson et al., 1956).

No. 4 Pocahontas coal (VA, WV). Coal mined at Pocahontas, Tazewell County, Virginia; in Pocahontas Formation between No. 5 Pocahontas coal (above) and Upper Pocahontas sandstone (Wanless, 1939).

No. 4A coal (OH). Extensively mined coal in southern Ohio underlying Obryan Member (formerly Ferriferous limestone of Stout [1916] in southern Ohio) in Allegheny Formation; coal probably correlative with Lower Kittanning coal in northern Ohio and generally miscorrelated with Clarion coal in other areas of Ohio (Rice, Kosanke, and Henry, this volume).

No. 5 and No. 6 coals (OH). Principal coals overlying Obryan Member in Allegheny Formation in southern Ohio (Rice, Kosanke, and Henry, this volume); names misused in other areas of Ohio for Lower and Middle Kittanning coals of northern Ohio and western Pennsylvania underlying Columbiana Member (Sturgeon and DeLong, 1964) and Washingtonville shale, respectively.

No. 5 coal (VA). Name applied in The Pocket, Lee County, Virginia, to coal more generally called Taggart coal; in Wise Formation between Low Splint (above) and Taggart Marker coals (Wanless, 1946).

No. 5A coal (OH). See Strasburg coal.

No. 5 Block and No. 6 Block coals (WV). Zone of several coals extensively mined in Kanawha River Valley; in Allegheny Formation between "Mahoning" sandstone (above) and Little No. 5 Block coal (Englund et al., 1979).

No. 5 Durham coal (GA). Coal extensively mined in Round Mountain at Durham, Walker County, Georgia, on Lookout Mountain plateau; previously in Walden Formation (Wanless, 1946); also called Durham coal; now assigned to Crab Orchard Mountains Group (Wilson et al., 1956).

No. 5 Pocahontas coal (VA, WV). Locally minable coal, exposed at Pocahontas, Tazewell County, Virginia; in Pocahontas Formation between Eckman sandstone (above) and No. 4 Pocahontas coal (Wanless, 1939).

No. 6 coal (TN). Name applied to coal at Soddy, Hamilton County, Tennessee; in upper part of Gizzard Group; probably equivalent to Angel coal (Wanless, 1946).

No. 6 coal (VA). Name applied to coal also known as "34-inch" or Low Splint E coal in Wise County, Virginia; in Wise Formation (Nolde et al., 1988).

No. 6A coal (OH). Name applied to coal also known as Lower Freeport coal in Ohio; near top of Allegheny Group (Sturgeon et al., 1958).

No. 6 Pocahontas coal (VA, WV). Coal locally mined at Pocahontas, Tazewell County, Virginia; in Pocahontas Formation between Pierpont (above) and Eckman sandstones (Wanless, 1939).

No. 7 coal (OH). Name applied locally to Upper Freeport coal at top of Allegheny Group in Ohio (Sturgeon et al., 1958).

No. 7 coal (TN). Name applied to extensively mined coal at Soddy, Hamilton County, Tennessee; in Whitwell Shale just above Sewanee Conglomerate; probably equivalent to Richland coal of Rhea County, Tennessee; same coal apparently known as No. 10 at Daisy, about 4 mi (6 km) south of Soddy (Wanless, 1946); Whitwell Shale and Sewanee Conglomerate now assigned to Crab Orchard Mountains Group (Wilson et al., 1956).

No. 7 coal (VA). Name locally applied to coal also known as Phillips coal, in Wise County, Virginia; in Wise Formation; equivalent to Fire Clay coal of Kentucky and West Virginia (Nolde et al., 1988).

No. 7 Pocahontas coal (VA, WV). Coal locally mined near Pocahontas, Tazewell County, Virginia; highest coal of Pocahontas Formation between Flattop Mountain Sandstone Member (above) and Pierpont sandstone (Wanless, 1939).

No. 8 coal (OH). Name applied to coal also known as Pittsburgh coal at base of Monongahela Group in Ohio (Sturgeon et al., 1958).

No. 8 coal (TN). Coal in upper part of Whitwell Shale of Crab Orchard Mountains Group at Soddy, Hamilton County, Tennessee; probably equivalent to Sewanee coal of Cumberland Plateau (Wanless, 1946; Wilson et al., 1956).

No. 8 coal (VA). Name applied in The Pocket coal field, Lee County, Virginia, to coal also known as Gin Creek coal; in Wise Formation between Wax (above) and Phillips coals (Wanless, 1946).

No. 8A coal (OH). Name applied to coal also known as Redstone or Pomeroy coal in Ohio; near base of Monongahela Group (Sturgeon et al., 1958).

No. 8 Pocahontas coal (VA, WV). Thin coal exposed at Pocahontas, Tazewell County, Virginia; near base of New River Formation between No. 9 Pocahontas coal (above) and Flattop Mountain Sandstone Member of Pocahontas Formation (Wanless, 1939).

No. 9 coal (TN). Stray coal in middle part of Whitwell Shale at Soddy, Hamilton County, Tennessee (Wilson et al., 1956).

No. 9 coal (OH) Name applied to coal also know as Sewickley coal in Monongahela Group in Ohio (Sturgeon et al., 1958).

No. 9 coal (VA). Name applied in The Pocket coal field, Lee County, Virginia; coal also known as Wax coal; in Wise Formation between Pardee (above) and Gin Creek coals (Wanless, 1946).

No. 9 Pocahontas coal (WV). Coal bed in lower part of New River Formation below Pineville Sandstone Member (Englund et al., 1979).

No. 10 coal (OH). Name applied to coal also known as Uniontown coal in Monongahela Group in Ohio (Sturgeon et al., 1958).

No. 10 coal (TN). Name applied to upper split of Sewanee coal in Whitwell Shale at Soddy, Hamilton County, Tennessee (Wilson et al., 1956).

No. 10 coal (VA). Name locally applied to coal also known as Pardee coal in Wise County, Virginia; in Wise Formation (Nolde et al., 1988).

No. 10 Pocahontas coal (WV). Name locally applied to coal also known as Little Fire Creek coal in New River Formation.

No. 11 coal (OH). Name applied to coal also known as Waynesburg coal at top of Monongahela Group in Ohio (Sturgeon et al., 1958).

No. 11 coal (VA). Name locally applied to coal also known as Morris coal in Wise Formation in Wise County, Virginia (Nolde et al., 1988).

No. 11 Pocahontas coal (WV). Name locally applied to coal also known as Fire Creek coal in New River Formation.

No. 12 coal (OH). Name applied to coal also known as Washington coal of Pennsylvanian or Permian age in Washington Group in Ohio (Sturgeon et al., 1958).

No. 12 coal (TN). Name applied near Soddy, Hamilton County, Tennessee, to a coal in Vandever Formation of Crab Orchard Mountains Group; possibly same as Lantana coal of Cumberland Plateau field (Wanless, 1946; Wilson et al., 1956).

No. 12 Pocahontas coal (WV). Name locally applied to coal also known as Beckley or War Creek coal in New River Formation.

No. 13 coal (VA). Name applied to coal about 100 ft (30 m) above High Splint coal in Harlan Formation on Black Mountain in Wise County, Virginia, and Harlan County, Kentucky (Nolde et al., 1988).

No. 13 Pocahontas coal (WV). Name locally applied to coal also known as Little Raleigh coal in middle part of New River Formation (Englund et al., 1979).

Noble limestone (OH). Unranked marine limestone as much as 5 ft (1.5 m) thick exposed in shale pit of Ava Brick Company, 7 mi (11 km) northwest of Caldwell, Noble County, Ohio; in Conemaugh Group between Harlem coal (above) and Ewing nonmarine limestone (Murphy, 1973).

Norman Pond coal (TN). Coal named for Norman Pond Knob, Fork Mountain 7.5-min quadrangle, in southern Scott County on county line with Morgan County, Tennessee; in lower part of Graves Gap Formation between Armes Gap Sandstone Member (above) and Jordan coal (Wilson et al., 1956, as revised by Hardeman et al., 1966).

North Coalburg coal (WV). Highest coal in bluff above Shrewsbury (formerly North Coalburg) in Cabin Creek district, Kanawha County, West Virginia; in Allegheny Formation; tentatively correlated with Middle Kittanning coal in Pennsylvania (Wanless, 1939).

North Fork coal (VA). Coal bed in middle part of Norton(?) Formation, Lee County, Virginia (Miller and Roen, 1973); correlatives unknown.

North Fork shale (WV). Unranked thin marine shale exposed at North Fork Station on Elkhorn Creek, McDowell County, West Virginia; in Pocahontas Formation between Keystone (above) and Simmons coals (Wanless, 1939).

North Jellico coal (KY). Formerly mined at Gray's Station on L & N Railroad, Whitley County, Kentucky; in Breathitt Formation; correlated with Jellico coal (Wanless, 1946).

Norton coal (VA). Mined at Norton, Wise County, Virginia; at top of Norton Formation directly below Gladeville Sandstone (Wanless, 1939).

Norton Formation (VA). Exposed just south of Norton, Wise County, Virginia; between Gladeville Sandstone (above) and Lee Formation; includes several named coals from Norton (at top) to Tiller coal (Wanless, 1946); formation estimated to be about 1,300 ft (396 m) thick (Miller, 1969).

Nosben coal (KY). Coal exposed near mouth of Nosben Branch of Long Fork of Big Creek, Pike County, Kentucky (Alvord and Trent, 1962); in Breathitt Formation; equivalent to coal in Upper Elkhorn No. 3 coal zone; probably Upper Elkhorn No. 3½ coal.

Nuttall Sandstone Member (WV). Massive conglomeratic sandstone forming upper wall of New River Canyon from Caperton to Deepwater, in part of Fayette County, West Virginia; named for nearby Nuttallburg (now Nuttall); unit as much as 110 ft (34 m) thick in type area; uppermost member of New River Formation where present (Keroher et al., 1966).

Oakdale coal (TN). See Catoosa coal.

Oak Hill clay (OH). Refractory clay produced at Oak Hill, Jackson County, southern Ohio; in lower part of Allegheny Formation above No. 5 coal in southern Ohio (Stout, 1916); probably miscorrelated with Lawrence clay in other areas (Rice, Kosanke, and Henry, this volume).

Oak Hill coal (TN). Name Oak Hill used by W. F. Pond and the late A. W. Evans, but not known to have been published; highest coal in Etna Mountain, Marion County, Tennessee; above Slate coal; commonly called Walker coal; probably in upper part of Whitwell Formation in Crab Orchard Mountains Group (Wanless, 1946; Wilson et al., 1956).

Oakley coal (KY). Coal named for mines along Oakley Creek in southern part of Salyersville South 7.5-min quadrangle, Magoffin County, Kentucky; in Breathitt Formation; probably equivalent to coal in lower part of Hazard No. 7 or Peach Orchard coal zones (Welch, 1958).

Obryan Member (KY, OH). Type section along a power line 1 mi (1.6 km) east of Obryan Cemetery (for which it is named), Ironton 7.5-min quadrangle, Greenup County, Kentucky; marine limestone in southern Ohio and northeastern Kentucky formerly miscorrelated with and called Vanport limestone; as much as 9 ft (3 m) thick; in Breathitt Formation in Kentucky between Hitchins clay bed (above) and Princess No. 5B coal and in Allegheny Formation in Ohio between Lawrence clay bed (above) and No. 4A coal; probably equivalent to Columbiana Member of Allegheny Formation in Pennsylvania and northern Ohio (Rice, Kosanke, and Henry, this volume).

Oceana limestone (WV). Unranked thin marine limestone named by Hennen and Gawthrop (1915) from exposures on Dry Branch, 0.5 mi

(0.8 km) north of Oceana, Oceana district, Wyoming County, West Virginia; in lower part of Kanawha Formation in dark shales above Glenalum Tunnel coal (Wanless, 1939); incorrectly correlated with Dorothy limestone and shale of Krebs and Teets (1916).

Ogan coal (OH). Locally thick coal exposed on Ogan farm northeast of McArthur, Vinton County, Ohio; in lower part of Allegheny Formation just above Putnam Hill Limestone Member (McArthur limestone of Stout, 1916) and just below Zaleski black flint; coal has very limited extent (Stout, 1927).

Olean Conglomerate (N.Y., PA). Named for Olean, Cattaraugus County, New York; at base of Pottsville Group in Pennsylvania (Keroher et al., 1966); equivalent to Sharon Conglomerate Member of Pottsville Formation.

Olive Hill flint clay (KY). Important refractory clay produced at Olive Hill, Carter County, Kentucky; in Breathitt Formation; equivalent to Sciotoville fire clay in Ohio; commonly at very base of Pennsylvanian (Wanless, 1939).

Oliver Springs coal (TN). Name locally applied to Poplar Creek coal near Oliver Springs in Anderson and Morgan Counties, Tennessee; previously in Briceville Formation (Wanless, 1946); now assigned to uppermost part of Crooked Fork Group above Wartburg Sandstone (Wilson et al., 1956, as revised by Hardeman et al., 1966).

Orme coal (TN). Coal extensively mined at Orme, Marion County, Tennessee; correlated with Battle Creek or Etna coal by some and considered a little lower (perhaps equivalent to Dade coal) by others; in Gizzard Group below Warren Point Sandstone (Wanless, 1946).

Paintrock coal (TN). One of the names applied in Scott County to Poplar Creek coal; named because it is mined in Paintrock Creek east of Huntsville; previously in Briceville Formation (Wanless, 1946); now assigned to uppermost part of Crooked Fork Group above Wartburg Sandstone (Wilson et al., 1956).

Panther sandstone (WV). Unranked massive conglomeratic sandstone exposed on Tug Fork between Panther and Douglas stations, Sandy River district, McDowell County, West Virginia; in uppermost part of New River Formation (Wanless, 1939).

Pardee coal (VA, KY). Coal mined at Pardee, on Potcamp Fork of Roaring Fork, Wise County, Virginia; in Wise Formation between Morris (above) and Phillips coals; correlated with Limestone and Taylor coals in eastern Kentucky (Wanless, 1939); locally, the Pardee and Limestone coals are separated by Jesse Sandstone Member of Breathitt Formation in Kentucky.

Parsons coal (zone) (KY). Local name for Limestone or Pardee coals in Breathitt Formation (Rice and Smith, 1980, Table 2).

Path Fork coal (zone) (KY). Coal named for Path Fork of Puckett Creek, Harlan County, Kentucky; in Breathitt Formation between Harlan (above) and Puckett Creek coals (Englund et al., 1963b); equivalent to Imboden coal.

Peach Orchard coal (zone) (KY). Coal mined at Peach Orchard, on Left Fork of Nats Creek, southern Lawrence County, Kentucky; in Breathitt Formation; equivalent to Francis coal (Wanless, 1939); Peach Orchard coal zone may include more than five coals, the lower of which are difficult to distinguish from coals of the Hazard zone.

Pebble seam (WV). See Aily coal.

Peerless coal (WV). Coal named for mining village of Peerless (now abandoned) east of Lewistown, Cabin Creek district, Kanawha County, West Virginia; in Kanawha Formation, commonly considered upper split of Campbell Creek coal, the No. 2 Gas coal being the lower split (Wanless, 1939).

Peerless sandstone (WV). Massive sandstone in Kanawha River Valley, Kanawha County, West Virginia; in Kanawha Formation between Williamson coal (above) and Campbell Creek limestone (Wanless, 1939).

Penn Lee coal (VA). Coal in Lee County, Virginia, indicated by Miller and Roen (1973) to be in upper part of Norton Formation but may be equivalent to Blair coal of Wise Formation.

Petree coal (TN). Coal named after Mr. L.J.A. Petree, who prospected it near Cross Mountain in Anderson County, Tennessee; called Frozenhead coal by Mr. Petree, but that name more generally applied to a lower coal equivalent to Sharp coal; previously at or near top of Scott Formation (Wanless, 1946); now assigned to Vowell Mountain Formation (Wilson et al., 1956, as revised by Hardeman et al., 1966).

Petros Sandstone Member (TN). Sandstone exposed at town of Petros, Morgan County, Tennessee; in middle part of Slatestone Formation; commonly divided into upper and lower members (Upper and Lower Petros Sandstone Members), Petros coal being in between (Wilson et al., 1956, as revised by Hardeman et al., 1966).

Pewee coal (TN). Coal so-named at mine of Sun Coal Company, Caryville, Campbell County, Tennessee; also known as Merwin or "X" coal; previously in Scott Formation between Split Seam (above) and Red Ash coals (Wanless, 1946); now assigned to uppermost part of Redoak Mountain Formation (Wilson et al., 1956, as revised by Hardeman et al., 1966).

Phillips coal (VA). Coal mined on Ambrose Phillips farm at head of South Fork of Pound River, Wise County, Virginia; in Wise Formation between Pardee (above) and Low Splint coals; coal contains flint clay (tonstein) parting and is equivalent to Fire Clay coal in eastern Kentucky (Wanless, 1939).

Phoebe coal (KY). Coal named for exposures on Phoebe Branch of Little Clear Creek, Bell County, Kentucky; in Breathitt Formation between Rich Mountain coal zone (above) and Betsie Shale Member (Rice and Maughan, 1978).

Piedmont coal (MD). Coal mined near Piedmont, Allegany County, Maryland; in lower part of Conemaugh Formation between Gallitzin (above) and Upper Freeport coals; locally called Six-Foot coal; probably equivalent to Mahoning coal of Pennsylvania (Swartz, 1922).

Pierpont sandstone (WV). Massive sandstone exposed at Pierpont station on Slab Fork of Guyandot River, Slab Fork district, Wyoming County, West Virginia; in Pocahontas Formation between No. 7 Pocahontas (above) and No. 5 Pocahontas coals (Wanless, 1939).

Pikeville unit (KY). Informal subdivision of Breathitt Formation between base of Kendrick Shale Member (above) and base of Betsie Shale Member (Chesnut, 1989); unit unranked and not otherwise described or defined.

Pilot Mountain Sandstone Member (TN). Originally named Pilot Knob sandstone by Glenn (1925), renamed Pilot Mountain Sandstone Member for Pilot Mountain, Duncan Flats 7.5-min quadrangle, Anderson County, Tennessee; in Vowell Mountain Formation; as much as 60 ft (18 m) thick (Wilson et al., 1956, as revised by Hardeman et al., 1966).

Pine Creek limestone (PA, WV). Unranked marine limestone exposed on Pine Creek, Allegheny County, West Virginia; in lower part of Conemaugh Formation; more generally known as Cambridge limestone from Ohio locality (Wanless, 1939); probably equivalent to Meyersdale limestone (Flint, 1965).

Pine Hill No. 1 coal (PA). See Redstone coal.

Pine Hill No. 2 coal (PA). See Pittsburgh coal.

Pineville Sandstone Member (WV). Massive sandstone exposed along Guyandot River at Pineville, Wyoming County, West Virginia; basal member of New River Formation where it eroded No. 8 Pocahontas coal (Englund et al., 1979); commonly in interval between Little Fire Creek (above) and No. 9 Pocahontas coals; unit at least 100 ft (30 m) thick in type area (Wanless, 1939).

Pineville coal (KY). See Straight Creek coal.

Pinhook coal (VA). Coal mined locally near Keokee and Mohawk, Lee County, Virginia; source of name not stated; in Wise Formation between Lower St. Charles (above) and Kelly coals (Wanless, 1946).

Pioneer Sandstone Member (TN) Prominently exposed as cap of bench at Old Pioneer, Campbell County, Tennessee; bench formed by this sandstone known as Braden Bench; previously topmost member of Jellico Formation between Jordan (above) and Joyner coals (Wanless, 1946); now assigned as top member of Indian Bluff Formation (Wilson et al., 1956, as revised by Hardeman et al., 1966).

Pittsburgh coal (PA, OH, WV, MD). Coal extensively mined at Pittsburgh, Allegheny County, Pennsylvania; basal unit of Monongahela Formation (Wanless, 1939); locally called Big Vein or Pine Hill No. 2 coal (Flint, 1965) and No. 8 coal in Ohio (Bownocker and Dean, 1929).

Pittsburgh cyclothems (OH). Name (Pittsburgh and Upper Pittsburgh cyclothems) given two cyclothems at base of Monongahela Group below Redstone cyclothem in Ohio (Sturgeon et al., 1958); equivalent to unnamed lower member of Pittsburgh Formation of Monongahela Group in Pennsylvania (Berryhill et al., 1971).

Pittsburgh Formation (PA). Formation includes mudstone, siltstone, sandstone, limestone, and minor coal between bases of Uniontown (above) and Pittsburgh coals in lower part of Monongahela Group in Pennsylvania; formation as much as 264 ft (80 m) thick in type area; includes Redstone, Fishpot, and Sewickley Members (Berryhill et al., 1971), which in Ohio are treated as cyclothems of Monongahela Group (Sturgeon et al., 1958).

Pittsburgh shale (PA, WV). Unranked red shale exposed near Pittsburgh, Allegheny County, Pennsylvania; in Conemaugh Formation between Ames (above) and Cambridge limestones; known as Round Knob shale in Ohio (Wanless, 1939).

Pocahontas coal (zone) (WV, VA). Coals in Tazewell County and adjacent areas of West Virginia numbered upward from 1 to 7 in Pocahontas Formation and from 8 to 13 in overlying New River Formation; No. 8 Pocahontas coal marks base of New River Formation (Englund et al., 1979); see also numbered coals (such as No. 1 Pocahontas coal).

Pocahontas Formation (VA, WV). Formation exposed at Pocahontas, Tazewell County, Virginia; extends from base of No. 8 Pocahontas coal or unconformity at base of Pineville Sandstone Member of New River Formation (above) to base of Pennsylvanian System; formation varies from 0 to about 700 ft (213 m) in thickness (Englund et al., 1979).

Pocahontas unit (KY). Informal member of Breathitt Formation of eastern Kentucky extending from top of improperly termed "Warren Point Sandstone Member of Lee Formation," where present, to top of Mississippian strata (Chesnut, 1989); unit unranked and not otherwise described or defined, presumed to be equivalent to Pocahontas Formation of West Virginia and Virginia; Pocahontas not a member of Lee Formation and incorrectly extended across Kentucky

Pomeroy coal (OH). Coal mined near Pomeroy, Meigs County, Ohio; in Monongahela Formation between Pomeroy sandstone (above) and Redstone limestone; equivalent to Redstone or No. 8A coal in Ohio, Pennsylvania, and West Virginia (Wanless, 1939).

Pond Creek coal (KY). Coal mined on Pond Creek, Pike County, Kentucky, south of Williamson, West Virginia; in Breathitt Formation between Upper Elkhorn No. 1 (above) and Bingham coals (Wanless, 1939); equivalent to Eagle coal in West Virginia (Blake et al., this volume).

Poplar Creek coal (TN). Coal mined along Poplar Creek, Anderson County, and near Oliver Springs and Coalfield, Morgan County, Tennessee; formerly correlated with Coal Creek coal but later recognized to belong 200 ft (61 m) lower in section; previously in Briceville Formation between Coal Creek (above) and Hooper coals; correlated with Glenmary and Helenwood coals of Scott County, Tennessee; also known as Murray coal in LaFollette region, Tennessee (Wanless, 1946); now assigned to uppermost part of Crooked Fork Group above Wartburg Sandstone (Wilson et al., 1956).

Poplar Lick coal (KY, TN). Important coal in Log Mountain area, Bell County, Kentucky; type locality not known; in Breathitt Formation; correlated with lower split of Amburgy coal in Kentucky and Williamson coal in West Virginia (Rice and Ping, 1989).

Portersville limestone (OH). Unranked marine limestone and calcareous shale exposed in an abandoned tunnel entrance near Portersville, Bearfield Township, Perry County, Ohio; in Conemaugh Formation, overlies Anderson coal (Wanless, 1939); probably equivalent to Friendsville shale of Maryland and Pennsylvania (Flint, 1965).

Potters Falls coal (TN). Thin coal named for waterfall of same name where coal is exposed on Crooked Fork, Morgan County, Tennessee; near top of Dorton Shale of Crooked Fork Group (Wilson et al., 1956).

Pottsville Formation (Group) (PA, MD, OH, WV). Formation exposed at Pottsville Gap, Schuylkill County, Pennsylvania (in Southern Anthracite field); commonly identified as lowest division of Pennsylvanian System in Appalachian basin; defined in Pennsylvania and Ohio to include beds to top of Homewood Sandstone (Pennsylvania) or Homewood Sandstone Member (Ohio) or base of Brookville coal; in southeastern West Virginia, equivalent to Pocahontas, New River, and Kanawha Formations; Pottsville Formation is used in northeastern West Virginia only where unable to identify Pocahontas, New River, or Kanawha Formations and extends to top of Roaring Creek sandstone (probably higher than Homewood according to Wanless, 1939).

Poverty Run limestone (OH). Unranked thin marine limestone exposed on Poverty Run, Hopewell Township, Muskingum County, Ohio; in Pottsville Formation between Lower Mercer (above) and Vandusen coals (Wanless, 1939); equivalent to Lowellville limestone.

Powellton coal (WV). Coal mined at Powellton, on Armstrong Creek, Kanawha district, Fayette County, West Virginia; in Kanawha Formation between No. 2 Gas coal (above) and Cannelton limestone of White (1885) (Wanless, 1939).

Powellton "A" coal (WV). Thin coal 15 to 20 ft (5 to 6 m) above Powellton coal at Matewan, Mingo County, West Virginia; in Kanawha Formation in West Virginia and Breathitt Formation in Kentucky, *not present at Powellton* (Wanless, 1939); miscorrelated from Kanawha district of West Virginia as one of several thin "Powellton" coals between Eagle or Pond Creek coal (above) and Betsie Shale Member in Tug Fork area in West Virginia and adjacent parts of Kentucky (Blake et al., this volume).

Prater coal (zone) (KY). Coal named for small abandoned mine on southern side of Prater Branch of Spring Fork, eastern Breathitt County, Kentucky, about 1.2 mi (2 km) from its mouth; in Breathitt Formation; equivalent to Hazard coal (Welch, 1958); zone includes Adele coal in lower part.

Prestonsburg coal (KY). Coal mined near Dwale and Prestonsburg, Floyd County, Kentucky; in Breathitt Formation; equivalent to Van Lear coal or Upper Elkhorn No. 3 coal; below marine Dwale shale (Wanless, 1939).

Princess No. 3 coal (KY). In northeastern Kentucky, a coal equivalent to coal in lower part of Peach Orchard coal zone; in Breathitt Formation (Huddle et al., 1963).

Princess No. 4 coal (KY). Nonpersistent coal probably equivalent to Torchlight coal (Huddle et al., 1963) or a number of poorly defined coals in Breathitt Formation of Princess coal district, northeastern Kentucky; equivalent to Broas coal zone or Hazard No. 9 coal zone (Rice and Smith, 1980).

Princess No. 5 coal (zone) (KY). Coal (coals) locally underlying Kilgore Flint Member of Breathitt Formation in northeastern Kentucky (Huddle et al., 1963); equivalent to Skyline or Richardson coal zones; equivalent to Newland or No. 4 coal of southern Ohio (Rice, Kosanke, and Henry, this volume).

Princess No. 5A coal (KY). Coal between Princess No. 5B (above) and Kilgore Flint Member of Breathitt Formation in northeastern Kentucky (Huddle et al., 1963).

Princess No. 5B coal (KY). Coal directly underlying Obryan Member of Breathitt Formation in northeastern Kentucky; probably equivalent to No. 4A coal of southern Ohio (Rice, Kosanke, and Henry, this volume).

Princess No. 6 coal (zone) (KY). Coal commonly a single bed a few tens of feet above Obryan Member of Breathitt Formation in northeastern Kentucky; equivalent to No. 5 coal of Ohio (Rice, Kosanke, and Henry, this volume); also known as Winslow coal in Kenova 30-min quadrangle (Wanless, 1939).

Princess No. 7 coal (KY). Coal extensively mined in Boyd and Carter Counties, Kentucky; in Breathitt Formation, about 65 ft (20 m) above Obryan Member; also known as Coalton coal, from Coalton, Boyd County, Kentucky; equivalent to No. 6 coal of Ohio (Rice, Kosanke, and Henry, this volume).

Princess No. 8 and No. 9 coals (KY). Coals found locally at top of Breathitt Formation in northeastern Kentucky (Huddle et al., 1963; Ward, 1978); lower split of Princess No. 9 tentatively correlated with Upper Freeport coal of Ohio.

Princess No. 10 coal (KY). Coal about 10 ft (3 m) below strata erroneously called Cambridge Limestone Member of Conemaugh Formation (probably Brush Creek Limestone Member of Conemaugh Formation) in northeastern Kentucky (Huddle et al., 1963).

Princess unit (KY). Topmost informal subdivision of Breathitt Formation in eastern Kentucky between base of Conemaugh Formation (above) and base of Stoney Fork Member (Chesnut, 1989); unit unranked and not otherwise described or defined; Stoney Fork Member discontinuous and does not occur in same parts of eastern Kentucky as Conemaugh Formation.

Puckett Creek (or Puckett) coal (KY). Coal mined at several places along Puckett Creek, Bell and Harlan Counties, Kentucky; in Breathitt Formation; probably equivalent to a coal in Path Fork coal zone (Wanless, 1946).

Puncheon Camp coal (VA). Coal named for Puncheon Camp Branch of Indian Creek, Buchanan County, Virginia; in Norton Formation between Lower Banner (above) and Kennedy rider coals (Miller and Meissner, 1977).

Putnam Hill Limestone Member (OH). Marine limestone exposed at Putnam Hill, Zanesville, Muskingum County, Ohio; in lower part of Allegheny Formation above Newland coal or coal erroneously called Brookville coal; correlated with McArthur limestone in southern Ohio (Wanless, 1939), Kilgore Flint Member of Breathitt Formation in Kentucky, and Kanawha black flint in West Virginia (Rice, Kosanke, and Henry, this volume).

Quakertown coal (PA, OH, MD). Coal locally mined on Quakertown Run, Mahoning Township, Lawrence County, Pennsylvania; in Pottsville Formation (Group) below Upper Connoquenessing Sandstone Member (Wanless, 1939); probably younger than erroneously identified Quakertown coal in southern Ohio (Slucher and Rice, this volume).

Quinnimont sandstone (WV). Unranked massive sandstone exposed in New River Canyon at Quinnimont, Fayette County, West Virginia; in New River Formation between Beckley (above) and Fire Creek coals (Wanless, 1939).

Quinnimont Shale Member (WV). Dark shale named for Quinnimont, Fayette County, West Virginia; in New River Formation between Quinnimont sandstone (above) and Fire Creek coal (Wanless, 1939); contains flora of Zone 5 of Read and Mamay (1964).

Raccoon Mountain Formation (TN). Named for Raccoon Mountain on border between Marion and Hamilton Counties, Tennessee; type section on mine road beginning 0.4 mi (0.6 km) northwest of Whiteside, Marion County, Tennessee, and leading to strip mines at head of Scratch Ankle Hollow, Shellmound 7.5-min quadrangle, Tennessee; includes all shale, sandstone, and coal beds from base of Warren Point Sandstone (above) to base of Pennsylvanian; formation commonly ranges from 0 to 260 ft (79 m) thick but is 353 ft (108 m) thick at type locality; in Gizzard Group (Wilson et al., 1956; thickness revised by Hardeman et al., 1966).

Raleigh Sandstone Member (WV). Named for exposures on road from Prince to Raleigh, Raleigh County, West Virginia; a conglomerate commonly composed of two units locally separated by Little Raleigh coal and shale, an upper unit as much as 75 ft (23 m) thick and a lower unit as much as 100 ft (30 m) thick in type area; in New River Formation between Sewell coal (above) and Pineville Sandstone Member (Englund et al., 1979).

Rattlesnake coal (GA, TN). Coal formerly mined at Rattlesnake mine near Cole City, Dade County, Georgia; in Gizzard Group between Dade (above) and Red Ash coals; name also used in Raccoon Mountains, Hamilton and Marion Counties, Tennessee (Wanless, 1946).

Raven coal (zone) (VA, KY). Coal mined at Red Ash, near Raven, western Tazewell County, Virginia; in Norton Formation between Kennedy (above) and Jawbone coals (Wanless, 1939).

Ravenscroft coal (TN). Coal formerly mined at Ravenscroft, White County, Tennessee; a short distance below Sewanee Conglomerate near top of Signal Point Shale; correlates with Angel and Wilder coals (Wilson et al., 1956).

Red Ash coal (GA, TN). Coal locally mined in Raccoon and Sand Mountains in Dade County, Georgia, and Marion and Hamilton Counties, Tennessee; in Gizzard Group between Rattlesnake (above) and Mill Creek coals (Wanless, 1946; Wilson et al., 1956).

Red Ash coal (northeastern Tennessee). Coal mined at Red Ash, Block, and Caryville, Campbell County, Tennessee; previously in Scott Formation between Braden Mountain (above) and Sharp or Frozenhead coals (Wanless, 1946; Englund, 1968b); now assigned to Redoak Mountain Formation (Wilson et al., 1956, as revised by Hardeman et al., 1966).

Red Ash coal (WV, VA). Name locally used for Douglas and Kennedy coals in lower part of Kanawha Formation in border areas of West Virginia and Virginia.

Red Kidney ore (OH). Nodular ironstone in shale above Strasburg coal in Allegheny Formation (Sturgeon and DeLong, 1964).

Redoak Mountain Formation (TN). Formation named for Redoak Mountain in southwestern part of Lake City 7.5-min quadrangle, Anderson County, Tennessee; strata previously assigned to Scott Formation (Wanless, 1946); includes all strata between top of Pewee coal (above) and top of Windrock coal; formation ranges in thickness from 340 to 420 ft (104 to 128 m); unit between Vowell Mountain (above) and Graves Gap Formations (Wilson et al., 1956, as revised by Hardeman et al., 1966).

Red Springs coal (KY, TN). Coal cropping out high in Log Mountains, near head of Stony Creek, Bell County, Kentucky; type locality not known; above Magoffin Member in Breathitt Formation; tentatively correlated with Hazard coal in Kentucky (McDowell et al., 1985; Rice and Smith, 1980).

Redstone coal (PA, WV, OH). Coal exposed along Redstone Creek, Fayette County, Pennsylvania; in lower part of Monongahela Formation between Sewickley (above) and Pittsburgh coals; locally called Pine Hill No. 1 coal in Pennsylvania (Flint, 1965); overlies Redstone limestone in Pennsylvania and West Virginia; correlated with Pomeroy or No. 8A coal in Ohio (Wanless, 1939); in southwestern Pennsylvania, forms base of Redstone Member of Pittsburgh Formation.

Redstone Member (PA, WV). Member is typically siltstone and mudstone overlain by persistent limestone between Fishpot Member (above) and Redstone coal at base; in Pittsburgh Formation (Monongahela Group) in Pennsylvania (Berryhill et al., 1971); in lower part of Monongahela Group in West Virginia; equivalent to Redstone cyclothem in Ohio, which overlies Upper Pittsburgh cyclothem at base of Monongahela Group; unit is commonly about 35 ft (11 m) thick (Sturgeon et al., 1958).

Rex coal (TN). Coal mined by slope at Rex mine near LaFollette, Campbell County, Tennessee; previously in lower part of Briceville Formation (Wanless, 1946); now assigned to Breathitt Formation below Murray coal on Cumberland overthrust sheet (Rice and Newell, 1990) and near base of Crooked Fork Group in other areas of Tennessee (Wilson et al., 1956).

Reynolds Sandstone Member (KY, VA). Massive, locally conglomeratic sandstone capping Reynolds Mountain, Harlan County, Kentucky; in Breathitt Formation (Kentucky) or Wise Formation (Virginia) between Morris coal (above) and Magoffin Member (Wanless, 1946).

Rich Mountain coal (zone) (TN, KY). Coal mined on Rich Mountain near Habersham, Campbell County, Tennessee; previously defined as basal bed of Briceville Formation; correlated with Blue Gem coal in Kentucky and Tennessee and Imboden coal in Virginia (Wanless, 1946); now assigned to Breathitt Formation between Mingo (above) and Phoebe coals (Rice and Newell, 1990).

Richardson coal (zone) (KY). Coal mined at Richardson on Levisa Fork, Big Sandy River, southeastern Lawrence County, Kentucky; in upper part of Breathitt Formation above Broas coal; correlated with Princess No. 5 and Skyline coals (Huddle et al., 1963).

Richland coal (TN). Coal mined along Richland Creek, near Morgantown, Rhea County, Tennessee; in Crab Orchard Mountains Group in Whitwell Shale (Formation) a few feet above Sewanee Conglomerate; equivalent to Coke Oven coal and Lower Sewanee coal of Cumberland Plateau district (Wanless, 1946; Wilson et al., 1956).

Ridge coal (KY). Coal exposed in ridge above Williams Branch of Elk Fork of Licking River, Morgan County, Kentucky (Johnston, 1962); underlies Limekiln limestone in Breathitt Formation; probably equivalent to Princess No. 5 coal.

Rim coal (KY). Local name for coal bed in Jellico coal zone in area of Pineville, Bell County, Kentucky; in Breathitt Formation between Kendrick Shale Member (above) and Straight Creek coal zone (Froelich and Tazelaar, 1974).

River Gem coal (KY). Coal at base of Betsie Shale Member of Breathitt Formation in McCreary and Whitley Counties, Kentucky; equivalent to Swamp Angel coal (Huddle et al., 1963).

Roach Creek Sandstone Member (TN). Sandstone exposed above Roach Creek near Dean, Block 7.5-min quadrangle, Campbell and Scott Counties, Tennessee, where it is 60 ft (18 m) thick and only about 10 ft (3 m) below Windrock coal; in Graves Gap Formation (Wilson et al., 1956, as revised by Hardeman et al., 1966).

Roaring Creek sandstone (WV). Unranked massive sandstone exposed on Roaring Creek, a branch of Tygart Valley River, Randolph County, West Virginia; possibly uppermost unit of Kanawha Formation but correlated with a sandstone in lower part of Allegheny Formation by Wanless (1939).

Roaring Run shale (PA). Unranked marine shale and siltstone, 4.3 ft (1.3 m) thick, exposed near Roaring Run, Fulton County, Pennsylvania; below pebbly or conglomeratic sandstones in basal part of Pottsville Formation (Edmunds, 1992); probably equivalent to marine zone above No. 2 ("Quakertown") coal in vicinity of Jackson, southern Ohio.

Robbins Chapel Sandstone Member (VA). Sandstone exposed in churchyard at Robbins Chapel in northwestern part of Keokee 7.5-min quadrangle, Lee County, Virginia; as much as 80 ft (24 m) thick; in lower part of Wise Formation, about 120 to 150 ft (37 to 46 m) above base (Miller, 1969).

Robbins coal (TN). Coal formerly mined in connection with brick manufacture at Robbins, Scott County, Tennessee; previously in Briceville Formation; correlated by Wanless with Poplar Creek coal (Wanless, 1946); now assigned to Crooked Fork Group (Wilson et al., 1956).

Rock House coal (TN). Coal exposed on Rock House Branch of Duskin Creek, Rhea County, Tennessee; in Vandever Formation of Crab Orchard Mountains Group just below Rockcastle Sandstone; correlates with Morgan Springs or Isoline coals (Wilson et al., 1956).

Rock Riffle limestone (OH, WV, PA). Unranked nonmarine limestone as much as 4.5 ft (1.4 m) thick exposed along channel of Rock Riffle Run, Athens Township, Athens County, Ohio; in Conemaugh Group associated with underclay of Harlem coal (Sturgeon et al., 1958).

Rock Spring coal (TN). Coal named for its outcrop in a wayside rock spring in gap between Straight Fork and Puncheon Camp Branch 2 mi (3 km) southwest of Turley, Campbell County, Tennessee; previously in Anderson Formation between Frozen Head Sandstone Member (above) and Pilot Mountain Sandstone Member (Wanless, 1946); now assigned to Vowell Mountain Formation (Wilson et al., 1956, as revised by Hardeman et al., 1966).

Rockcastle Conglomerate or Sandstone (Member, Conglomerate Member, or Sandstone Tongue) (KY, TN). Massive conglomeratic sandstone in Lee Formation exposed along Rockcastle River, Rockcastle and Laurel Counties, Kentucky; best exposed at Cumberland Falls, Whitley County, Kentucky; about 150 ft (46 m) below Corbin Sandstone Member; ranges in thickness from 0 to more than 200 ft (61 ft); member or tongue of Lee Formation in Kentucky (Rice, 1984a); name later used more extensively in Tennessee for a Lower Pennsylvanian conglomerate capping broad areas of Cumberland Plateau, mainly Cumberland, Fentress, Overton, and Putnam Counties, Tennessee; the Tennessee Rockcastle Conglomerate is typically exposed at Rockcastle Cove southwest of Jamestown, Fentress County, Tennessee (Wanless, 1946), between Dorton Shale (above) and Vandever Formation; unit now assigned as uppermost formation of Crab Orchard Mountains Group; formation ranges from 150 to 220 ft (46 to 67 m) in thickness (Wilson et al., 1956; thickness revised by Hardeman et al., 1966).

Rockwood coal (TN). Coal mined at Rockwood, Roane County, Tennessee; in Whitwell Shale; correlated with Richland or Lower Sewanee coal in Tennessee (Wanless, 1946).

Rocky Fork coal (VA). Thin coal exposed on Rocky Fork of Guest River north of Addington, Wise County, Virginia; in Wise Formation between Imboden (above) and Addington coals (Wanless, 1939).

Round Knob shale (OH). Unranked red shale exposed on Round Knob, Madison Township, Columbiana County, Ohio; in Conemaugh Formation between Harlem (above) and Barton coals; equivalent to Pittsburgh red shale in Pennsylvania (Wanless, 1939).

Royal shale (WV). Unranked marine shale exposed at Royal, on New River, Shady Spring district, Raleigh County, West Virginia; in Pocahontas Formation, above No. 6 Pocahontas coal (Wanless, 1939).

Ruffner fire clay (WV). Refractory clay mined at South Ruffner, near Charleston, Kanawha County, West Virginia; in Allegheny Formation (Wanless, 1939); equivalent to Elk fire clay above No. 6 Block coal (Englund et al., 1979); probably equivalent to Mount Savage fire clay of Maryland (Swartz, 1922).

Sale Creek coals (TN). Coals formerly mined at Sale Creek, northeast of Chattanooga, Hamilton County, Tennessee; in lower part of Raccoon Mountain Formation of Gizzard Group in Walden Ridge; Sale Creek mines are in a detached fault block separated from the main Walden Ridge escarpment and their correlation with coals in the main Walden Ridge field is uncertain (Wanless, 1946; Wilson et al., 1956).

Salt Lick beds (KY). Unranked thin marine limestone named by Morse (1931) exposed in gap between Salt Lick and Quicksand Creeks, Knott County, Kentucky; in Breathitt Formation (Wanless, 1946); beds represent upper fossiliferous zone in Magoffin Member.

Saltsburg Sandstone Member (PA, MD, OH, WV). Exposed along Conemaugh and Loyalhanna Rivers near Saltsburg, Indiana County, Pennsylvania; in Conemaugh Formation between Harlem coal or Pittsburgh redbeds (above) and Upper Bakerstown coal (Keroher et al., 1966).

Sand Gap Sandstone Member (TN). Sandstone exposed in Sand Gap, a short distance north of Elk Valley, Pioneer 7.5-min quadrangle, Campbell and Scott Counties, Tennessee; in upper part of Slatestone Formation; locally may be 100 ft (30 m) thick (Wilson et al., 1956, as revised by Hardeman et al., 1966); member correlated by Englund (1968b) with Newcomb Sandstone Member as defined by Wilson et al. (1956) and revised by Hardeman et al. (1966).

Sand Block ore (OH). Unranked siliceous iron ore characterized by blocky structure, formerly mined in southern Ohio; in Pottsville Formation between Upper Mercer (above) and Bedford coals; ore contains a few marine fossils (Wanless, 1939).

Sandstone Parting coal (KY, TN). Coal mined in Log Mountains, Bell County, Kentucky, and Claiborne County, Tennessee, containing a sandstone parting 2 to 6 in. (5 to 15 cm) thick; in Breathitt Formation between Poplar Lick (above) and Mingo coals; also called Jack Rock coal because of tough parting; equivalent to Buckeye Spring or Upper Elkhorn No. 3 coals (Rice and Maughan, 1978).

Sciotoville fire clay (OH). Unranked very refractory clay produced near Sciotoville, Scioto County, Ohio; occurs locally at unconformity at base of Pennsylvanian (Pottsville Formation) but locally as much as 100 ft (30 m) above base between Anthony (above) and Sharon coals in southern Ohio; equivalent to Olive Hill flint clay in Kentucky (Wanless, 1939).

Scott Formation (or Shale) (TN). *Name herein abandoned.* Formation consisting largely of shale exposed in mountains of Scott County and adjacent counties, Tennessee; included strata from base of Pilot

Knob Sandstone Member at top to top of Pioneer Sandstone Member of Jellico Formation, including Petree, Pewee, "Walnut Mountain," Red Ash, Sharp, Beech Grove, Big Mary, Windrock, Upper Pioneer, Lower Pioneer, and Jordan coals (Wanless, 1946); strata now assigned to Graves Gap Formation, Redoak Mountain Formation, and lower part of Vowell Mountain Formation (Wilson et al., 1956, as revised by Hardeman et al., 1966).

Scrubgrass coal (PA, OH). Thin coal exposed on Scrubgrass Creek, Clinton Township, Venango County, Pennsylvania; in lower part of Allegheny Formation between Vanport Limestone Member (above) and Clarion coal; may be a split of Clarion coal (Wanless, 1939).

Sebastian coal (KY). Coal named for small mine on Carl Sebastian farm at head of third left hollow of Webb Branch of Caney Creek, Morgan County, Kentucky; in Breathitt Formation (Adkison, 1957); in upper part of Peach Orchard coal zone between Broas (above) and Nickell coals.

Seeber Flats Sandstone Member (TN). Named for Seeber Flats, formed by this sandstone; exposed in Briceville-Norman School section, Lake City 7.5-min quadrangle, Anderson County, Tennessee; near base of Indian Bluff Formation; averages 25 ft (8 m) in thickness (Wilson et al., 1956, as revised by Hardeman et al., 1966).

Seth limestone (WV). Unranked brackish to marine limestone and shale named by Krebs and Teets (1915) from exposures on Coal River at Seth, Sherman district, Boone County, West Virginia; in Kanawha Formation; equivalent to Campbell Creek limestone of White (1885) and Elkins Fork shale in Kentucky (Blake et al., this volume).

Sewanee coal (TN). Coal formerly mined in ridge 2 mi (3 km) east of Sewanee, Franklin County, Tennessee; in Whitwell Shale, about 30 to 50 ft (9 to 15 m) above Sewanee Conglomerate; most extensively mined Lower Pennsylvanian coal in Tennessee; approximately equivalent to Barren Fork coal in Kentucky, Sharon coal in Ohio and Pennsylvania, and Sewell coal in West Virginia (Wanless, 1946).

Sewanee Conglomerate (or Formation) (TN). Massive conglomerate sandstone exposed on campus of University of the South at Sewanee, Franklin County, Tennessee; formerly called Upper Conglomerate of Lookout Sandstone; between Whitwell Shale (above) and Gizzard Group (Wanless, 1946); formation as much as 200 ft (61 m) thick; now assigned as basal formation of Crab Orchard Mountains Group (Wilson et al., 1956; thickness revised by Hardeman et al., 1966).

Sewell coal (WV). Important coal mined on Sewell Mountain on New River, Fayette County, West Virginia; in New River Formation between Hartridge shale (above, where present) and Welch sandstone; may be equivalent to Lower Seaboard coal in Virginia and Sharon coal in Ohio and Pennsylvania (Wanless, 1939).

Sewell "A" coal (WV). Minor coal 30 to 50 ft (9 to 15 m) above Sewell coal in Fayette County, West Virginia; in New River Formation between Sewell "B" coal (above) and Lower Guyandot sandstone (Wanless, 1939).

Sewell "B" coal (WV). Minor coal 75 to 100 ft (23 to 30 m) above Sewell coal; in New River Formation between Guyandot sandstone (above) and Sewell "A" coal (Wanless, 1939).

Sewickley coal (PA, OH, WV). Coal locally mined at Sewickley, Allegheny County, Pennsylvania; in Pittsburgh Formation of Monongahela Group between Lower Uniontown (above) and Lower Sewickley coals; equivalent to Meigs Creek or No. 9 coals in Ohio (Wanless, 1939) and Upper (Borden) and Lower (Tyson) Sewickley coals in Maryland (Swartz, 1922).

Sewickley Member (PA). Member exposed near Sewickley, Allegheny County, Pennsylvania; in Pittsburgh Formation of Monongahela Group between unnamed upper member (above) and Fishpot Member in Pennsylvania (Berryhill et al., 1971); in Ohio equivalent to Sewickley cyclothem (Sturgeon et al., 1958).

Sharon coal (PA, OH). Coal extensively mined near Sharon, Mercer County, Pennsylvania; in Pottsville Formation between Lower Connoquenessing sandstone (above) and Sharon conglomerate; may be equivalent to Sewell coal in West Virginia (Wanless, 1939).

Sharon Conglomerate Member (PA, OH, MD). Massive conglomeratic sandstone exposed near Sharon, Mercer County, western Pennsylvania; basal member of Pottsville Formation through most of western Pennsylvania, eastern Ohio, and locally identified in Maryland (Swartz, 1922; Flint, 1965); locally called Olean Conglomerate in New York and western Pennsylvania; probably equivalent to Raleigh Sandstone Member of New River Formation in West Virginia (Wanless, 1939) and Livingston Conglomerate Member of Lee Formation in eastern Kentucky (Rice and Schwietering, 1988).

Sharon ore (PA, OH). Unranked marine siliceous iron ore above Sharon coal, particularly in southern Ohio; in lower part of Pottsville Formation (Wanless, 1939).

Sharon shale (MD, NY, OH, WV). Thin unranked shale named for Sharon, Mercer County, Pennsylvania; in lower part of Pottsville Formation; overlies Sharon coal or Sharon Conglomerate Member or its equivalent, Olean Conglomerate; contains plant flora of Zone 6 of Read and Mamay (1964).

Sharp coal (TN). Coal prospected and mined by Mr. L.J.A. Petree on Beech Fork of Coal Creek, Anderson County, Tennessee; also known as Frozenhead coal; previously in Scott Formation between Red Ash (above) and Beech Grove coals (Wanless, 1946); now assigned to middle of Redoak Mountain Formation (Wilson et al., 1956, as revised by Hardeman et al., 1966); underlies Magoffin Member; equivalent to Copland or Taylor coal in Kentucky.

Shawnee limestone (OH). Unranked nonmarine limestone exposed near Shawnee, Perry County, Ohio; in Allegheny Group between Bolivar clay (above) and Lower Freeport coal (DeLong and White, 1963).

Sidney coal (KY). Coal exposed in and near Sidney, Belfry 7.5-min quadrangle, northern Pike County, Kentucky; in Breathitt Formation (Alvord and Trent, 1962); equivalent to main bed of Upper Elkhorn No. 3 coal.

Signal Point Shale (TN). Shale exposed on Tennessee State Highway 8 just east of Signal Point and just south of town of Signal Mountain, Chattanooga 7.5-min quadrangle, Hamilton County, Tennessee; top unit of Gizzard Group; shale ranges from 0 to 180 ft (55 m) thick and averages about 60 ft (18 m) thick (Wilson et al., 1956; thickness revised by Hardeman et al., 1966).

Silvey Gap Sandstone Member (TN). Sandstone exposed in Silvey Gap above Windrock, Windrock 7.5-min quadrangle, Anderson County, Tennessee; 60 ft (18 m) thick in type area; near top of Red-

Oak Mountain Formation (Wilson et al., 1956, as revised by Hardeman et al., 1966).

Simmons coal (WV, VA). Thin coal exposed at Simmons Station, near Bramwell, Rock district, Mercer County, West Virginia; in Pocahontas Formation between North Fork shale (above) and Squire Jim coal; in many places, Simmons is used as a synonym for Squire Jim (Wanless, 1939).

Six-Foot coal (MD). See Piedmont coal.

Skelley limestone (OH). Unranked impure marine limestone exposed at Skelley Station, Wayne Township, Jefferson County, Ohio; in Conemaugh Formation between Morgantown sandstone (above) and Ames limestone member; youngest marine horizon in Appalachian coal field (Wanless, 1939).

Skyline coal (zone) (KY). Coal named for Skyline mine on ridge just north on Spring Fork in central part of Tiptop 7.5-min quadrangle, Breathitt County, Kentucky (Welch, 1958); in Breathitt Formation between Flint Ridge flint of Morse (1931) (above) and Stoney Fork Member; coal splits into as many as seven beds in a zone 200 ft (61 m) thick; equivalent to Richardson and Princess No. 5 coal zones (not including Princess Nos. 5A and 5B coals).

Slate coal (TN). Middle of three coals exposed in Etna Mountain, Marion County, Tennessee; in Whitwell Shale; probably equivalent to Sewanee coal (Wanless, 1946).

Slatestone Formation (TN). Formation named for town of Slatestone, Anderson County, Tennessee; includes all strata between top of Jellico (above) and top of Poplar Creek coals; type section begins at track and road at The Wye and extends to Jellico coal at 1,445 ft (440 m) elevation on Militia Hill on Cross Mountain, Lake City 7.5-min quadrangle, Anderson and Campbell Counties, Tennessee; formation ranges from 500 to 720 ft (152 to 219 m) in thickness (Wilson et al., 1956, as revised by Hardeman et al., 1966).

Smith 11-foot coal or Smith coal (zone) (KY). Thick coal exposed on Granville Smith property at head of Lee Fork of Puckett Creek, Harlan County, Kentucky; in Breathitt Formation; probably equivalent to Fire Clay rider coal (Wanless, 1946).

Splash Dam coal (VA, KY). Coal mined at splash dam in Russell Fork of Big Sandy River north of mouth of Pound River, Dickenson County, Virginia; in Norton (Virginia) and Breathitt (Kentucky) Formations between Hagy (above) and Upper Banner coals (Wanless, 1939).

Splint coal (TN). Name locally applied to Jordan coal in vicinity of Elk Valley, Campbell County, Tennessee; previously assigned to position near base of Scott Formation (Wanless, 1946); now assigned to basal part of Graves Gap Formation (Wilson et al., 1956, as revised by Hardeman et al., 1966).

Split Seam coal (TN). Coal commonly split by a thick clay band, in upper levels of mountains of Anderson and Campbell Counties, Tennessee; previously in Scott Formation between Petree (above) and Pewee or Merwin coals (Wanless, 1946); now assigned to Vowell Mountain Formation (Wilson et al., 1956, as revised by Hardeman et al., 1966).

Splitseam coal (KY). Coal mined locally in upper part of Brownies Creek, Varilla 7.5-min quadrangle, Bell County, Kentucky (Englund et al., 1963a); in Breathitt Formation; between Mason (above) and Clear Fork coals.

Springfield coal (KY). Local name for Fire Clay coal in Big Sandy district, eastern Kentucky; in Breathitt Formation (Rice and Smith, 1980).

Squire Jim coal (WV). Thin coal exposed on North Fork of Big Creek, east of Squire Jim, Big Creek district, McDowell County, West Virginia; lowest coal in Pocahontas Formation and probably lowest Pennsylvanian coal in Appalachian coal field; below Simmons coal but locally inadvertently called Simmons (Wanless, 1939).

Stamper coal (KY). Coal exposed on old Stamper land near head of Turkey Creek, a tributary of Line Fork in western Letcher County, Kentucky; in Breathitt Formation (Wanless, 1939); equivalent to Helton coal of Middle Fork of Kentucky River.

Standiford coals (VA). Two coals about 20 ft (9 m) apart, formerly mined by Mr. Standiford near head of South Fork of Pound River, Wise County, Virginia; in Wise Formation between Taggart Marker (above) and Kelly coals; correlated with Harlan coal in Kentucky but considered probably somewhat higher than Harlan coal by Wanless (1946).

Starns coal (VA). Thin coal named for locality on Stony Creek, Scott County, Virginia; in Lee Formation between Milner coal (above) and Bald Rock Conglomerate Member (Wanless, 1939).

State coal (TN). Coal mined by State of Tennessee at Brushy Mountain State Prison, Petros, Morgan County, Tennessee; previously in Jellico Formation; also called Brushy Mountain coal; equivalent to Jellico coal in Tennessee and Kentucky (Wanless, 1946); now assigned to top of Slatestone Formation in Tennessee (Wilson et al., 1956, as revised by Hardeman et al., 1966).

Stearns coal zone (KY). Coals mined west of community of Stearns, McCreary County, Kentucky; in Breathitt Formation at base of Pennsylvanian in that area; includes (upward) Stearns Nos. 1, 2, and 3 (Rice and Smith, 1980) and commonly Barren Fork coal.

Stephens Sandstone Member (TN). Sandstone exposed along Tennessee State Highway 62 immediately south of Stephens siding, Petros 7.5-min quadrangle, eastern Morgan County, Tennessee; in lower part of Slatestone Formation; sandstone is about 40 ft (12 m) thick at type locality (Wilson et al., 1956, as revised by Hardeman et al., 1966).

Sterling coal (TN, KY). Coal mined at Sterling mine near Manring in Log Mountain coal field, Claiborne County, Tennessee; also known as Klondike coal (Wanless, 1946); in Breathitt Formation between Kendrick Shale Member (above) and Poplar Lick coal (Rice and Maughan, 1978); equivalent to upper coal of Amburgy zone.

Stock Creek coal (VA). Coal exposed on Stock Creek, Scott County, Virginia; in Lee Formation between Little Fire Creek (above) and Cove Creek coals; equivalent to Egan coal (Henika, 1988).

Stockstill coal (TN). Coal found in shale overlying Stockstill Sandstone Member of Indian Bluff Formation in Morgan County, Tennessee (Wilson et al., 1956, as revised by Hardeman et al., 1966).

Stockstill Sandstone Member (TN). Sandstone named for Stockstill Creek and exposed on Tennessee State Highway 116 between Petros and Armes Gap, Morgan County, Tennessee; in middle of Indian Bluff Formation; sandstone is 40 ft (12 m) thick at type section (Wilson et al., 1956, as revised by Hardeman et al., 1966).

Stockton coal (WV). Coal formerly mined by Aaron Stockton in hills north of Montgomery, Cabin Creek district, Kanawha County, West Virginia; in upper part of Kanawha Formation between Kanawha black flint (above) and Coalburg coal; also known as Lewiston coal (Wanless, 1939).

Stockton "A" coal (WV). Thin coal locally mined near Eagle, Fayette County, West Virginia; in lowermost Allegheny Formation above Kanawha black flint of White (1891) (Wanless, 1939).

Stoney Fork Member (KY). Marine limestone and shale unit, 20 to 40 ft (6 to 12 m) thick, named for settlement of Stoney Fork, 5 mi (8 km) south of type section in Balkan 7.5-min quadrangle, Bell County, Kentucky; in Breathitt Formation above Hindman (Hazard No. 9) coal; name replaces Lost Creek limestone; equivalent to Boggs (Blunt Run) limestone of Pottsville Formation in Ohio (Ping and Rice, 1979).

Straight Creek coal (KY). Coal mined extensively along Straight Creek, north of Pine Mountain, Bell County, Kentucky; in Breathitt Formation between Moss (above) and Little Blue Gem coal; also known as Pineville coal (Wanless, 1946); equivalent to Blue Gem coal.

Strasburg coal (OH). Coal exposed near Strasburg, Tuscarawas County, Ohio; in Allegheny Group, believed to be between Middle Kittanning (above) and Lower Kittanning coals, locally called No. 5A coal (Sturgeon et al., 1958); locally overlain by Tuscarawas shale with sparse marine fauna (Sturgeon and DeLong, 1964); stratigraphic position uncertain (see Rice, Kosanke, and Henry, this volume).

Stray coal (zone) (KY). Group of as many as 10, commonly thin coals, between Hignite coal (above) and Kendrick Shale Member of Breathitt Formation in Log Mountain area, Bell County, Kentucky (Rice and Maughan, 1978); equivalent to Whitesburg coal zone.

Summerfield limestone (OH). See Lower Pittsburgh limestone.

Swamp Angel coal (TN, KY). Name applied to lowest minable coal near Jellico and Newcomb, Campbell Counties, Tennessee; previously in Briceville Formation below Dixie coal; equivalent to Lily or Williamsburg coal in Kentucky (Wanless, 1946); now assigned to top of Crooked Fork Group in Tennessee (Wilson et al., 1956, as revised by Hardeman et al., 1966).

Tacus coal (VA). See Cove Creek coal.

Taggart coal (VA, KY). Coal named for locality near Dunbar on Roaring Fork of Powell River, Wise County, Virginia; extensively mined; in Wise Formation between Low Splint (above) and Taggart Marker coals (Wanless, 1946); equivalent to Upper Elkhorn No. 3 coal in Letcher County, Kentucky, and Campbell Creek coal in West Virginia.

Taggart Marker coal (VA, KY). Coal mined in Wise and Lee Counties, Virginia; in Wise Formation 30 to 50 ft (9 to 15 m) below Taggart coal (Wanless, 1939).

Tattlers coal (KY). Coal exposed in Tattlers Hollow, Jackson County, Kentucky (Weir and Mumma, 1973); in Breathitt Formation stratigraphically below Corbin Sandstone Tongue of Lee Formation; approximately equivalent to Beattyville coal.

Tatum coal (GA). Coal named after Dr. Robert Tatum, who once prospected this seam in Rock Creek gulch, north of Durham, Walker County, Georgia; also called No. 4 Durham coal; previously assigned to Walden Formation (Wanless, 1946); now assigned to Crab Orchard Mountains Group (Wilson et al., 1956).

Taylor coal (KY). Coal in Breathitt Formation directly underlying Magoffin Member in Pike County, Kentucky; equivalent to Copland and Limestone coals (Rice and Smith, 1980); equivalent to Sharp coal in Tennessee.

Terry Creek coal (TN). Coal named for mine on southern side of Terry Creek, northeastern Campbell County and northern Scott County, Tennessee; in upper part of Slatestone Formation (Wilson et al., 1956, as revised by Hardeman et al., 1966).

Thacker coal (zone) (WV, KY). Coals named for mines in Thacker Branch of Tug Fork of Big Sandy River, Mingo County, West Virginia, and Pike County, Kentucky; generally equivalent to Upper Elkhorn No. 3 coal zone (Rice et al., 1977); in Kanawha Formation (West Virginia) or Breathitt Formation (Kentucky); the name Thacker also may have been used locally for other coals in West Virginia such as Campbell Creek, Peerless, and No. 2 Gas coals.

"34-inch" coal (VA). Name for a coal about 140 ft (43 m) above Taggart coal, also called Low Splint E or No. 6 (or incorrectly, Cedar Grove); in Wise Formation in Wise County, Virginia (Nolde et al., 1988); equivalent to coal in Amburgy coal zone in Kentucky

Thomas coal (MD). Coal named for mines at Thomas, southeastern Garrett County, Maryland; in Conemaugh Formation, equivalent to Lower Bakerstown coal; also known locally as Honeycomb coal (Swartz, 1922).

Tightwad coal (KY). Local name for Manchester coal in Breathitt Formation of eastern Kentucky (Rice and Smith, 1980, Table 2).

Tiller coal (VA). Coal mined by S. J. Tiller of Duty, Virginia, near head of Indian Creek, Dickenson County, Virginia; in Norton Formation between Jawbone coal (above) and Bee Rock Sandstone Member of Lee Formation (Wanless, 1939); may correlate with Lower Iaeger coal.

Tionesta coal (PA, OH). Coal crops out near Tionesta, Forrest County, Pennsylvania; formerly correlated with Brookville coal; basal member of Allegheny Formation; now considered to underlie Brookville coal in Pottsville Formation (Wanless, 1939; see also Rice, Kosanke, and Henry, this volume); also referred to as No. 3B coal in Ohio.

Tiptop coal (KY). Coal mined extensively in Tiptop 7.5-min quadrangle, Breathitt, Magoffin, and Knott Counties, Kentucky; in Breathitt Formation approximately 50 ft (18 m) below basal bed of Skyline coal zone (Danilchik, 1977); coal sometimes referred to as "A" seam; probably part of Skyline coal zone in other areas.

Tom Cooper coal (KY). Coal exposed on land of Tom Cooper on Lick Creek, in northern Magoffin County, Kentucky; in Breathitt Formation between Gun Creek (above) and Lacey Creek coals; equivalent to Upper Elkhorn No. 3 coal (Wanless, 1939).

Torchlight coal (zone) (KY). Coals mined at Torchlight, on Levisa Fork of Big Sandy River, Lawrence County, Kentucky; in Breathitt Formation (Wanless, 1939); probably equivalent to coals in Princess No. 4 coal zone.

Trace Fork coal (KY). Thin coal exposed on Trace Fork, southeastern Magoffin County, Kentucky; in Breathitt Formation between Haddix coal (above) and Magoffin Member (Wanless, 1939).

Travellers Rest coal (KY). Thin coal exposed near Travellers Rest, Owsley County, Kentucky; in Breathitt Formation between Manchester (above) and Beattyville coals; now called Gray Hawk coal (Weir, 1978).

Tub Spring Sandstone Member (TN). Sandstone exposed at top of Frozen Head Mountain and spring on northern side of mountain above Petros, Morgan County, Tennessee, where 50 ft (15 m) of sandstone is preserved; in Cross Mountain Formation (Wilson et al., 1956, as revised by Hardeman et al., 1966).

Tunnel coal (zone) (KY). Thin coals commonly found in Hensley Member of Lee Formation along Pine Mountain in southeastern Kentucky; commonly exposed at sites of tunnels cut through Pine Mountain (Rice and Smith, 1980).

Turkey Pen coal (KY). Discontinuous coal about 100 ft (30 m) above Rim coal in Knox and Bell Counties, Kentucky (Froelich and Tazelaar, 1974); probably equivalent to split of Upper Elkhorn No. 3 coal.

Turner coal (KY). Coal at Turner mines on Lane Branch of Yellow Creek just southwest of Middlesboro, Bell County, Kentucky; in Breathitt Formation 200 ft (61 m) below Bennetts Fork coal (Wanless, 1946); equivalent to Mason or Chenoa coal (Rice and Ping, 1989).

Tuscarawas shale (OH). Name given a marine shale between Red Kidney ironstone (above) and Strasburg coal; in Allegheny Group in eastern Ohio (Sturgeon and DeLong, 1964).

Twin seam (MD). See Wellersburg coal.

Twomile limestone (WV). Unranked nonmarine limestone named by White (1885) for Twomile Run below Charlestown, Kanawha County, West Virginia; in Conemaugh Formation between Morgantown sandstone (above) and Bakerstown coal (Keroher et al., 1966); placed directly below unnamed lower sandstone member of Conemaugh Formation by Englund et al. (1979).

Tyson coal (MD). See Sewickley coal.

Uniontown coal (PA, OH, WV). Coal mined near Uniontown, Fayette County, Pennsylvania; in Monongahela Formation between Waynesburg (above) and Sewickley coals (Wanless, 1939); also called No. 10 coal in Ohio.

Uniontown Formation (PA). Unit exposed at Uniontown, Fayette County, Pennsylvania, including strata from base of Waynesburg coal (above) to base of Uniontown coal; upper of two formations of Monongahela Group; formation varies in thickness from 32 to 93 ft (10 to 28 m) in type area (Berryhill et al., 1971); in Ohio equivalent to (descending) Waynesburg, Little Waynesburg, Uniontown, Lower Uniontown, Arnoldsburg, and Benwood cyclothems of Monongahela Group (Sturgeon et al., 1958).

Upper Bakerstown coal (MD, PA, WV). Upper of two coals exposed at Bakerstown, Allegheny County, Pennsylvania; in Glenshaw Formation of Conemaugh Group above Ewing limestone; locally called Maynadier coal in Maryland (Flint, 1965).

Upper Banner coal (VA, KY). Upper of two coals mined at Banner, on Toms Creek, eastern Wise County, Virginia; in Norton Formation between Splash Dam (above) and Lower Banner coals; may be equivalent to Gilbert coal in West Virginia (Wanless, 1939); identified near base of Breathitt Formation in border areas of Kentucky; in Virginia, contains a distinctive tonstein parting of volcanic origin, commonly described as a sandstone parting 0.1 ft (3 cm) thick.

Upper Bolling coal (VA). Upper of two coals formerly mined by Bolling family in southwestern part of Pound quadrangle, Wise County, Virginia; in Wise Formation (Wanless, 1939).

Upper Bon Air coal (TN). See No. 2 Bon Air coal.

Upper Brush Creek limestone (OH, KY). Exposed in Cranberry Township, Butler County, Pennsylvania; upper of two unranked fossiliferous horizons in southern Ohio and northeastern Kentucky corresponding with Brush Creek limestone; cherty limestone in Conemaugh Formation between Cambridge limestone (above and Lower Brush Creek limestone (Wanless, 1939).

Upper Cliff coal (GA). Coal between unnamed upper and lower conglomerates in Sand and Lookout Mountains, Dade County, Georgia; in Lookout Formation; possibly equivalent to Angel coal in Tennessee (Wanless, 1946); probably equivalent to Bluff seam in a split in Warren Point Sandstone (E. T. Luther, written communication, 1993).

Upper Elkhorn No. 1 coal (KY). Lowest of three coals constituting Upper Elkhorn coal zone; exposed on Elkhorn Creek, southern Pike County, Kentucky; in Breathitt Formation between Upper Elkhorn No. 2 (above) and Lower Elkhorn coals (Wanless, 1946).

Upper Elkhorn Nos. 1 and No. 2 coal zone (KY). Lowest coals constituting Upper Elkhorn coal zone; named for outcrops on Elkhorn Creek, southern Pike County, Kentucky; in Breathitt Formation between Upper Elkhorn No. 3 coal zone (above) and Lower Elkhorn coals; coals locally form a single bed equivalent to Alma coal in West Virginia (Rice et al., 1977).

Upper Elkhorn No. 2 coal (KY). Thick, extensively mined coal near head of Elkhorn Creek, Letcher County, Kentucky; in Breathitt Formation (Wanless, 1939); probably equivalent to No. 2 Gas coal (West Virginia).

Upper Elkhorn No. 3 coal (zone) (KY). Principal Elkhorn coal of Letcher and Pike Counties, Kentucky; equivalent to "Cedar Grove" coal in southwestern West Virginia (Wanless, 1939), which is miscorrelated from Kanawha County, central West Virginia (see Blake et al., this volume); in Breathitt Formation between Elkins Fork shale of Morse (1931) (above) and Upper Elkhorn No. 2 coal.

Upper Elkhorn No. 3½ coal (KY). Coal in Pike County, Kentucky, locally at base of Elkins Fork shale of Morse (1931) in upper part of Upper Elkhorn No. 3 coal zone; in Breathitt Formation; locally called Nosben.

Upper Freeport coal (PA, OH, WV, MD, KY). Coal mined at Freeport, South Buffalo Township, Armstrong County, Pennsylvania; uppermost bed of Allegheny Formation (Wanless, 1939); also locally called "E" or Lemon coal in Pennsylvania, Bleachy coal in Maryland, Davis coal in West Virginia (Flint, 1965), and No. 7 coal in Ohio; tentatively described as uppermost bed of Breathitt Formation in Kentucky by Huddle et al. (1963).

Upper Freeport limestone (PA, OH, WV). Unranked nonmarine limestone below Upper Freeport coal; in upper part of Allegheny Formation (Wanless, 1939).

Upper Grassy Spring coal (TN). Upper of two coals in lower part of Cross Mountain Formation, Campbell County, Tennessee (Wilson et al., 1956, as revised by Hardeman et al., 1966).

Upper Horsepen coal (VA). Upper of three coals mined near Horsepen Post Office, on upper part of Horsepen Creek, Tazewell County, Virginia; in Lee Formation between Lower Seaboard (above) and Middle Horsepen coals; equivalent to Welch coal in West Virginia (Wanless, 1939).

Upper Kittanning coal (PA, OH, WV). Thin coal exposed along Allegheny River near Kittanning, Armstrong County, Pennsylvania; in Allegheny Formation between Lower Freeport (above) and Middle Kittanning coals (Wanless, 1939); in Ohio, may be an upper split of Middle Kittanning coal; equivalent to Montell coal of Maryland (Swartz, 1922).

Upper Mahoning sandstone (PA, OH). Upper of two unranked sandstones exposed on Mahoning Creek, Armstrong County, Pennsylvania; in lower Conemaugh Formation between Humbert or Mason coal (above) and Mahoning coal in Pennsylvania or Ohio, respectively (Flint, 1965).

Upper Mercer coal (PA, OH). Coal below Upper Mercer limestone in Mercer County, Pennsylvania; in Pottsville Formation; in Ohio, coal occurs between Bedford coal (above) and Lower Mercer limestone (Wanless, 1939); also called No. 3A coal in Ohio.

Upper Mercer limestone (PA, OH). Unranked marine limestone and black flint exposed in Mercer County, Pennsylvania; in upper part of Pottsville Formation between Tionesta (above) and Upper Mercer coals; in Ohio, Bedford coal intervenes between it and Upper Mercer coal (Wanless, 1939).

Upper Nuttall sandstone (WV). Unranked massive conglomeratic sandstone forming upper wall of New River Canyon in part of Fayette County, West Virginia; uppermost unit of New River Formation; equivalent to Panther sandstone (Wanless, 1939).

Upper Petros Sandstone Member (TN). Sandstone named for town of Petros, Morgan County, Tennessee; in middle of Slatestone Formation above Petros coal; thinner of two benches of sandstone as much as 60 ft (18 m) thick at type section (Wilson et al., 1956, as revised by Hardeman et al., 1966).

Upper Pine Bald coal (TN) Coal between Rock Spring (above) and Lower Pine Bald coals (source of name unknown), Campbell County, Tennessee; in Vowell Mountain Formation (Wilson et al., 1956, as revised by Hardeman et al., 1966).

Upper Pioneer coal (TN, KY). Upper of two coals formerly mined at Pioneer Gap, Campbell County, Tennessee; previously in Scott Formation between Windrock (above) and Lower Pioneer coals; coal is 68 ft (21 m) above Lower Pioneer coal at type locality; equivalent to Whitesburg coal in Kentucky (Wanless, 1946); now assigned to upper part of Graves Gap Formation (Wilson et al., 1956, as revised by Hardeman et al., 1966).

Upper Pocahontas sandstone (WV). Unranked massive sandstone quarried at Pocahontas, Tazewell County, Virginia; in Pocahontas Formation between No. 4 Pocahontas (above) and No. 3 Pocahontas coals (Wanless, 1939).

Upper Productive Measures (Appalachian coal field). Name formerly applied to Monongahela Formation between Upper Barren Measures (Dunkard Group) above and Lower Barren Measures (Conemaugh Formation) (Wanless, 1939).

Upper Raleigh sandstone (WV). Unranked massive sandstone exposed at top of New River Canyon between Sewell and Prince, Raleigh County, West Virginia; in New River Formation between Welch coal (above) and Little Raleigh coal; equivalent to Rockcastle Conglomerate Member of Lee Formation (eastern Kentucky) and Sharon Conglomerate Member of Pottsville Formation (Ohio) (Wanless, 1939).

Upper St. Charles coal (VA). Thin coal cropping out just above level of main street of St. Charles, Lee County, Virginia; also called No. 2A coal; in Wise Formation between misidentified Harlan (above) and Lower St. Charles coal (Wanless, 1946).

Upper Standiford coal (VA). Upper of two thin coals locally mined on Standiford farm, near head of South Fork of Pound River, Wise County, Virginia; in Wise Formation between Taggart Marker (above) and Lower Standiford coals (Wanless, 1939).

Upper Whitesburg coal (KY). Uppermost coal of Whitesburg coal zone in Letcher, Knott, Floyd, and Pike Counties, Kentucky; in Breathitt Formation; commonly overlain by a brackish-water or marine shale; called Little Fire Clay coal where locally contains thin flint clay (tonstein) parting (Huddle et al., 1963); equivalent to Cedar Grove coal in Kanawha County, West Virginia (Blake et al., this volume).

Van Cleve coal (KY). Coal named for mines at Van Cleave, Breathitt County, Kentucky (Hansen and Johnston, 1963); in Breathitt Formation; probably equivalent to Zachariah or Manchester coals or lower split of those coals.

Van Lear coal (KY). Coal mined extensively on Miller Creek at Van Lear, Johnson County, Kentucky; in Breathitt Formation (Wanless, 1939); equivalent to Upper Elkhorn No. 3 coal in Pike County, Kentucky.

Vanderpool coal (KY). Coal named for coals mined on Vanderpool Mountain, Saxton 7.5-min quadrangle, Whitley County, Kentucky; in Breathitt Formation; probably equivalent to coal in Fire Clay or Fire Clay rider coal zones; in Bell County, name Vanderpool used for coal 60 to 100 ft (18 to 30 m) below Fire Clay coal (Froelich and Tazelaar, 1974), probably a coal in Whitesburg coal zone.

Vandever Formation (or Shale) (TN). Formation consisting mainly of shale exposed at Vandever, Cumberland County, Tennessee; previously in Lee Group between Rockcastle (above) and Bon Air (Newton) Sandstones; includes Lantana, Morgan Springs, and Isoline coals (Wanless, 1946); formation as much as 450 ft (137 m) thick; strata now assigned to Crab Orchard Mountains Group (Wilson et al., 1956).

Vandusen coal (OH). Coal mined on land of Marion Van Dusen, Hamilton Township, Jackson County, Ohio; in upper part of Pottsville Formation between Lower Mercer (above) and Bear Run coals; commonly overlain by Poverty Run limestone (Wanless, 1939).

Vanport limestone (PA, OH). Unranked marine limestone exposed in railroad cut near Vanport, along Ohio River, in Brighton district,

Beaver County, Pennsylvania; in lower part of Allegheny Formation between Lower Kittanning (above) and Clarion coals (Wanless, 1939); formal Vanport Limestone Member of Allegheny Formation in Pennsylvania (Flint, 1965).

Vires coal (KY). Coal named for Vires coal mine at Jackson, Breathitt County, Kentucky (Prichard and Johnston, 1963); in Breathitt Formation between Grassy coal (above) and Frozen Sandstone Member; probably equivalent to Lower Elkhorn coal.

Vivian sandstone (WV). Unranked sandstone quarried for building stone near East Vivian station, Browns Creek district, McDowell County, West Virginia; in Pocahontas Formation between No. 2 Pocahontas (above) and No. 1 Pocahontas coals (Wanless, 1939).

Vowell Mountain Formation (TN). Formation named for Vowell Mountain, a spur on eastern side of Cross Mountain, Campbell County, Tennessee; includes all strata between top of Frozen Head Sandstone Member (above) to top of Pewee coal; formation ranges in thickness from 230 to 375 ft (70 to 114 m); unit between Cross Mountain (above) and Redoak Mountain Formations (Wilson et al., 1956, as revised by Hardeman et al., 1966).

Walden Sandstone (GA, TN). *Name herein abandoned.* Exposed on Walden Ridge, Marion and Hamilton Counties, Tennessee; above Lookout Sandstone (Wanless, 1946); in Tennessee, strata now assigned to Whitwell Shale, Newton Sandstone, Vandever Formation, and Rockcastle Conglomerate of Crab Orchard Mountains Group of Tennessee.

Walker coal (TN). See Oak Hill coal.

Wallins Creek coal (zone) (KY). Coal exposed at head of Wallins Creek, Harlan County, Kentucky; previously at base of Catron Formation (Wanless, 1946); now in Breathitt Formation (McDowell et al., 1985); correlated with Fire Clay or Dean coal in Kentucky and Windrock coal in Tennessee because of its flint clay parting.

Walnut Mountain coal (TN). Coal exposed near summit of Walnut Mountain, near LaFollette, Campbell County, Tennessee; in Breathitt Formation; previously misplaced in Hignite Formation by Englund (1968b); equivalent to Wallins Creek and Fire Clay coals in Kentucky and Windrock coal in Tennessee; generally incorrectly correlated on autochthonous plate (as distinct from Cumberland overthrust sheet) with Braden Mountain or related coals in upper part of Redoak Mountain Formation (Rice, 1984b).

War Creek coal (WV, VA). Coal mined along War Creek, Big Creek district, McDowell County, West Virginia; in New River Formation between Lower Raleigh sandstone (above) and Quinnimont sandstone; now generally correlated with Beckley coal (Wanless, 1939).

Warfield coal (KY, WV). Important coal mined on Tug Fork at Warfield, eastern Martin County, Kentucky; in Breathitt Formation; equivalent to Alma coal in Pike County, Kentucky (Huddle and Englund, 1966).

Warm Fork coal (KY). Coal named for mines along Warm Fork of Cold Cave Creek, Menifee County, Kentucky; in Breathitt Formation but underlies Corbin Sandstone Tongue of Lee Formation (Pipiringos et al., 1968).

Warren Point Sandstone (TN). Sandstone exposed on Warren Point 0.5 mi (0.8 km) north of Monteagle, Grundy County, Tennessee, where sandstone forms bluff 65 ft (20 m) high; formation ranges from 0 to 300 ft (91 m) thick; in upper part of Gizzard Group between Angel (above) and Battle Creek coals (Wanless, 1946; Wilson et al., 1956; thickness revised by Hardeman et al., 1966).

Wartburg Sandstone (TN). Massive sandstone capping an extensive bench in Morgan County, Tennessee, on which town of Wartburg is built; strata previously in Briceville Formation below Poplar Creek coal (Wanless, 1946); sandstone ranges from 0 to 50 ft (15 m) in thickness; now assigned to upper part of Crooked Fork Group (Wilson et al., 1956; thickness revised by Hardeman et al., 1966).

Washington coal zone (OH, WV). Thin, generally very impure coals exposed in highlands of Washington County, Pennsylvania; in Dunkard Group at base of Washington Formation (Late Pennsylvanian or Early Permian in age); directly overlies Waynesburg Formation (Englund et al., 1979); also called No. 12 coal; locally contains parting with brackish-water fossils.

Washingtonville shale (OH). Unranked marine shale exposed near Washingtonville, Salem Township, Columbiana County, Ohio; in Allegheny Formation immediately above Middle Kittanning coal (Wanless, 1939), commonly called No. 6 coal in northern Ohio.

Wax coal (VA). Coal named for Wax prospect in valley of Straight Creek, Lee County, Virginia; in Wise Formation between Pardee (above) and Gin Creek coals; formerly known as No. 9 coal in The Pocket (Wanless, 1946).

Wayland coal (KY). Coal mined at Wayland, southern Floyd County, Kentucky; in Breathitt Formation; equivalent to Upper Elkhorn No. 1 coal (Rice and Smith, 1980).

Waynesburg coal (PA, OH, WV, MD). Coal exposed near Waynesburg, Greene County, Pennsylvania; coal is uppermost unit of Monongahela Formation or Group; in Ohio, equivalent to No. 11 coal (Sturgeon et al., 1958); in Pennsylvania, locally called Koontz coal (Swartz, 1922).

Waynesburg "A" coal zone (PA, OH, WVA). Several thin coals occurring in middle of Waynesburg Formation, basal formation of Dunkard Group (Late Pennsylvanian and Early Permian age); exposed near Waynesburg, Greene County, Pennsylvania; overlies Waynesburg sandstone of Englund et al. (1979).

Waynesburg cyclothems (OH). Names (Waynesburg and Little Waynesburg) given strata at top of Monongahela Group in southeastern Ohio (Sturgeon et al., 1958).

Waynesburg Formation (PA). Formation named for Waynesburg, Greene County, Pennsylvania; includes strata from base of Washington coal (above) to base of Waynesburg coal; formation of Late Pennsylvanian and Early Permian age; lowest formation of Dunkard Group; unit ranges in thickness from 112 to 172 ft (34 to 52 m) in type area (Berryhill et al., 1971).

Welch coal (WV). Coal mined at Welch, McDowell County, West Virginia; in New River Formation between Sewell coal (above) and Upper Raleigh sandstone; equivalent to Upper Horsepen coal in Virginia (Wanless, 1939).

Welch "A" coal (WV). Thin coal, an upper split, a few feet above Welch coal, in southern West Virginia; in New River Formation (Wanless, 1939).

Welch sandstone (WV). Unranked massive sandstone exposed at Welch, McDowell County, West Virginia; in New River Formation between Sewell (above) and Welch coals (Wanless, 1939).

Wellersburg coal (PA). Coal mined near Wellersburg, southeastern Somerset County, Pennsylvania; in middle of Casselman Formation of Conemaugh Group below Morgantown sandstone and above Wellersburg nonmarine limestone (Flint, 1965); locally known as Twin seam in Maryland (Swartz, 1922) and Elk Lick coal in West Virginia.

Wellston coal (OH). Coal mined near Wellston, Jackson County, Ohio; in Pottsville Formation; equivalent to No. 2 or "Quakertown" coal of southern Ohio (Stout, 1916).

Wheelersburg coal (KY). Coal locally mined along Mine Fork at Wheelersburg, northern Magoffin County, Kentucky; in lower part of Breathitt Formation between Howard coal (above) and Corbin Sandstone Tongue of Lee Formation (Wanless, 1939).

Whetstone Creek shale (KY). *Name herein abandoned.* Unranked thick dark gray shale exposed along Cumberland River near mouth of Whetstone Creek about 9 mi (14 km) east of Williamsburg, Whitley County, Kentucky; in Breathitt Formation (Wanless, 1946) above Lily, Manchester, or River Gem coals; strata now assigned to Betsie Shale Member of Breathitt Formation (Rice et al., 1987).

White Oak coal (TN). Coal named for White Oak Creek in Fentress County, east of Jamestown, Tennessee; in middle of Raccoon Mountain Formation of Gizzard Group (Wilson et al., 1956).

White Rocks Sandstone Member (VA, KY). Sandstone exposed at White Rocks, a conspicuous south-facing cliff at crest of Cumberland Mountain, Lee County, Virginia; in Lee Formation between Dark Ridge Member (above) and Chadwell Member; as much as 340 ft (104 m) thick; placed in Mississippian by Englund (1964a); now placed in Pennsylvanian by Englund et al. (1985).

Whitesburg coal (zone) (KY). Coal mined near Whitesburg, Letcher County, Kentucky; in Breathitt Formation between Fire Clay (above) and Amburgy coals (Wanless, 1946); zone may include more than five coals; correlated with Hernshaw and Cedar Grove coals in Kanawha River valley in West Virginia.

Whitwell Formation (or Shale) (TN). Formation exposed at Whitwell mines, Marion County, Tennessee; includes strata between Newton Sandstone (above) and Sewanee Conglomerate; includes Sewanee and Richland coals (Wanless, 1946); formation is as much as 220 ft (67 m) thick; in Crab Orchard Mountains Group (Wilson et al., 1956; thickness revised by Hardeman et al., 1966).

Wild Cat coal (TN). Coal named for a prospect close beneath Wild Cat cliffs above Cold Gap at head of Beech Fork of Coal Creek, Anderson County, Tennessee; previously in Anderson Formation between Bald Knob cannel coal (above) and Cold Gap coal (Wanless, 1946); now assigned to Cross Mountain Formation (Wilson et al., 1956, as revised by Hardeman et al., 1966).

Wilder coal (TN). Coal mined extensively at Wilder, Fentress County, Tennessee; coal in upper part of Signal Point Shale in upper part of Gizzard Group (Wilson et al, 1956).

Wilgus coal (OH). Coal exposed near Wilgus, Lawrence County, Ohio; in Conemaugh Group between Cambridge limestone (above) and Brush Creek limestone (Sturgeon et al., 1958).

Williamsburg coal (KY). Local name applied to Lily coal near Williamsburg, Whitley County, Kentucky; in Breathitt Formation between Whetstone Creek shale (now called Betsie Shale Member) (above) and Corbin Sandstone Tongue of Lee Formation; equivalent to Lily and Manchester coals (Wanless, 1946).

Williamson coal (zone) (WV, KY). Coal mined on Williamson Creek at Williamson, Mingo County, West Virginia; in Kanawha Formation between Dingess Shale Member (above) and Campbell Creek limestone of White (1885); equivalent to Amburgy coal in Kentucky and Low Splint coal in Virginia (Wanless, 1939); lower part or bench locally contains (in Kentucky) thin flint clay (tonstein) parting (Huddle and Englund, 1966).

Wilson coal (VA). Coal in Wise Formation of Lee County, Virginia; equivalent to Harlan coal in adjacent areas of Kentucky (Giles, 1925).

Windrock coal (TN). Coal formerly mined at Upper Windrock, Anderson County, Tennessee; previously in Scott Formation between Big Mary (above) and Upper Pioneer coals; correlated (because of flint clay bed on which it rests at many places) with Fire Clay, Dean, or Wallins Creek coals in Kentucky and Phillips coal in Virginia (Wanless, 1946); now assigned to top of Graves Gap Formation (Wilson et al., 1956, as revised by Hardeman et al., 1966.)

Winifrede coal (WV). Coal mined extensively at Winifrede, Cabin Creek district, Kanawha County, West Virginia; in Kanawha Formation between Coalburg coal (above) and Chilton "A" coals (Wanless, 1939); commonly called Dorothy coal owing to early mining near Dorothy, Raleigh County, West Virginia; possibly correlated with a slightly older coal by the same name in Pike County, Kentucky, and reported by Froelich and Stone (1973) to be misused in eastern Harlan County, Kentucky, for Limestone or Pardee coal.

Winifrede Shale Member (WV). Marine shale and limestone exposed on South Hollow, a branch of Fields Creek, south of Winifrede, Cabin Creek district, Kanawha County, West Virginia; in Kanawha Formation between Chilton "A" (above) and Chilton rider coals; equivalent to Magoffin Member of Breathitt Formation (Blake et al., this volume).

Winslow coal (KY). Coal mined extensively at Winslow, Boyd County, Kentucky; in Breathitt Formation; also known as Princess No. 6 coal (Wanless, 1939).

Winters coal (OH). Coal mined on Winters farm southwest of McArthur, Vinton County, Ohio; in Allegheny Formation between No. 4A coal (above) and Zaleski black flint (Wanless, 1939).

Wise Formation (VA). Formation exposed in Wise County, Virginia; division of the Pennsylvanian including principal coals mined in Wise County between Harlan Sandstone (above) and Gladeville Sandstone; includes beds from High Splint coal (near top) to Dorchester coal (Wanless, 1939); thickness of formation estimated at 2,440 ft (744 m) (Miller, 1969).

Wolf Creek coal (KY). Coal named for exposures in Wolf Creek near Leon, Carter County, Kentucky; in Breathitt Formation, overlies Grayson sandstone bed (Whittington and Ferm, 1967).

Wolf Creek coal (KY). Coal in Breathitt Formation mined on Wolf Creek, Rockcastle County, Kentucky (Gualtieri, 1968); probably equivalent to New Livingston or No. 2 Lee coal.

Woods Run limestone (shale) (PA, MD). Unranked marine limestone or shale named for Woods Run, Pittsburgh area, Allegheny County, Pennsylvania; in Conemaugh Formation between Upper and Lower Bakerstown coals; equivalent to Friendsville shale (Flint, 1965).

Worthington Sandstone Member (PA, MD). Sandstone exposed near Worthington, Armstrong County, Pennsylvania; in Allegheny Formation between Upper and Lower Kittanning coals; commonly subdivided into lower and upper parts by Middle Kittanning coals (Flint, 1965).

Wyatt coal (KY). Coal named for Wyatt's opening on Wills Branch of Grays Fork of Little Goose Creek, Clay County, Kentucky; in Breathitt Formation between Burns (above) and Manchester coals (Wanless, 1946).

"X" coal (TN). Coal prospected and named by H. J. Merwin near Beech Grove, Anderson County, Tennessee; renamed Merwin coal and more widely known as Pewee coal; previously in Scott Formation between Split Seam (above) and Red Ash coals (Wanless, 1946); now assigned to top of Redoak Mountain Formation (Wilson et al., 1956, as revised by Hardeman et al., 1966).

Yellow Creek coal (KY). Local name for Mingo coal in Bell County, Kentucky; in Breathitt Formation (Rice and Smith, 1980, Table 2).

Young coal (KY). Coal mined by William Young on Elklick Fork of Youngs Fork of Lots Creek, Knott County, Kentucky; in Breathitt Formation between Hazard (above) and Haddix coals (Wanless, 1939).

Zachariah coal (zone) (KY). Coal named for mines near village of Zachariah, Lee and Wolfe Counties, Kentucky (Black, 1978); in Breathitt Formation, overlies Corbin Sandstone Tongue of Lee Formation; probably equivalent to Manchester coal.

Zaleski black flint (OH). Unranked marine black flint of limited extent exposed west of Zaleski, Madison Township, Vinton County, Ohio; in Allegheny Formation between Winters (above) and Ogan coals (Wanless, 1939); may be equivalent to Vanport limestone of central Ohio (Rice, Kosanke, and Henry, this volume).

Zelda coal (KY). Coal mined near Zelda, on Big Sandy River, northern Lawrence County, Kentucky; also called No. 9 coal; equivalent to Upper Freeport coal, which is uppermost coal of Breathitt Formation in Kentucky and of Allegheny Formation in Ohio (Wanless, 1939).

ACKNOWLEDGMENTS

We thank all those personnel of the West Virginia Geological and Economic Survey, the Ohio Geological Survey, and the Kentucky Geological Survey who reviewed versions of this manuscript. Jack E. Nolde, the Virginia Geological Survey, and Edward T. Luther, Tennessee Division of Geology, painstakingly reviewed the final version and deserve our special thanks. Edward Luther, in particular, was extremely helpful in updating the stratigraphic usages for the southern part of the central Appalachian basin.

REFERENCES CITED

Adkison, W. L., 1957, Coal geology of the White Oak quadrangle, Magoffin and Morgan Counties, Kentucky: U.S. Geological Survey Bulletin 1047-A, 23 p., map scale 1:24,000.

Alvord, D. C., 1971, Geologic map of the Hellier quadrangle, Kentucky-Virginia and part of the Clintwood quadrangle, Pike County, Kentucky: U.S. Geological Survey Geologic Quadrangle Map GQ-950, scale 1:24,000.

Alvord, D. C., and Miller, R. L., 1972, Geologic map of the Elkhorn City quadrangle, Kentucky-Virginia, and part of the Harman quadrangle, Pike County, Kentucky: U.S. Geological Survey Geologic Quadrangle Map GQ-951, scale 1:24,000.

Alvord, D. C., and Trent, V. A., 1962, Geology of the Williamson quadrangle in Kentucky: U.S. Geological Survey Geologic Quadrangle Map GQ-187, scale 1:24,000.

Ashley, G. H., and Glenn, L. C., 1906, Geology and mineral resources of the Cumberland Gap coal field, Kentucky: U.S. Geological Survey Professional Paper 49, 239 p., map scale 1:45,000.

Berryhill, H. L., Jr., Schweinfurth, S. P., and Kent, B. H., 1971, Coal-bearing Upper Pennsylvanian and Lower Permian rocks, Washington area, Pennsylvania: U.S. Geological Survey Professional Paper 621, 47 p.

Black, D.F.B., 1977, Geologic map of the Heidelberg quadrangle, east-central Kentucky: U.S. Geological Survey Geologic Quadrangle Map GQ-1340, scale 1:24,000.

Black, D.F.B., 1978, Geologic map of the Zachariah quadrangle, east-central Kentucky: U.S. Geological Survey Geologic Quadrangle Map GQ-1452, scale 1:24,000.

Bownocker, J. A., and Dean, E. S., 1929, Analysis of the coals of Ohio: Ohio Division of Geological Survey Bulletin 34, 360 p.

Brown, A., Berryhill, H. L., Jr., Taylor, D. A., and Trumbull, J.V.A., 1952, Coal resources of Virginia: U.S. Geological Survey Circular 171, 57 p.

Brown, W. R., 1977, Geologic map of the Willard quadrangle, eastern Kentucky: U.S. Geological Survey Geologic Quadrangle Map GQ-1387, scale 1:24,000.

Brown, W. R., and Osolnik, M. J., 1974, Geologic map of the Livingston quadrangle, southeastern Kentucky; U.S. Geological Survey Geologic Quadrangle Map GQ-1179, scale 1:24,000.

Chesnut, D. R., Jr., 1989, Pennsylvanian rocks of the Eastern Kentucky Coal Field, in Cecil, C. B., and Eble, C. F., eds., Carboniferous geology of the eastern United States, St. Louis, Missouri, to Washington, D.C., June 28–July 8, 1989, Field trip guidebook T143 for the 28th International Geological Congress: Washington D.C., American Geophysical Union, p. 57–63.

Danilchik, W., 1977, Geologic map of the Tiptop quadrangle, eastern Kentucky: U.S. Geological Survey Geologic Quadrangle Map GQ-1410, scale 1:24,000.

DeLaney, A. O., and Englund, K. J., 1973, Geologic map of the Ault quadrangle, northeastern Kentucky: U.S. Geological Survey Geologic Quadrangle Map GQ-1066, scale 1:24,000.

DeLong, R. M., and White, G. W., 1963, Geology of Stark County: Ohio Division of Geological Survey Bulletin 61, 209 p., map scale 1:62,500.

Diffenbach, R. N., 1989, Geology of the Nora quadrangle, Virginia: Virginia Division of Mineral Resources Publication 92, scale 1:24,000.

Dutcher, R. R., Ferm, J. C., Flint, N. K., and Williams, E. G., 1959, The Pennsylvanian of western Pennsylvania (Geological Society of America guidebook, field trip 2): New York, New York: Geological Society of America, p. 61–114.

Edmunds, W. E., 1992, Early Pennsylvanian (middle Morrowan) marine transgression in south-central Pennsylvania (Northeastern Geology, v. 14, no. 2, p. 225–231.

Englund, K. J., 1955, Geology of the Cannel City area, Kentucky: U.S. Geological Survey Bulletin 1020-A, 21 p., map scale 1:24,000.

Englund, K. J., 1964a, Stratigraphy of the Lee Formation in the Cumberland

Mountains of southeastern Kentucky, *in* Geological Survey Research 1964: U.S. Geological Survey Professional Paper 501-B, p. B30–B38.

Englund, K. J., 1964b, Geology of the Middlesboro South quadrangle, Tennessee, Kentucky, and Virginia: U.S. Geological Survey Geologic Quadrangle Map GQ-301, scale 1:24,000.

Englund, K. J., 1968a, Geologic map of the Bramwell quadrangle, West Virginia–Virginia: U.S. Geological Survey Geologic Quadrangle Map GQ-745, scale 1:24,000.

Englund, K. J., 1968b, Geology and coal resources of the Elk Valley area, Tennessee and Kentucky: U.S. Geological Survey Professional Paper 572, 59 p., map scale 1:24,000.

Englund, K. J., 1981, Geologic map of the Jewell Ridge quadrangle, Buchanan and Tazewell Counties, Virginia: U.S. Geological Survey Geologic Quadrangle Map GQ-1550, scale 1:24,000.

Englund, K. J., and DeLaney, A. O., 1966, Geologic map of the Bruin quadrangle, Elliott and Carter Counties, Kentucky: U.S. Geological Survey Geologic Quadrangle Map GQ-522, scale 1:24,000.

Englund, K. J., and Goett, H. J., 1968, Occurrence of refractory clay in Randolph County, West Virginia: U.S. Geological Survey Professional Paper 600-C, p. C1–C3.

Englund, K. J., Landis, E. R., and Smith, H. L., 1963a, Geology of the Varilla quadrangle, Kentucky-Virginia: U.S. Geological Survey Geologic Quadrangle Map GQ-190, scale 1:24,000.

Englund, K. J., Smith, H. L., Harris, L. D., and Stephens, J. G., 1963b, Geology of the Ewing quadrangle, Kentucky and Virginia: U.S. Geological Survey Bulletin 1142-B, p. B1-B23, map scale 1:24,000.

Englund, K. J., Arndt, H. H., and Henry, T. W., eds., 1979, Proposed Pennsylvanian System stratotype, Virginia and West Virginia: American Geological Institute Selected Guidebook Series, no. 1, 138 p.

Englund, K. J., Gillespie, W. H., Cecil, C. B., Windolph, J. F., Jr., and Crawford, T. J., 1985, Characteristics of the Mississippian-Pennsylvanian boundary and associated coal-bearing rocks in the southern Appalachians: U.S. Geological Survey Open-File Report 85-577, 83 p.

Flint, N. K., 1965, Geology and mineral resources of southern Somerset County, Pennsylvania: Pennsylvania Geological Survey (4th ser.) County Report C56A, 267 p.

Froelich, A. J., and Stone, B. D., 1973, Geologic map of parts of the Benham and Appalachia quadrangles, Harlan and Letcher Counties, Kentucky: U.S. Geological Survey Geologic Quadrangle Map GQ-1059, scale 1:24,000.

Froelich, A. J., and Tazelaar, J. F., 1974, Geologic map of the Pineville quadrangle, Bell and Knox Counties, Kentucky: U.S. Geological Survey Geologic Quadrangle Map GQ-1129, scale 1:24,000.

Giles, A. W., 1925, The geology and coal resources of the coal-bearing portion of Lee County, Virginia: Virginia Geological Survey Bulletin 26, 216 p.

Glenn, L. C., 1925, The northern Tennessee coal field: Tennessee Division of Geology Bulletin 33-B, 478 p.

Gualtieri, J. L., 1968, Geologic map of the Wilde quadrangle, Garrard and Rockcastle Counties, Kentucky: U.S. Geological Survey Geologic Quadrangle Map GQ-684, scale 1:24,000.

Hansen, W. R., and Johnston, J. E., 1963, Geology of the Landsaw quadrangle, Kentucky: U.S. Geological Survey Geologic Quadrangle Map GQ-201, scale 1:24,000.

Hardeman, W. D., and 6 others, 1966, Geologic map of Tennessee, east-central sheet: Nashville, Tennessee Division of Geology, scale 1:250,000.

Hatch, N. L., Jr., 1963a, Geology of the Bernstadt quadrangle, Kentucky: U.S. Geological Survey Geologic Quadrangle Map GQ-202, scale 1:24,000.

Hatch, N. L., Jr., 1963b, Geology of the Billows quadrangle, Kentucky: U.S. Geological Survey Geologic Quadrangle Map GQ-228, scale 1:24,000.

Henika, W. S., 1988, Geology of the East Stone Gap quadrangle, Virginia: Virginia Division of Mineral Resources Publication 79, scale 1:24,000.

Hennen, R. V., and Gawthrop, R. M., 1915, Wyoming and McDowell Counties: West Virginia Geological and Economic Survey [County Report], 783 p., 1 map, scale 1:62,500.

Hinds, H., 1918, The geology and coal resources of Buchanan County, Virginia: Virginia Geological Survey Bulletin 18, 278 p.

Huddle, J. W., and Englund, K. J., 1966, Geology and coal reserves of the Kermit and Varney area, Kentucky: U.S. Geological Survey Professional Paper 507, 83 p., scale 1:24,000.

Huddle, J. W., Lyons, E. J., Smith, H. L., and Ferm, J. C., 1963, Coal reserves of eastern Kentucky: U.S. Geological Survey Bulletin 1120, 247 p.

Hunt, C. B., Briggs, G. H., Jr., Munyan, A. C., and Wesley, G. R., 1937, Coal deposits of Pike County, Kentucky: U.S. Geological Survey Bulletin 876, 92 p.

Johnston, J. E., 1962, Geology of the Lenox quadrangle, Kentucky: U.S. Geological Survey Geologic Quadrangle Map GQ-181, scale 1:24,000.

Keroher, G. C., and 14 others, 1966, Lexicon of geologic names of the United States for 1936-1960: U.S. Geological Survey Bulletin 1200, 4341 p.

Krebs, C. E., and Teets, D. D., Jr., 1914, Kanawha County: West Virginia Geological and Economic Survey [County Report], 679 p., scale 1:62,500.

Krebs, C. E., and Teets, D. D., Jr., 1915, Boone County: West Virginia Geological and Economic Survey [County Report], 648 p., scale 1:62,500.

Krebs, C. E., and Teets, D. D., Jr., 1916, Raleigh County and the western part of Mercer and Summers Counties: West Virginia Geological and Economic Survey [County Report], 648 p., scale 1:62,500.

McDowell, R. C., Rice, C. L., and Newell, W. L., 1985, Revision of Lower and Middle Pennsylvanian nomenclature in the Cumberland overthrust sheet of southeastern Kentucky and eastern Tennessee, *in* Stratigraphic notes, 1984: U.S. Geological Survey Bulletin 1605-A, p. A35–A43.

Miller, A. M., 1910, Coals of the Lower Measures along the western border of the eastern coal field: Kentucky Geological Survey Bulletin 12, 83 p.

Miller, R. L., 1965, Geologic map of the Big Stone Gap quadrangle, Virginia: U.S. Geological Survey Geologic Quadrangle Map GQ-424, scale 1:24,000.

Miller, R. L., 1969, Pennsylvanian formations of southwest Virginia: U.S. Geological Survey Bulletin 1280, 62 p.

Miller, R. L., and Meissner, C. R., Jr., 1977, Geologic map of the Big A Mountain quadrangle, Buchanan and Russell Counties, Virginia: U.S. Geological Survey Geologic Quadrangle Map GQ-1350, scale 1:24,000.

Miller, R. L., and Roen, J. B., 1973, Geologic map of the Pennington Gap quadrangle, Lee County, Virginia, and Harlan County, Kentucky: U.S. Geological Survey Geologic Quadrangle Map GQ-1098, scale 1:24,000.

Morse, W. C., 1931, Pennsylvanian invertebrate fauna: Kentucky Geological Survey, ser. 6, v. 36, p. 293–348.

Murphy, J. L., 1973, The Noble Limestone Member (Conemaugh Group, Pennsylvanian)—New occurrence in Noble and Guernsey Counties, Ohio: Ohio Journal of Science, v. 73, no. 1, p. 42–46.

Nelson, W. A., 1925, The southern Tennessee coal field: Tennessee Division of Geology Bulletin 33-A, 239 p.

Nolde, J. E., and Diffenbach, R. N., 1988, Geology of the Coeburn quadrangle and the coal-bearing portion of the Dungannon quadrangle, Virginia: Virginia Division of Mineral Resources Publication 81, scale 1:24,000.

Nolde, J. E., Whitlock, W. W., and Lovett, J. A., 1988, Geology of Virginia portion of the Flat Gap quadrangle: Virginia Division of Mineral Resources Publication 71, scale 1:24,000.

North American Commission on Stratigraphic Nomenclature, 1983, North American stratigraphic code: American Association of Petroleum Geologists Bulletin, v. 67, p. 841–875.

Outerbridge, W. F., 1964, Geology of the Offutt quadrangle, Kentucky: U.S. Geological Survey Geologic Quadrangle Map GQ-348, scale 1:24,000.

Outerbridge, W. F., 1967, Geologic map of the Oil Springs quadrangle, eastern Kentucky: U.S. Geological Survey Geologic Quadrangle Map GQ-586, scale 1:24,000.

Outerbridge, W. F., 1976, The Magoffin Member of the Breathitt Formation, *in* Cohee, G. V., and Wright, W. B., eds., Changes in stratigraphic nomenclature by the U.S. Geological Survey, 1975: U.S. Geological Survey Bulletin 1422-A, p. A64, A65.

Outerbridge, W. F., 1978, Geologic map of the Cowcreek quadrangle, Owsley and Breathitt Counties, Kentucky: U.S. Geological Survey Geologic Quadrangle Map GQ-1448, scale 1:24,000.

Ping, R. G., and Rice, C. L., 1979, Stoney Fork Member (new name) of the Breathitt Formation in southeastern Kentucky, in Sohl, N. F., and Wright, W. B., eds., Changes in stratigraphic nomenclature by the U.S. Geological Survey 1978: U.S. Geological Survey Bulletin 1482-A, p. A70–A76.

Pipiringos, G. N., Bergman, S. C., and Trent, V. A., 1968, Geologic map of the Ezel quadrangle, Morgan and Menifee Counties, Kentucky: U.S. Geological Survey Geologic Quadrangle Map GQ-721, scale 1:24,000.

Prichard, G. E., and Johnston, J. E., 1963, Geology of the Jackson quadrangle, Kentucky: U.S. Geological Survey Geologic Quadrangle Map GQ-205, scale 1:24,000.

Prostka, H. J., and Seiders, V. M., 1968, Geologic map of the Leatherwood quadrangle, southeastern Kentucky: U.S. Geological Survey Geologic Quadrangle Map GQ-723, scale 1:24,000.

Read, C. B., and Mamay, S. H., 1964, Upper Paleozoic flora zones and floral provinces: U.S. Geological Survey Professional Paper 454-K, 35 p., 19 plates.

Reeves, R. G., 1964, Geology of the Hima quadrangle, Kentucky: U.S. Geological Survey Geologic Quadrangle Map GQ-319, scale 1:24,000.

Rice, C. L., 1969, Geologic map of the Ivyton quadrangle, eastern Kentucky: U.S. Geological Survey Geologic Quadrangle Map GQ-801, scale 1:24,000.

Rice, C. L., 1980, Kendrick Shale Member of the Breathitt Formation in eastern Kentucky, in Sohl, N. F., and Wright, W. B., eds., Changes in stratigraphic nomenclature by the U.S. Geological Survey 1979: U.S. Geological Survey Bulletin 1502-A, p. A117–A122.

Rice, C. L., 1984a, Sandstone units of the Lee Formation and related strata in eastern Kentucky: U.S. Geological Survey Professional Paper 1151-G, 53 p.

Rice, C. L., 1984b, Stratigraphic framework and nomenclatural problems in the Pennsylvanian of the Cumberland overthrust sheet, Kentucky and Tennessee: Geological Society of America Bulletin, v. 95, p. 1475–1481.

Rice, C. L., and Maughan, E. K., 1978, Geologic map of the Kayjay quadrangle and part of the Fork Ridge quadrangle, Bell and Knott Counties, Kentucky: U.S. Geological Survey Geologic Quadrangle Map GQ-1505, scale 1:24,000.

Rice, C. L., and Newell, W. L., 1990, Geologic map of part of the Jellico East quadrangle, Campbell and Claiborne Counties, Tennessee: U.S. Geological Survey Geologic Quadrangle Map GQ-1674, scale 1:24,000.

Rice, C .L., and Ping, R. G., 1989, Geologic map of the Middlesboro North quadrangle, Bell County, Kentucky: U.S. Geological Survey Geologic Quadrangle Map GQ-1663, scale 1:24,000.

Rice, C. L., and Schwietering, J. F., 1988, Fluvial deposition in the central Appalachians during the Early Pennsylvanian: U.S. Geological Survey Bulletin 1839-B, p. B1–B10.

Rice, C. L., Ping, R. G., and Barr, J. L., 1977, Geologic map of the Belfry quadrangle, Pike County, Kentucky: U.S. Geological Survey Geologic Quadrangle Map GQ-1369, scale 1:24,000.

Rice, C. L., Sable, E. G., Dever, G. R., and Kehn, T. M., 1979, The Mississippian and Pennsylvanian (Carboniferous) Systems—Kentucky: U.S. Geological Survey Professional Paper 1110-Г, 32 p.

Rice, C. L., and Smith, J. H., 1980, Correlation of coal beds, coal zones, and key stratigraphic units in the Pennsylvanian rocks of eastern Kentucky: U.S. Geological Survey Miscellaneous Field Studies Map MF-1188.

Rice, C. L., Currens, J. C., Henderson, J. A., Jr., and Nolde, J. E., 1987, The Betsie Shale Member—A datum for exploration and stratigraphic analysis of the lower part of the Pennsylvanian in the central Appalachian basin: U.S. Geological Survey Bulletin 1834, 17 p.

Rice, C. L., Henry, T. W., and Chesnut, D. R., Jr., 1993, The distribution and biostratigraphy of the Crummies Member (new name) of the Breathitt Formation in Pennsylvanian rocks of eastern Kentucky, in Sando, W. J., ed., Shorter contributions to paleontology and stratigraphy: U.S. Geological Survey Bulletin 2073-A, p. A1–A9.

Rice, D. D., 1975, Geologic map of the Helton quadrangle, southeastern Kentucky: U.S. Geological Survey Geologic Quadrangle Map GQ-1227, scale 1:24,000.

Rogers, H. D., 1858, The geology of Pennsylvania: Philadelphia, Lippincott, v. 1, 586 p.; v. 2, 1,046 p.

Russell, P. G., 1918, The coals of Sexton Creek and the tributaries of South Fork on the right between the mouth of Redbird Creek and the mouth of Sexton Creek: Kentucky Geological Survey, ser. 4, v. 4, pt. 3, p. 185–260.

Stager, H. K., 1963, Geology of the London SW quadrangle, Kentucky: U.S. Geological Survey Geologic Quadrangle Map GQ-195, scale 1:24,000.

Stout, W., 1916, Geology of southern Ohio: Ohio Division of Geological Survey Bulletin 20, 723 p.

Stout, W., 1927, Geology of Vinton County: Ohio Division of Geological Survey Bulletin 31, 402 p.

Sturgeon, M. T., and DeLong, R. M., 1964, Revision of some stratigraphic names between the Lower and Middle Kittanning coals of eastern Ohio: Ohio Journal of Science 64, no. 1, p. 41–43.

Sturgeon, M. T., and Merrill, W. M., 1949, An additional fossiliferous member in the Allegheny Formation (Pennsylvanian) of Ohio: Ohio Journal of Science, v. 49, p. 1–11.

Sturgeon, M. T., and 15 others, 1958, Geology and mineral resources of Athens County, Ohio: Ohio Division of Geological Survey Bulletin 57, 600 p.

Swartz, C. K., 1922, Distribution and stratigraphy of the coal measures of Maryland, in Swartz, C. M., and Baker., W. A., Jr., Second report on the coals of Maryland, pt. 1: Baltimore, Maryland Geological Survey, v. 11, p. 27–79.

Wanless, H. R., 1939, Pennsylvanian correlations in the Eastern Interior and Appalachian coal fields: Geological Society of America Special Paper 17, 130 p.

Wanless, H. R., 1946, Pennsylvanian geology of a part of the southern Appalachian coal field: Geological Society of America Memoir 13, 162 p.

Wanless, H. R., 1975, Appalachian region, in McKee, E. D., and Crosby, E. J., coord., Introduction and regional analyses of the Pennsylvanian System, pt. 1, of Paleotectonic investigations of the Pennsylvanian System in the United States: U.S. Geological Survey Professional Paper 853, p. 69–96.

Ward, D. E., 1978, Geologic map of the Adams quadrangle, Lawrence County, Kentucky: U.S. Geological Survey Geologic Quadrangle Map GQ-1489, scale 1:24,000.

Weir, G. W., 1978, Geologic map of the Sturgeon quadrangle, east-central Kentucky: U.S. Geological Survey Geologic Quadrangle Map GQ-1455, scale 1:24,000.

Weir, G. W., and Mumma, M. D., 1973, Geologic map of the McKee quadrangle, Jackson and Owsley Counties, Kentucky: U.S. Geological Survey Geologic Quadrangle Map GQ-1125, scale 1:24,000.

Welch, S. W., 1958, Geology and coal resources of the Tiptop quadrangle, Kentucky: U.S. Geological Survey Bulletin 1042-P, p. P585–P612, map scale 1:24,000.

White, I. C., 1885, Resume of the work of the U.S. Geological Survey in the Great Kanawha Valley during the summer of 1884: The Virginias, v. 6, p. 7–16.

White, I. C.,1891, Stratigraphy of the bituminous coal field of Pennsylvania, Ohio, and West Virginia: U.S. Geological Survey Bulletin 65, 212 p.

Whittington, C. L., and Ferm, J. C., 1965, Geology of the Oldtown quadrangle, Kentucky: U.S. Geological Survey Geologic Quadrangle Map GQ-353, scale 1:24,000.

Whittington, C. L., and Ferm, J. C., 1967, Geologic map of the Grayson quadrangle, Carter County, Kentucky: U.S. Geological Survey Geologic Quadrangle Map GQ-640, scale 1:24,000.

Williamson, A. D., and Adkison, W. L., 1953, Principal coal beds in the Troublesome quadrangle, Breathitt, Knott, and Perry Counties, Kentucky: U.S. Geological Survey Coal Investigation Map C-18, scale 1:62,500.

Wilson, C. W., Jr., Jewell, J. W., and Luther, E. T., 1956, Pennsylvanian geology of the Cumberland Plateau: Nashville, Tennessee Division of Geology, unnumbered folio, 21 p., scale 1:190,000.

MANUSCRIPT ACCEPTED BY THE SOCIETY FEBRUARY 1, 1994

Printed in U.S.A.

Typeset by WESType Publishing Services, Inc., Boulder, Colorado
Printed in U.S.A. by Johnson Printing, Boulder, Colorado